29 Springer Series in Solid-State Sciences

Edited by Peter Fulde

Springer Series in Solid-State Sciences

Editors: M. Cardona P. Fulde H.-J. Queisser

1 **Principles of Magnetic Resonance**
2nd Edition 2nd Printing
By C. P. Slichter

2 **Introduction to Solid-State Theory**
2nd Printing
By O. Madelung

3 **Dynamical Scattering of X-Rays in Crystals** By Z. G. Pinsker

4 **Inelastic Electron Tunneling Spectroscopy**
Editor: T. Wolfram

5 **Fundamentals of Crystal Growth I**
Macroscopic Equilibrium and Transport Concepts. 2nd Printing
By F. Rosenberger

6 **Magnetic Flux Structures in Superconductors** By R. P. Huebener

7 **Green's Functions in Quantum Physics**
By E. N. Economou

8 **Solitons and Condensed Matter Physics**
2nd Printing
Editors: A. R. Bishop and T. Schneider

9 **Photoferroelectrics**
By. V. M. Fridkin

10 **Phonon Dispersion Relations in Insulators** By H. Bilz and W. Kress

11 **Electron Transport in Compound Semiconductors** By B. R. Nag

12 **The Physics of Elementary Excitations**
By S. Nakajima, Y. Toyozawa, and R. Abe

13 **The Physics of Selenium and Tellurium**
Editors: E. Gerlach and P. Grosse

14 **Magnetic Bubble Technology** 2nd Edition
By A. H. Eschenfelder

15 **Modern Crystallography I**
Symmetry of Crystals
Methods of Structural Crystallography
By B. K. Vainshtein

16 **Organic Molecular Crystals**
Their Electronic States
By E. Silinsh

17 **The Theory of Magnetism I**
Statics and Dynamics By D. C. Mattis

18 **Relaxation of Elementary Excitations**
Editors: R. Kubo and E. Hanamura

19 **Solitons,** Mathematical Methods for Physicists By G. Eilenberger

20 **Theory of Nonlinear Lattices**
By M. Toda

21 **Modern Crystallography II**
Structure of Crystals
By B. K. Vainshtein, V. M. Fridkin, and V. L. Indenbom

22 **Point Defects in Semiconductors I**
Theoretical Aspects
By M. Lannoo and J. Bourgoin

23 **Physics in One Dimension**
Editors: J. Bernasconi, T. Schneider

24 **Physics in High Magnetic Fields**
Editors: S. Chikazumi and N. Miura

25 **Fundamental Physics of Amorphous Semiconductors**
Editor: F. Yonezawa

26 **Theory of Elastic Media with Microstructure I** By I. A. Kunin

27 **Superconductivity in Transition Metals and Compounds**
By S. Vonsovsky, Yu. A. Isyumov, and E. Z. Kurmaev

28 **The Structure and Properties of Matter**
Editor: T. Matsubara

29 **Electron Correlation and Magnetism in Narrow-Band Systems**
Editor: T. Moriya

30 **Statistical Physics I**
By M. Toda and R. Kubo

31 **Statistical Physics II**
By R. Kubo and M. Toda

32 **Quantum Theory of Magnetism**
By R. M. White

33 **Mixed Crystals**
By A. I. Kitaigorodsky

34 **Phonons: Theory and Experiments I**
Lattice Dynamics and Models of Interatomic Forces
By P. Brüesch

35 **Point Defects in Semiconductors II**
Experimental Aspects
By M. Lannoo and J. Bourgoin

36 **Modern Crystallography III**
Formation of Crystals
By A. A. Chernov et al.

37 **Modern Crystallography IV**
Physical Properties of Crystals
By L. A. Shuvalov et al.

Electron Correlation and Magnetism in Narrow-Band Systems

Proceedings of the
Third Taniguchi International Symposium,
Mount Fuji, Japan, November 1–5, 1980

Editor: T. Moriya

With 99 Figures

Springer-Verlag Berlin Heidelberg New York 1981

Professor *Tôru Moriya*
The Institute for Solid-State Physics, University of Tokyo,
Roppongi, Minato-ku, Tokyo 106, Japan

ISBN 3-540-10767-3 Springer-Verlag Berlin Heidelberg New York
ISBN 0-387-10767-3 Springer-Verlag New York Heidelberg Berlin

Offset printing: Beltz Offsetdruck, 6944 Hemsbach/Bergstr. Bookbinding: J. Schäffer oHG, 6718 Grünstadt.
2153/3130-543210

Dedicated to the Memory of

John Hubbard

Susono, November, 1980

Preface

This volume is the product of the Third Taniguchi International Symposium on the Theory of Condensed Matter, which was held at the Fuji Institute of Education and Training in Susono-shi (at the foot of Mt. Fuji) in Japan during the period of 1-5 November 1980. The topic of the Symposium was "Electron Correlation and Magnetism in Narrow-Band Systems", one of the central issues in present-day condensed-matter physics. Special emphases were put on the ferro- and antiferromagnetism of transition metals, alloys and compounds – in particular, various approaches toward resolution of the long-standing controversy between itenerant versus localized pictures – and on the valence fluctuations in rare-earth compounds.

In the former area marked progress has been made in recent years owing to investigations of spin fluctuations from a very general point of view, as discussed by many contributors in Part I of this volume. Although there are still controversies as to the relative importance of the short-range order, etc. in each specific substance such as iron or nickel, and although the studies of dynamical effects are far from satisfactory, I believe that we are now on a promising track. Part II deals with the ground-state properties of d metals and alloys, including the possible importance of the many-body effects in nickel and its alloys and the importance of the antisymmetric exchange interaction as an origin of the long-period helical spin structure in certain metallic compounds. Various theoretical aspects of the narrow-band problems such as spin glasses, low dimensionality, and some exact results for simplified models are discussed in Part III. The problem of mixed valence in 4f compounds is a relatively new and hardly settled subject, where the effects of electron correlation are expected to be stronger than in transition metals and the many-body effect seems to be essential even in the ground state. Some of the fundamental problems in this area, as viewed from the standpoint of narrow-band theory, are discussed by several contributors in Part IV.

A limited number of active researchers were invited to participate in the Taniguchi Symposium in line with the philosophy and policy of the Taniguchi Foundation as described in Mr. Taniguchi's speech delivered at the reception, which may be found in the Foreword. Although this set up was useful in order to facilitate intense communication among the participants, we feel obliged to share the information obtained in the symposium with scientists throughout the world. We are therefore grateful to Springer-Verlag for their assistance in reaching this goal.

Finally, on behalf of the participants, I would like to dedicate this volume to the memory of Dr. John Hubbard, who passed away shortly after the Symposium. We are indebted to him for invaluable contributions to the symposium both in his enlightening talks and very stimulating discussions, not to mention his outstanding contributions in the field of narrow-band problems in the course of his brilliant career.

February 1981

Tôru Moriya

Foreword

Speech by Toyosaburo Taniguchi

Welcome my friends to the Third International Symposium, Division on the Theory of Condensed Matter, of the Taniguchi Foundation.

The need is now greater than ever for Japan to consider how to strengthen and foster international understanding between nations, peoples and societies, and how to contribute towards the establishment of peace and prosperity in the world.

For more than twenty years, I have been supporting a symposium on mathematics in which distinguished scholars from all over the world have engaged in free discussions. In this symposium, all the participants live together in community style. I have heard from members of some of these study groups that this type of setup has helped to strengthen their ties and relationships with their colleagues on a personal basis. What developed in the mathematics group led me to reorganize and strengthen the Taniguchi Foundation only a few years ago through additional funding.

In order to effectively translate the objectives of the Foundation into action with the funds available, it becomes necessary to select those fields which are not necessarily in the limelight of popular interest, which means those fields which, I am afraid, are low in funding. I would rather choose for the Foundation, projects from modest unimpressive academic fields than those that stand out in gaudy, gorgeous popular acclaim.

I would desire to see the promising younger groups of scholars throughout the world assume an even more active role in our symposia. With world-shaking events occurring daily, deep-rooted changes are taking place. It is really the younder generation who will lead us into the promise of tomorrow. Their experiences in the symposia will be instrumental in forging the bonds of mutual understanding between peoples, between societies and between nations.

I would like to add that the number of the participants is being preferably limited to about 15 or 20 – and no more – because of my desire to see a situation of living and studying together in a close-knit community setup. The more the number of participants increases, the more the substance of the symposium becomes vague and the perfume is gone; therefore I beg your best understanding of our basic philosophies and thoughts.

Today the Foundation extends its activities to nearly ten divisions. In addition to the Theory of Condensed Matter, we hold symposia in Mathematics, Business History, Biophysics, Medical History, Brain Sciences, Ethnology and Neurobiology in Vision. Whenever I meet with the many distinguished scholars participating in our symposia, I am impressed by their hones dedication to their mission of pursuing the vocation they are in. Words are simply inadequate to express my sincere respect for these unselfish and dedicated efforts of theirs, which cannot help but make a major contribution to the well-being of mankind. It is my hope that these symposia will continue for ten years or more under the guidance of Mr. H. Shima, President of the Foundation, who designed and brought to completion the "Shin-Kansen", the world-famous bullet train of Japan.

Kyoto, the 1200-year-old capital of Japan, is widely known as a veritable treasure house of cultural and historical interest in Japan. It is in truth a city sitting in a nest of great natural beauty. You might be interested to know that three out of five Japanese Nobel Laureates, Dr. Hideki Yukawa, Dr. Leona Esaki and the late Dr. Shin-ichiro Tomonaga, lived in this beautiful environment during their student days. They schooled in the same institution located here in Kyoto. While I realize that your time in the Kyoto area is limited, I hope that you will be enjoying yourself as they did in their youth since there should be a meeting of the hearts and of experiences common to all the scholastic minded.

In conclusion, I would like to end these words of welcome with an earnest prayer for the success of the symposium throughout all sessions, and also sincerely hope that all the participants, both Japanese and foreign, will return, bearing heartwarming memories of the symposium and our enjoyable time together this evening.

Thank you.

Participants in the Taniguchi International Symposium, November 1 - 5, 1980

(photographed with Mt. Fuji in the background)

Seated (left to right)

T. Moriya A. Yanase J. A. Hertz J. Hubbard R. Kubo Mrs. Prange
D. M. Edwards K. Yosida T. Kasuya K. Yamada

Standing (left to right)

O. Nakanishi K. Ueda K. Kubo J. Igarashi E. Müller-Hartmann
F. D. M. Haldane C. P. Enz J. Kanamori R. E. Prange H. Hasegawa
Y. Takahashi K. Usami

Contents

Part I. **d Metals and Compounds – Finite Temperature Properties**

Part II. **d Metals and Compounds – Ground-State Properties**

Part I

d Metals and Compounds – Finite Temperature Properties

Spin Fluctuations and a Unified Picture of Magnetism

Tôru Moriya

Institute for Solid State Physics, University of Tokyo,
Roppongi, Tokyo

1. Introduction — What Have Been the Problems? — Historical Development

Ferro- and antiferromagnetism of transition metals and their compounds have been a controversial subject of continued interest for half a century since the modern theoretical research of magnetism has started with the advent of quantum mechanics. As is well known, in the major part of the modern theory of solids the mean-field approximation or the one-electron picture has been extremely successful. While in the problem of magnetism, especially at finite temperatures, the consideration of electron-electron correlation going beyond the mean-field approximation was indispensable even for qualitative understanding of the subject. This may have been one of the reasons why the controversy lasted for such long time.

Recently substantial progress has been made in this area and we are having a unified picture, though crude, of magnetism including both metals and insulators. This picture has emerged out of recent investigations of spin density fluctuations from a very general point of view. The purpose of the present note is to discuss on this unified picture after tracing very briefly the modern theoretical developments in the field of magnetism.

1.1 Local Moment Hypothesis and the Curie-Weiss Law

In the end of the 19th century the Curie law for the magnetic susceptibility: $\chi=C/T$, where C is the Curie constant and T the temperature, was discovered. Also discovered was the Curie temperature T_C above which ferromagnetism disappeared. As is well known these phenomena have been explained by the hypothesis of local magnetic moments by LANGEVIN [1] and the additional hypothesis by WEISS [2] of the molecular field which is exerted on each local moment and is proportional to the total magnetization. The Curie law was revised and the magnetic susceptibility above T_C was found to obey the Curie-Weiss law:

$$\chi = C/(T - \theta), \tag{1.1}$$

with $\theta=T_C$ in the Weiss theory.

1.2 Heisenberg Model and Its Developments

The quantum mechanics has given justification for the hypothesis of local magnetic moments associated with atoms and the quantum-mechanical theory of the magnetic susceptibility of atoms has been developed by VAN VLECK [3]. The origin of the coupling between the neighboring local moments, or the origin of the Weiss molecular field has been attributed by HEISENBERG [4] to the quantum-mechanical exchange interaction and the Heisenberg model has been set out.

Subsequent investigations of magnetism based on the Heisenberg model or its modifications have been remarkably successful and our understandings of magnetism and of phase transition have been advanced enormously. For example various magnetic structures such as antiferromagnetism, ferrimagnetism, helimagnetism, etc. have been discovered by using this model and all the possible types of magnetic structures have now been exhausted. The new concepts such as spin waves and magnetic resonances have been brought forth with the use of this model. Also in the development of the theory of the second-order phase transition this model was indispensable. To sum, the Heisenberg model has been an extremely fruitful model in the fields of magnetism and of statistical mechanics of phase transitions.

However, it has been pointed out for 50 years that this model is not appropriate for the description of ferromagnetism of transition metals such as iron, cobalt and nickel where the d electrons which are responsible for magnetism are considered to be itinerant or Bloch electrons. It has been shown later that the Heisenberg model is the basic model for the description of the magnetism of insulator compounds and its theoretical foundation was established by ANDERSON [5] on the basis of the Mott concept of electron localization by strong correlation [6].

1.3 Itinerant Electron Model -- Stoner Theory

Another main stream of the quantum-mechanical theory of magnetism has been based on the Bloch band theory of electronic states in solids. After the Bloch theory of ferromagnetism of an electron gas [7], STONER [8] and SLATER [9] in 1930's initiated the theory of itinerant (collective) electron ferromagnetism with the use of the mean-field approximation for the electron-electron interaction. As one can easily see, this type of theory can be consistent with the non-integral Bohr magneton number per atom of the observed spontaneous magnetizations of iron, cobalt and nickel, their large cohesive energies and large coefficients of the T-linear terms in the low-temperature specific heat, etc., which are not compatible with the local moment picture.

However, the finite temperature properties of ferromagnetic d metals were not explained by the Stoner theory based on the mean-field approximation. For example the Curie-Weiss susceptibility observed in all ferromagnetic metals, the large specific-heat anomaly around T_C and the temperature dependence of magnetization were not explained consistently by this theory. These properties were better described by the localized (Heisenberg) model which has been used widely in the analysis of experimental data of magnetic d metals and their alloys. Also, the Curie temperature calculated by using the parameters consistent with the experimental results at T=0 K were generally too high compared with experiment.

1.4 Controversy Between Localized and Itinerant Models

It is interesting to point out here that the above-mentioned two models have mutually opposite starting points; the localized model starts with the electronic states localized in the real space while the itinerant model starts with those localized in the reciprocal (momentum) space. Neither one of these pictures by itself could explain all the physical properties of ferromagnetic metals in any systematic way and it was necessary for this purpose to have recourse to the two types of concepts associated respectively with these models. Thus the controversy over these models has lasted for half a century.

As a matter of fact recent theoretical efforts have been concentrated on finding a way of reconciling these mutually opposite pictures into a unified one and for this purpose the effect of electron correlation was invoked. There were two main streams of theoretical investigations in this general direction. One was the improvement of the Stoner theory to take account of the electron correlation and the other was the study of local moments in metals.

1.5 Local Moment in Metals -- A Reconciliation

In early 1950's VAN VLECK [10] has stressed the importance of the electron correlation in narrow d bands and suggested the possibility of approximate validity of the local moment picture in d metals. In 1961 ANDERSON [11] has presented his model of local moment in metals on the basis of the Friedel virtual bound-state picture in dilute alloys [12]. The local moment here is nothing but the slow large local spin fluctuation in the itinerant electron systems. Anderson derived the condition for the appearance of the local moment with the use of the Hartree-Fock approximation and related it with the Curie-Weiss susceptibility. Thus in this theory the local moment as the local spin fluctuation was treated within an adiabatic approximation. Subsequently WOLFF [13] presented a tight binding model for the description of local moments in metals. The interaction between a pair of local moments and the spin polarization due to the local moment have been investigated subsequently and the results were applied successfully to interpret magnetic properties of various transition metals and their alloys. The theoretical approach along this line has been advanced further in recent years as will be discussed in §§2 and 5.

1.6 Weakly Ferromagnetic Metals — *A New Limiting Case*

The concept of local moments in metals, though successful in interpreting magnetic properties of many of d metals and alloys, soon turned out to be only a partial picture of magnetism of metals. Around 1960 MATTHIAS et al. discovered weakly ferromagnetic metals, $ZrZn_2$ [14] and Sc_3In [15], which had very low T_C and very small saturation moments at T=0 K. The magnetic properties of these substances were not compatible with the local moment model. Although there was an expectation that the Stoner theory might be valid in the weakly ferromagnetic metals [16], the observations in $ZrZn_2$ and Sc_3In of very good Curie-Weiss susceptibilities up to $T \sim 10T_C$ suggested the contrary, i.e., they indicated the necessity of a theory going beyond the Hartree-Fock approximation even in this case.

One might think that the studies in 1950's and 60's of spin-density fluctuations including spin waves [17] and paramagnons [18] by using the dynamical Hartree-Fock or random phase approximation (RPA) may be an improvement of the Stoner theory. However, they were simply extensions of the Stoner theory to include the description of fluctuations around the Stoner equilibrium state and the latter was not modified at all. Thus the major difficulties associated with the Stoner theory remained without any improvement. In early 1970's the self-consistent renormalization (SCR) theory of spin fluctuations going one step beyond the Hartree-Fock-RPA theory has been put forth and various physical properties including the Curie-Weiss susceptibility of weakly ferromagnetic metals have been explained consistently [19,20].

It was pointed out that *weakly ferro- and antiferromagnetic metals constitute an important limiting case opposite to the local moment case.* In this case the long-wave components of spin fluctuations or those with the wave vectors around the antiferromagnetic vector Q play a predominant role. In other words, *the spin fluctuations in weakly ferro- and antiferromagnetic limit are localized in reciprocal or q space in contrast with the local moment case where the spin fluctuations are localized in real space.* The Curie-Weiss law arises from a new mechanism entirely different from the well-known local moment mechanism, i.e., through the coupling between different modes of spin fluctuations. As will be discussed in §3, intensive and extensive theoretical and experimental investigations of this class of materials in mid-1970's have convinced us the soundness of the above-mentioned picture and the SCR theory [21] for weakly ferro- and antiferromagnetic metals.

1.7 A Unified Picture

As an important consequence of the success of the SCR theory of weakly ferro- and antiferromagnetic metals, it was realized that many of the physical properties of magnets so far regarded as specific to the local moment system were also exhibited by the mutually interacting long-wave spin-fluctuation modes in weakly ferro- and antiferromagnetic metals. This indicates that we have to make a very general study of spin fluctuations encompassing the weakly ferromagnetic limit and the local moment limit in order to get a full and general understanding of magnetism in metals. In other words, we can get a unified picture of magnetism by interpolating between the weakly ferro- and antiferromagnetic limit and the local moment limit. A simple theory of this kind has been put forth recently by MORIYA and TAKAHASHI [22].

In what follows we discuss about the unified theory and various important consequences of it [21], after brief sketch of the developments in the local moment approach (§2) and that of the approaches from the weakly ferro- and antiferromagnetic limit (§3).

2. Local Moment in Metals and Ferro- and Antiferromagnetism

The concept of a local magnetic moment in metals has been used quite fruitfully not only in the theory of dilute magnetic alloys but also in the theory of magnetism in metals. The theoretical approach to the latter problem has been initiated by the study of pairs of local moments with the use of the Anderson model as well as the tight-binding model by Wolff and Clogston. The interaction between a neighboring pair of local moments was first treated by ALEXANDER and ANDERSON [23] on the basis of the Anderson model and the theory was generalized by MORIYA [24] who pointed out that *the sign of the interaction between the neighboring local moments is primarily determined by the occupied fraction of the localized orbitals; when each atomic d shell is nearly half-filled the coupling between the local moments is antiferromagnetic, while the coupling is ferromagnetic when the occupied or empty fraction of each d shell is small.* The sign of the induced spin polarization is also governed by the corresponding rules. These simple rules have been applied successfully in interpreting magnetic properties of a number of magnetic metals and alloys. The interaction between a widely separated pair of magnetic atoms has been treated by CAROLI [25]. The same problem has been treated on the basis of the Wolff-Clogston model by INOUE and MORIYA [26] including both neighboring and separated pairs within the same theoretical framework. The results were qualitatively the same as those for the Anderson model.

The discussion of magnetism of transition metals and their alloys by using these results have naturally been qualitative, since one has necessarily had recourse to the intuitive use of the Heisenberg model with the above-mentioned interaction. Nevertheless good orders of magnitude have been obtained for the Curie temperature of ferromagnetic transition metals by simply evaluating the necessary parameters such as the width of the virtual bound-state energy and the transfer integral from the known band width.

This line of approach has been advanced by SCHRIEFFER et al. [27] with the use of the Stratonovitch-Hubbard technique in the functional integral formalism [28]. They have indicated a mathematical method of calculating the thermodynamical properties including the Curie susceptibility, thus giving a sound basis for the above arguments based on the adiabatic and local Hartree-Fock approximations. They have also discussed the interaction between a pair of local moments with half-filled orbitals and shown that the static and saddle point approximation for the functional integral leads to the above-mentioned results based on the local Hartree-Fock approximation [29].

It may be noteworthy that there has been a different line of attempts toward this type of theory, starting from the band structure for the entire crystal. FRIEDEL et al. [30] proposed a picture of interacting local moments in magnetic metals, interpreting very roughly the Stoner condition for ferromagnetism as the condition for the appearance of local moments. They suggested the possible relation between the size of the Fermi radii and the sign of the coupling between a pair of local moments, leading to the rules similar to the above-mentioned ones. BLANDIN and LEDERER [31] extended this idea and indicated a possible relation between the Fourier transform of the inversed wave vector-dependent susceptibility of non-interacting electrons and the interaction between the local moments. However, these arguments remained within the linear theory, i.e., the theory of non-local susceptibilities, and the non-linearity with respect to the amplitude of the local moment was not taken into account.

More recently CYROT [32] applied the functional integral method to this problem, in particular to the half-filled Hubbard model or the regular array of the Wolff models and discussed the metal-insulator (Mott) transition within the static and saddle point approximation. He made use of the single-site coherent potential approximation (CPA) in calculating the free energy functional for a system with randomly oriented local moments and interpreted the occurrence of the energy gap around the Fermi level as that of the Mott insulator state. This theory made significant improvements over previous theories of the Mott transition in various respects. LACOUR-GAYET and CYROT [33] extended the Inoue-Moriya type theory of pair interaction, though limited to the case of a half-filled band, to include the effect of local spin fluctuations in the surrounding medium through CPA. Thus the functional integral method and possibly the coherent potential approximation have given a sounder basis to the local moment approach to magnetic metals. Along a similar line CAPELLMANN [34] has proposed a generalized Hartree-Fock scheme which retains rotational symmetry in dealing with the local spin fluctuations. Although this theory reproduces an effective Heisenberg model with the correct coupling constant in the insulator (strong correlation) limit, the theory in general makes use of a long-wave approximation with fixed amplitude of the local spin density.

Quite recently, this line of approach has been advanced to the stage of dealing with quantitative values for the Curie temperature of Fe and Ni. This will be discussed in §5.

3. Weakly Ferro- and Antiferromagnetic Metals — A New Limiting Case and New Curie-Weiss Law

As was discussed already in §1, the approach from the opposite limit to the local moment theory should be the improvement of the conventional HF-RPA theory. The necessity of such a theory has been realized strongly by the experimental observations of various properties of weakly ferromagnetic metals which are compatible with neither the HF-RPA theory nor the local moment picture.

It was well known that some of the drawbacks of the Stoner theory were remedied by considering the spin density fluctuations such as spin waves and paramagnons. These thermal excitations have been treated in 1950's and 1960's mainly within RPA [17,18] which is an extension of HFA to include the description of fluctuations but was not an improvement of HFA as was stressed already. In other words, within the HF-RPA theory the spin fluctuations are still localized strictly in reciprocal space and the interaction between different wave vector components of spin fluctuations is not taken into account.

Several attempts have been made in 1960's to go beyond the HF-RPA theory by renormalizing the thermal equilibrium state. Although many of these attempts remained at the formal stage of the theory, physically meaningful results were obtained in nearly ferromagnetic metals in the low-temperature limit. DONIACH and ENGELSBERG [35] as well as BERK and SCHRIEFFER [36] pointed out the logarithmic enhancement of the coefficient of the T-linear term of low-temperature specific heat due to the paramagnons, with the use of RPA for spin fluctuations, while MORIYA [37] showed a weaker enhancement or the absence of logarithmic enhancement in nearly antiferromagnetic metals. BEAL-MONOD, MA and FREDKIN [38] pointed out that the same renormalization effect gave enhancement of the coefficient of the T^2 term in the low-temperature magnetic susceptibility of a nearly ferromagnetic Fermi liquid, i.e., the renormalized susceptibility decreases with increasing temperature much faster than the HF result.

In these theories the thermal equilibrium state is renormalized by taking account of the RPA spin fluctuations, or the thermal excitations are assumed to be described by RPA spin fluctuations. Thus these theories have a serious inconsistency in themselves except for the limit of $T\to 0$; the renormalized equilibrium state at finite temperature should be different from that of HFA while the spin fluctuations are treated by RPA as those around the HFA equilibrium state. The seriousness of this inconsistency may be apparent when we consider ferromagnets. The renormalized Curie temperature T_C should be lower than its HFA value T_C^{HF} and the RPA spin fluctuations between T_C and T_C^{HF} show the characteristics of those in ferromagnetic state even though the state in actuality is paramagnetic. Thus it is necessary to develop a theory where the equilibrium state and the spin fluctuations are renormalized at the same time in a self-consistent fashion.

Such a theory was developed by MORIYA and KAWABATA [19]. It was remarkable that the self-consistent renormalization (SCR) theory has given qualitatively new results for various physical properties among which most striking was a new explanation for the Curie-Weiss magnetic susceptibility without local magnetic moment. MURATA and DONIACH [20] have independently developed a phenomenological mode-mode coupling theory of spin fluctuations in weakly ferromagnetic metals and obtained qualitatively the same conclusions as the SCR theory. It was shown later that this theory corresponds to the high-temperature (classical) approximation to the Moriya-Kawabata theory.

We will not go into details of the SCR theory and its consequences here and refer to [19] and [21] for interested readers. We only note some characteristic features of the SCR theory as applied to weakly ferro- and antiferromagnetic metals, where only long-wave components of spin fluctuations or the components with the wave vectors near the antiferromagnetic vector Q are strongly exchange enhanced and are important. Thus we need to treat the spin fluctuation components only in a relatively small part of q space. Furthermore we can make an expansion in terms of the amplitude of spin fluctuations, since the latter is generally small in weakly ferro- and antiferromagnetic metals. Both of these make the theory significantly simple and *we can express various physical quantities in terms of relatively small number of parameters deduced from the band structure around the Fermi surface.* In particular, *we can derive various general results independent of the band structure as to the temperature and magnetic-field dependences of physical quantities.*

We discuss here very briefly the new physical origin of the Curie-Weiss (CW) law in weakly ferromagnetic metals. The free energy may be expressed in terms of spatially varying spin density $\vec{S}(r)$ and its Fourier transform $\vec{S}_q$ as follows:

$$F = \sum_q \frac{1}{2\chi_{0q}} |\vec{S}_q|^2 - N_0 I v_0 \int d\vec{r}[\vec{S}(r)]^2 + \frac{\gamma_0 v_0^3}{4} \int d\vec{r}[\vec{S}(r)]^4, \tag{3.1}$$

where v_0 is the atomic volume, N_0 the number of atoms in the crystal, χ_{0q} the magnetic susceptibility for the non-interacting electron system, I the intra-atomic exchange energy J divided by N_0, and γ_0 the constat. The second term is the exchange energy which is assumed to be local. The first and the third terms are the increase in the kinetic energy due to the excitation of spin fluctuations and we take terms up to the fourth order in the spin density, which is approximated to be local and gives the mode-mode coupling. The magnetic susceptibility as expressed in units of $g^2\mu_B^2$, g being the gyromagnetic ratio and μ_B the Bohr magneton, is given from (3.1) as follows:

$$1/\chi_q = (\partial^2 F/\partial S_{q\alpha}^2) = \frac{1}{\chi_{0q}} - 2I + \frac{5\gamma_0}{3N_0} \langle S_{loc}^2 \rangle,$$

with

$$\langle S_{loc}^2 \rangle = N_0^{-2} \sum_q \langle |\vec{S}_q|^2 \rangle. \tag{3.2}$$

Since we have the following relation from the fluctuation-dissipation theorem:

$$\langle S_{loc}^2 \rangle = 3TN_0^{-2} \sum_q \chi_q, \tag{3.3}$$

we can obtain $\langle S_{loc}^2 \rangle$ and χ_q by solving (3.2) and (3.3), provided the q dependence of χ_{0q} is given from the band structure.

The uniform susceptibility can be expressed as follows considering its divergence at $T=T_C$:

$$1/\chi_0 = (5\gamma_0/3N_0)(\langle S_{loc}^2 \rangle - \langle S_{loc}^2 \rangle_C), \tag{3.4}$$

where $\langle\ \rangle_C$ means the statistical average at T_C. Eq.(3.4) indicates that we have the CW law when $\langle S_{loc}^2 \rangle$ is proportional to T. This is possible in weakly ferromagnetic metals where $1/\chi_{0q}-2I$ is very small for $q\to 0$ and increases relatively rapidly with increasing q. In such a case χ_q for not so small q is almost temperature independent, since the first two terms in (3.2) dominate over the last term, and thus the corresponding $<|S_q|^2>$ increases linearly with T, giving rise to the linear increase in T of $<S_{loc}^2>$. As was stressed previously, *the Curie-Weiss law in weakly ferromagnetic metals is obeyed by* χ_q *with small q only in contrast with the local moment case where all the q components obey the CW law. Correspondingly in weakly antiferromagnetic metals with the antiferromagnetic wave vector Q, the CW law is obeyed by* χ_{Q+q} *with small q only. The important point is that the CW law in this case arises from the linear increase in T of* $<S_{loc}^2>$ *in contrast with the local moment mechanism where* $<S_{loc}^2>$ *is constant.* The SCR theory deals with this physical mechanism on the basis of the quantum statistical mechanics.

In addition to the magnetic properties the specific heat, nuclear magnetic relaxation rate $1/T_1$, electrical and thermal resistivities, and their magnetic-field dependences have been studied both theoretically and experimentally. Many of the theoretical predictions were qualitatively new and all of these predictions have been confirmed by experimental studies [21]. This fact is enough to convince us the soundness of the SCR theory in the weakly ferro- and antiferromagnetic limit.

The SCR theory have been extended to cover the helical spin structure and the coexistence of and phase transition between ferro- and antiferromagnetic phases have also been treated [21].

It is also noteworthy that various other theoretical approaches to the weakly ferromagnetic metals have been developed in mid 1970's and almost all of them reduced essentially to the SCR results. HERTZ and KLENIN [39] have applied a functional integral method to this problem and discussed an extension of the theory to the case of larger amplitude of the spin fluctuation within a long-wave approximation.

Extension of the SCR theory to include shorter wave components and the case of larger amplitude has also been made to some extent. In particular MORIYA [40] has derived integral equations for the dynamical susceptibility corresponding to the interacting modes of spin fluctuations with the use of the temperature Green's function method and the equations of motion method. This equation has been solved first in the weakly ferromagnetic limit, leading to the SCR theory. The relatively short-wave components were also treated assuming that the thermal spin fluctuations are of long-wave character and the sloppy spin waves above T_C was described. HERTZ and KLENIN [41] have also discussed the sloppy spin waves with the use of their functional integral method.

4. A Unified Picture of Magnetism

4.1 General Considerations

As may be seen from the discussions in the preceding sections the thermodynamical properties of ferro- and antiferromagnetic metals are governed by the exchange-enhanced and mutually interacting modes of spin density fluctuations in a very general sense, encompassing between the local moment limit and the weakly ferro- and antiferromagnetic limit as mutually opposite extremes. In the weakly ferro- and antiferromagnetic limit the spin fluctuation is localized in reciprocal space and its amplitude is small. In the local moment case the spin fluctuation is localized in real space and its local amplitude is

(a) SPIN FLUCTUATIONS

local in q space — *local in real space*

Amplitude: *small* (top) to *saturated* (bottom)

Weakly Ferromagnetic(a) & Antiferromagnetic(b) Limit

MnSi

Au_4V α-Mn

Cr CrB_2

γ-Mn   MnP CoS_2

Ni   Co   Fe

Local Moment Limit(c)

(a) Sc_3In, $ZrZn_2$, Ni_3Al, $Fe_{.5}Co_{.5}Si$, $Ni_{.43}Pt_{.57}$, etc...

(b) β-Mn, V_3Se_4, V_3S_4, V_5Se_8, etc...

(c) Magnetic Insulator Compounds, 4f Metals, Heusler Alloys.

(b) THEORETICAL APPROACHES

local in q space — *local in real space*

Amplitude: *small* (top) to *saturated* (bottom)

Self-consistent Renormalization Theory(1)

Extension to Larger Amplitude(2)

Interpolation Theory(3) *general amplitude non-local character*

Interacting Virtual Bound States Picture; Local Spin Fluctuation Approach(4)

Long Wave Approximation with Fixed Amplitude(5)

Local Moment Theory(6)

(1) Murata & Doniach[20]; Moriya & Kawabata[19] (ferromagnetism); Hasegawa & Moriya[42] (antiferromagnetism).
(2) Hertz & Klenin[39].
(3) Moriya & Takahashi[22]; Usami & Moriya[57]; Moriya & Hasegawa[54].
(4) Alexander & Anderson[23]; Moriya[24]; Inoue & Moriya[26]; Schrieffer, Evenson & Wang[27]; Cyrot[32]; Weller[43]; Liu[44]: Hasegawa[55,59]; Hubbard[56,58].
(5) Capellmann[34,52]; Korenman & Prange[53].
(6) Anderson[5] (magnetic insulators); Vonovsky; Zener; RKKY (4f metals).

Fig.1 (a) A classification of magnetic substances according to the nature of spin fluctuations (b) Theoretical approaches as classified by the ways of treating the spin fluctuations large and fixed. We show in Fig.1 the relation between the nature of spin fluctuations and magnetic properties. The positions of various substances in the intermediate range in Fig.1a are only by impression from experimental results. Note also that the local amplitude of the spin fluctuation is generally temperature dependent. In Fig.1b are shown the corresponding theoretical approaches.

There are two important points in developing a unified theory. First the amplitude of the local spin fluctuation can take any value up to saturation and the non-linearity with respect to the amplitude should fully be taken into account. The amplitude can generally vary with temperature as was seen in the case of weakly ferro- and antiferromagnetic limit. Next the non-local character of spin fluctuations either in real or reciprocal space have to be taken into account. Since we know the two opposite limiting cases and corresponding substances as shown in Fig.1a, it is necessary to interpolate between them in order to get a unified picture of magnetism. As we go to the left-hand side of this diagram, the degree of short-range order increases. There are various reasons to believe that in metals the spin fluctuations are generally more delocalized than in insulators.

To meet with the above requirements has been the basic idea of the theory by MORIYA and TAKAHASHI (MT) [22]. They introduced $\langle S_{loc}^2 \rangle$, the average squared amplitude of the local spin density, as an important physical variable of the theory, which is to be determined as a function of temperature in a self-consistent fashion taking account of the effect of short-range order.

A well-known method of dealing with the fluctuations with large amplitudes is the functional integral approach. In particular the Stratonovitch-Hubbard technique gives a formally accurate starting point for the quantum statistical theory and is useful in the interpolation theory. The physical idea of the theory may be represented by the following phenomenological expression for the free energy F:

$$e^{-F/T} = \int d\vec{S}(r)\, e^{-\mathcal{F}[S(r)]/T} = \int \prod_j d\vec{S}_j\, e^{-\mathcal{F}[S]/T}, \tag{4.1}$$

where $\vec{S}(r)$ is the spin density and we use a tight-binding model, denoting the spin of the jth atom as $\vec{S}_j$. We use the units $k_B=\hbar=1$ here and in what follows. The free energy functional <u>for the paramagnetic phase</u>, or for the case of $\langle\vec{S}_q\rangle\to 0$, is approximated by

$$\mathcal{F}[S] = -\frac{1}{2}\sum_{j,\ell} \mathcal{J}_{j\ell}(S_L^2)(\vec{S}_j\cdot\vec{S}_\ell) + \frac{1}{2}N_0 L(S_L^2)$$
$$= -\frac{1}{2}\sum_q \mathcal{J}_q(S_L^2)|\vec{S}_q|^2 + \frac{1}{2}N_0 L(S_L^2),$$

with

$$S_L^2 = N_0^{-1}\sum_j \langle|\vec{S}_j|^2\rangle = N_0^{-1}\sum_q \langle|\vec{S}_q|^2\rangle. \tag{4.2}$$

$L(S_L^2)$ and $\mathcal{J}_q(S_L^2)$ are assumed to be functions of S_L^2 only and can be calculated from a given band structure as will be discussed later. Eq.(4.2) has some similarity to the Heisenberg model but S_L^2 is variable here and should be determined so as to minimize F.

The model functional (4.2) is an extremely simplified one where (1) a local approximation is employed for the non-linear terms with respect to the squared amplitude of the spin density and (2) as for the non-local terms only bilinear terms with respect to the relative orientation of the local spin densities are taken into account. The range of interaction, however, is taken to be large enough.

Although we can generalize this functional further when necessary, we should note that this functional form leads to correct results in both the local moment limit and the weakly ferro- and antiferromagnetic limit. When $L(S_L^2)$ has a very sharp minimum the value of S_L^2 is fixed and (4.3) leads to the Heisenberg model. On the other hand, when S_L^2 is small we can expand $L(S_L^2)$ and $\mathcal{J}_q(S_L^2)$ in powers of S_L^2. Taking necessary lower order terms we get the results of the classical approximation to the SCR theory.

In actuality MT made use of the Stratonovitch-Hubbard method where the functional integral is expressed by spin and charge field variables which are conjugate to the density variables. We refer to [21] for details and only note here that within the static approximation we need to calculate the following expression corresponding to (4.2):

$$\psi_1[\xi,\eta] = -\,T\,\mathrm{Tr}\,\log\,(1 - g\cdot V), \tag{4.3}$$

where g and V are the matrices whose elements are given by

$$(k,n,\sigma|g|k',n',\sigma') = \delta_{kk'}\delta_{nn'}\delta_{\sigma\sigma'}g_k(i\omega_n), \quad g_k(i\omega_n) = (i\omega_n+\mu-\varepsilon_k)^{-1},$$

$$(j,n|V|\ell,n\text{-}m) = \delta_{j\ell}\delta_{m0}V_j, \quad V_j = \bar{c}_1\begin{pmatrix} \xi_j^z & \xi_j^- \\ \xi_j^+ & -\xi_j^z \end{pmatrix} + \bar{c}_2\eta_j,$$

$$\bar{c}_1 = (\pi JT)^{1/2}, \quad \bar{c}_2 = (-\pi UT)^{1/2}, \tag{4.4}$$

where $\omega_n=(2n+1)\pi T$, n being an integer, μ the chemical potential, ε_k the energy dispersion of the band, U and J the intra-atomic Coulomb and exchange energies, respectively. ψ_1 is nothing but the free energy of non-interacting electrons under the influence of the space-varying magnetic field $\{\bar{c}_1\vec{\xi}_j\}$ and the electro-static potential $\{\bar{c}_2\eta_j\}$. The partition function can be calculated by integrating $\exp[-\pi\sum_j(\xi_j^2+\eta_j^2)-\psi_1(\xi,\eta)/T]$ over ξ_j and η_j and taking average over direction of $\vec{\xi}_j$. Our model functional form as shown below corresponds to a generalization of (4.2) to include the ferromagnetic phase ($T<T_C$):

$$\psi_1[\xi,\eta] = \pi T\sum_q{}' \sum_\alpha[X_q^\alpha\xi_{q\alpha}\xi_{-q\alpha} + 2iZ_q\xi_{qz}\eta_{-q} + Y_q\eta_q\eta_{-q}] + \pi N_0L[\xi_0;x,y,z],$$

with

$$x_\alpha = (T/N_0)\sum_q{}'\xi_{q\alpha}\xi_{-q\alpha}, \; y = (T/N_0)\sum_q{}'\eta_q\eta_{-q}, \; z = (iT/N_0)\sum_q{}'\xi_{qz}\eta_{-q}, \tag{4.5}$$

where not only L but X_q^α, Y_q and Z_q are functions of ξ_0, x,y and z. If we confine ourselves to the paramagnetic phase ($T>T_C$), or the range of small ξ_0, and make a saddle point approximation for the integral over η_j, (4.5) reduces to a simple form:

$$\psi_1(\xi) = \pi T \sum_q{}' \sum_\alpha \chi_q^\alpha \xi_{q\alpha}\xi_{-q\alpha} + \pi N_0 L[x], \tag{4.6}$$

corresponding to (4.2).

In actually evaluating the functional integral, however, some approximations are necessary in order to get any tangible results. Here we will not go into details of the functional integral approach, for which we refer to [21], but give alternative arguments [45] to lead to the same results.

4.2 Spin Fluctuations and Magnetic Susceptibility

For brevity we confine ourselves here to the paramagnetic phase or above T_C. We first note that the space-time correlation function of the spin density is related with the dynamical susceptibility through the fluctuation-dissipation theorem as follows:

$$S_q^{\alpha\beta}(\omega) = \int_{-\infty}^{\infty} dt\; e^{i\omega t} \langle S_q^\alpha(t+t') S_{-q}^\beta(t')\rangle = \frac{2}{1-e^{-\omega/T}}\, Im\chi_q^{\alpha\beta}(\omega+is), \quad (s\to+0),$$

with

$$S_q = \sum_j e^{-iq\cdot R_j} S_j, \tag{4.7}$$

where α and β denote the x,y,z components. We also have

$$\langle (S_j^\alpha)^2\rangle = \frac{1}{2\pi N_0^2}\int_{-\infty}^{\infty} d\omega\, \coth\left(\frac{\omega}{2T}\right) \sum_q Im\chi_q^{\alpha\alpha}(\omega+is) = \frac{T}{N_0^2}\sum_m\sum_q \chi_q^{\alpha\alpha}(i\omega_m), \tag{4.8}$$

where $\omega_m=2\pi mT$, m being an integer. These relations should be satisfied by any correct theoretical results.

Next we consider to express the dynamical susceptibility as a function of spin correlation functions. We may write

$$\chi(Q,\Omega) = \bar{\chi}[Q,\Omega;\{S_q(\omega)\}]/\{1 - 2I\bar{\chi}[Q,\Omega;\{S_q(\omega)\}]\}, \tag{4.9}$$

where $\bar{\chi}(Q,\Omega)$ is defined by the irreducible diagrams in the temperature Green's function formalism and is calculated approximately by choosing appropriate diagrams or by a method of equations of motion. In the latter approach $\bar{\chi}$ is calculated as the dynamical susceptibility of non-interacting electrons under the influence of the fluctuating fields produced by the spin fluctuations $\{S_q(\omega)\}$. This type of theories have been developed previously and an integral equation for $\chi(q,\omega)$ has been obtained and solved for some special cases [40].

Since it is very hard to solve this equation in general, we here make a drastic simplification of assuming $\bar{\chi}(q,\omega)$ to be a function of S_L^2 only. This, as a matter of fact, is in line with the simplification of choosing the model functional (4.2) and is considered to be reasonable at least in the two limiting cases, i.e., when the spin fluctuations are localized in real and reciprocal spaces, respectively. Thus for the system with cubic symmetry we have

$$S_L^2 = \frac{3T}{N_0^2}\sum_m\sum_q \frac{\bar{\chi}(q,i\omega_m;S_L^2)}{1-2I\bar{\chi}(q,i\omega_m;S_L^2)}\,. \tag{4.10}$$

When expressions for $\bar{\chi}$ is given we can solve (4.10) for $S_L{}^2$ and then calculate $\chi(q,\omega)$ from (4.9) as functions of temperature. Here we simplify (4.10) further taking the terms with m=0 only, or making a static or high temperature approximation. We have

$$S_L^2 = \frac{3T}{2IN_0}\left\langle\frac{1-\sigma}{\delta+\sigma}\right\rangle,$$

$$\delta = 1/2I\chi(0,0) = [1/2I\bar{\chi}(0,0;S_L^2)] - 1,$$

$$\sigma_q(S_L^2) = 1 - [\bar{\chi}(q,0;S_L^2)/\bar{\chi}(0,0;S_L^2)], \tag{4.11}$$

where < > means the average over q.

The problem is now reduced to calculating $\bar{\chi}(q,0;S_L^2)$ from the band structure and solve (4.11) for $S_L{}^2$ and δ as functions of temperature and then calculate $\chi(q,0)$. Various approximation methods for calculating $\bar{\chi}(q,0;S_L{}^2)$ or the free energy functional corresponding to (4.2) or (4.5) have been developed and will be discussed later.

Before going into details of the band structure we can derive some general results of significance which will be discussed in what follows.

4.3 General Expression for the Curie Temperature

The Curie temperature is given by setting δ=0 in (4.11). We have

$$T_C = \frac{2IN_0S_L^2(T_C)}{3}\left\langle\frac{1-\sigma}{\sigma}\right\rangle_C^{-1} = \frac{2JS_L^2(0)}{3}\eta(T_C)\left\langle\frac{1-\sigma}{\sigma}\right\rangle_C^{-1},$$

$$1 - 2I\bar{\chi}[00;S_L^2(T_C)] = 0, \tag{4.12}$$

with

$$\eta(T) = S_L^2(T)/S_L^2(0). \tag{4.13}$$

This expression is shown to reproduce the results of the static approximation to the SCR theory in weakly ferromagnetic limit and the results of the spherical model in the local moment limit and thus is expected to be reasonable as a general expression.

It was shown that we have $\eta(T_C)$=3/5 in the weakly ferromagnetic limit [46] and we naturally have $\eta(T)$=1 in the local moment limit. In the intermediate range $\eta(T_C)$ should depend on the band structure and has to be calculated numerically.

It is easy to show the relation

$$\sigma_q - \langle\sigma\rangle \simeq - \mathcal{J}_q(S_L^2)/2J, \tag{4.14}$$

and in the local moment limit $\mathcal{J}_q$ reduces to the Fourier transform of the Heisenberg exchange interaction and (4.12) reduces to a familiar expression for the spherical model. Incidentally, if we replace $\langle(1-\sigma)/\sigma\rangle_C^{-1}$ in (4.12) by $\langle\sigma\rangle_C$ we get the result of the molecular field approximation. Since (4.12) takes account of the effect of short-range order the value for T_C is usually lower than the molecular field value. It was also shown that in the case of half-filled band in the strong coupling limit the corresponding expression for T_N reduced to the correct one for the Heisenberg model with the Anderson superexchange [21].

4.4 Magnetic Susceptibility — Unified View for the Two Mechanisms for the Curie-Weiss Law

The magnetic susceptibility can generally be calculated from (4.11) which may be rewritten as

$$S_L^2(T) - S_L^2(T_C) = \frac{3T}{2J}\left\langle\frac{1-\sigma}{\delta+\sigma}\right\rangle - \frac{3T_C}{2J}\left\langle\frac{1-\sigma}{\sigma}\right\rangle_C ,$$

$$\bar{\chi}[0,0;T,S_L^2(T)] - \bar{\chi}[0,0;T_C;S_L^2(T_C)] = -\frac{1}{2I}\frac{\delta}{1+\delta} . \qquad (4.15)$$

We consider here the case where the temperature variation of $S_L{}^2$ is not very large so that the second expression of (4.15) may be expanded in $S_L{}^2(T)-S_L{}^2(T_C)$. The opposite case will be discussed in the following subsection. Defining

$$-2I[\partial\bar{\chi}(0,0;T,S_L^2)/\partial S_L^2]_C = \gamma/2J = 2J/3T_0 ,$$

$$-[\partial\bar{\chi}(0,0;T,S_L^2)/\partial T]_C = \nu/2I , \qquad (4.16)$$

we get the following equation for the reduced inverse susceptibility δ:

$$\frac{\delta}{1+\delta} = \frac{T}{T_0}\left\langle\frac{1-\sigma}{\delta+\sigma}\right\rangle - \frac{T_C}{T_0}\left\langle\frac{1-\sigma}{\sigma}\right\rangle_C + \nu(T-T_C). \qquad (4.17)$$

Here $\gamma=4J^2/3T_0=\gamma_L-(\partial\mathcal{J}_Q/\partial S_L{}^2)$ is the mode-mode coupling constant between the uniform and the local spin fluctuations and is of the same order as the *longitudinal stiffness constant* γ_L for the spin density which measures the stiffness against the change in amplitude of the local spin fluctuation. In what follows we will call γ also the longitudinal stiffness constant.

This equation exhibits the above-mentioned two types of the CW law in the mutually opposite two extremes. First in the weakly ferromagnetic limit, where δ is small and T_0 is not small in general, we have for $\delta \ll \sigma$ and $\delta \ll 1$

$$\delta = (2J/3T_0)[S_L^2(T) - S_L^2(T_C)] + \nu(T-T_C)$$

$$= \left[\left\langle\frac{1-\sigma}{\sigma}\right\rangle T_0^{-1} + \nu\right](T-T_C). \qquad (4.18)$$

Since ν is usually small, the Curie constant is inversely proportional to the longitudinal stiffness constant.

In the local moment limit, we have $\gamma\to\infty$ or $T_0\to 0$ and S_L is fixed. Then (4.17) becomes

$$\left\langle\frac{1-\sigma}{\delta+\sigma}\right\rangle = \frac{T_C}{T}\left\langle\frac{1-\sigma}{\sigma}\right\rangle_C = \frac{2JS_L^2}{3T} . \qquad (4.19)$$

Expanding the left-hand side of (4.19) in terms of σ/δ for $T \gg T_C$, we get

$$\chi = N_0 S_L^2/3(1 - \langle\sigma\rangle)(T - T_C'),$$

with

$$T_C' = 2JS_L^2\langle\sigma(1 - \sigma)\rangle/3(1 - \langle\sigma\rangle)^2. \tag{4.20}$$

Note that the Curie constant is modified by the factor $1/(1-\langle\sigma\rangle)$. $\langle\sigma\rangle$ is usually quite small in magnetic insulators, where we have $\langle(1-\sigma)/\sigma\rangle^{-1}(1-\langle\sigma\rangle) = T_C/2IC$, C being the Curie constant. In magnetic metals $\langle\sigma\rangle$ is usually expected to be larger than in insulators.

Thus (4.17) is considered to be a general expression interpolating between the two limiting cases. In order to see the magnetic susceptibility in the intermediate range more explicitly we show as an example the results of calculation for the following distribution function $P(\sigma)$ for σ:

$$P(\sigma) = (2/\pi\langle\sigma\rangle^2)[\sigma(2\langle\sigma\rangle - \sigma)]^{1/2}. \tag{4.21}$$

We have

$$\frac{z}{1 + \langle\sigma\rangle z} = t(1 + \langle\sigma\rangle z)\{(1 + \langle\sigma\rangle z)(1 + z - \sqrt{z^2 + 2z}) - \frac{\langle\sigma\rangle}{2}\}$$
$$- t_C(1 - \frac{\langle\sigma\rangle}{2}),$$

with

$$z = \delta/\langle\sigma\rangle, \quad t = 2T/\langle\sigma\rangle^2 T_0, \quad t_C = 2T_C/\langle\sigma\rangle^2 T_0. \tag{4.22}$$

In particular, for small $\langle\sigma\rangle$ we have

$$z = t(1 + z - \sqrt{z^2 + 2z}) - t_C, \quad (\langle\sigma\rangle \to 0). \tag{4.23}$$

The important parameter here is t_C. When t_C is large we have local moment-like situation while the weakly ferromagnetic limit is realized for $t_C \to 0$. The results of calculation is shown in Fig.2. We see that the CW law is best obeyed both in the weakly ferromagnetic limit (t_C:small) and in the local moment limit ($t_C \to \infty$). In the intermediate range the linearity of $1/\chi$-T curves is generally worse as is naturally expected from entirely different origins of the CW law in the opposite ends.

In a rough sense, however, the CW law is obeyed even in the intermediate range. This is particularly true for relatively large $\langle\sigma\rangle$ as may be seen from Fig.2. The Curie constant is very roughly the sum of two contributions arising from the two mechanisms. This trend may easily be seen from Fig.2. If we assume that T_C is fixed, $S_L^2(T_C)$ is fixed for a given value of $\langle\sigma\rangle$. Thus the change in t_C from ∞ to 0 corresponds to that in γ from ∞ to 0 (or T_0 from 0 to ∞). Fig.2 clearly shows that the decrease in the longitudinal stiffness constant γ tends to increase the Curie constant. This is due to the increase of $S_L^2(T)$ with temperature.

In order to visualize this situation we show in Fig.3 the values of $\langle|S_q|^2\rangle$ as functions of temperature for various values of q or $\sigma_q/\langle\sigma\rangle$ for several

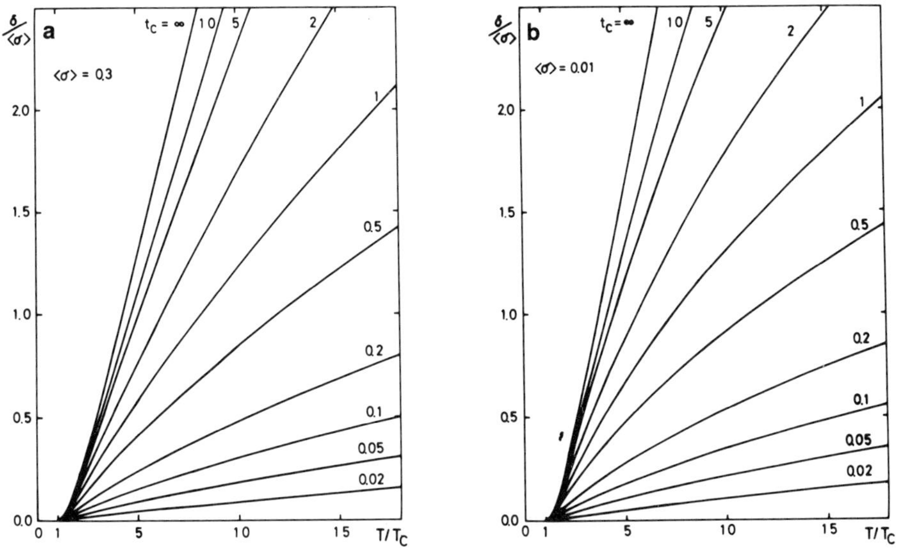

Fig.2 Reduced inverse susceptibility $z=\delta/\langle\sigma\rangle$, against reduced temperature for various values of $t_C=\langle 1/\sigma\rangle^2 T_C/2T_0$. $t_C\to\infty$ corresponds to the local moment limit and $t_C\to 0$ the weakly ferromagnetic limit (see text).

typical examples. $S_L{}^2(T)$ are also indicated in the figures by dashed lines. Fig.3a represents the weakly ferromagnetic case where small q components only are temperature independent at high temperatures and $S_L{}^2$ increases linearly with T. Fig.3d and f are examples for the local moment limit where $S_L{}^2$ is constant. All the q components become almost temperature independent at high temperatures. Fig.3b represents a typical intermediate case and Fig.3c and e are examples closer to the local moment limit. Note that the Curie constant is rather sensitive to the longitudinal stiffness constant even near the local moment limit. It should be mentioned that this effect might be somewhat exaggerated in the present approach, where the local moment condition is satisfied only on an average even in the limit of $\gamma\to\infty$, since the approximation in this limit corresponds to the spherical model rather than the Heisenberg model. It is generally important to consider the $S_L{}^2$ dependence of σ and the effect of charge density fluctuations as has been discussed in recent investigations (see §5), although general qualitative conclusions as discussed here are not altered.

We note here that the molecular field theory is derived from the present theory by simply assuming a δ function for $P(\sigma)$. In other words, it is essential to calculate $P(\sigma)$ from the band structure in order to get any information on the short-range order in actual substances, which is expected to be sensitive to the form of $P(\sigma)$ or $\{\sigma_q\}$.

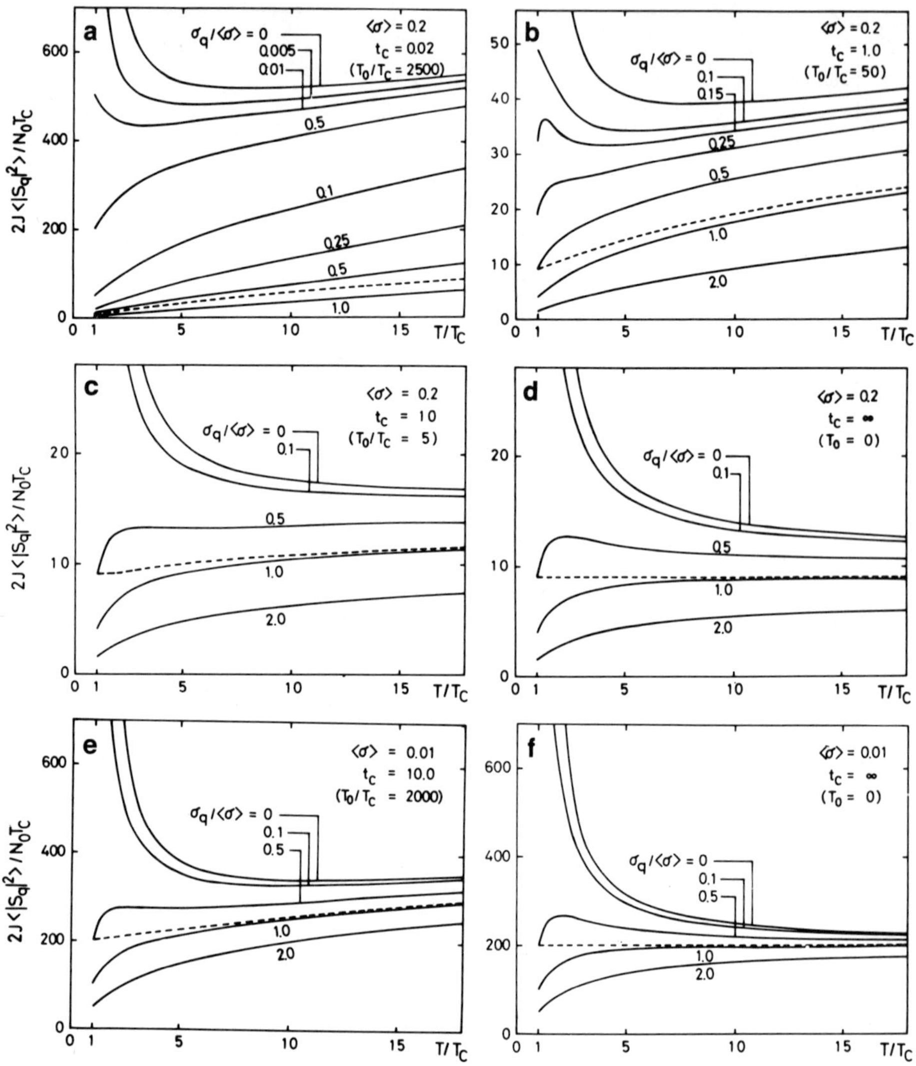

Fig.3 Temperature dependence of $<|S_q|^2>$ corresponding to the example shown in Fig.2 for various values of $<\sigma>$ and t_C. The dashed lines indicate the values for the local spin fluctuation $<S_{loc}^2>$.

4.5 Temperature-Induced Local Moments

Let us now consider the case of very small longitudinal stiffness constant where $S_L{}^2$ can increase rapidly with temperature. Since there is an upper bound $S_M{}^2$ for the value of $S_L{}^2$ associated with the number of electrons per atom, the increase with T of $S_L{}^2$ will be saturated at around certain temperature T*. We have

$$S_L^2(T) = \frac{3T}{2J} \langle\frac{1-\sigma}{\delta+\sigma}\rangle \leq S_M^2 = \frac{3T^*}{2J} \langle\frac{1-\sigma}{\delta^*+\sigma}\rangle , \tag{4.24}$$

and

$$T^* \simeq \frac{2}{3} J\, S_M^2 \, [(2I\bar{\chi}_0)^{-1} - 1 + \langle\sigma\rangle]. \tag{4.25}$$

Above T* we expect the magnetic susceptibility to obey the CW law of local moment type as described by (4.19) and (4.20). Below T* the temperature dependence of the susceptibility is expected to be weaker owing to the increase of $S_L{}^2$ with T. Thus we expect a change in slope of the $1/\chi$-T plot at around T*. Since the spin fluctuations above T* may be regarded as a set of local moments, we call this as temperature-induced local moments. It should be kept in mind, however, that the short-range order in this case is expected to be stronger than in usual magnetic insulator compounds. Also T* is only a measure of the temperature of saturation and does not indicate any phase transition. Order estimation for T* from (4.25) shows that it can be fairly low or at around room temperature in favorable cases [47].

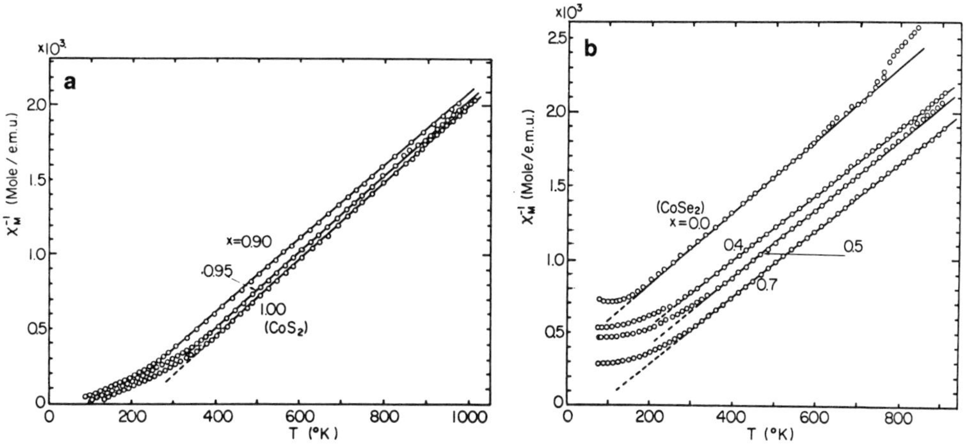

Fig.4 Inverse magnetic susceptibility for $Co(S_xSe_{1-x})_2$; (a) x>0.88, (b) x< 0.88 (from [47])

This phenomenon is considered to be observed in CoS_2, a ferromagnetic metal with T_C=120 K, and $Co(S_xSe_{1-x})_2$ system with pyrite type crystal structures [48]. Experimental data for their magnetic susceptibilities are shown in Fig. 4. In CoS_2 and $Co(S_xSe_{1-x})_2$ with x>0.88 the inversed magnetic susceptibility $1/\chi$ above T_C is linear in T and shows a break at around 300-350 K, above which $1/\chi$ is linear in T but with a larger slope corresponding to the Curie constant of the local moment system with $1\mu_B$ per Co atom. Since the electronic structure of this system is understood as having one electron per Co atom in the e_g bands with 2-fold orbital degeneracy, we may interpret this phenomenon as the temperature-induced local moments with T* at around the break in $1/\chi$-T plot.

In $CoSe_2$ and $Co(S_xSe_{1-x})_2$ with x<0.88, this phenomenon appears more dramatically. These substances are exchange-enhanced Pauli paramagnets at low temperatures. The temperature dependence of $1/\chi$ is very weak or even negative at low temperatures and above around T*≃200–300 K χ obeys a very good CW law with almost the same Curie constant as CoS_2 above T^*. Thus even in paramagnetic metals we can have temperature-induced local moments. These behaviors of the susceptibilities of $Co(S_xSe_{1-x})_2$ system had been a mystery for some time and were explained by this new concept which is one of the most interesting outcomes of the unified theory. Recently YASUOKA et al. [49] have measured the nuclear spin-lattice relaxation rate $1/T_1$ of ^{59}Co in $CoSe_2$ which increases linearly with temperature at low temperatures and then tends to saturate above around T^* as is shown in Fig.5. This is another strong indication for the occurrence of temperature-induced local moments. This mechanism may also be related with the CW susceptibilities in metallic phases of V_2O_3 and $(Ti_xV_{1-x})_2O_3$ system and possibly in Pd. YASUOKA et al [49] have also observed a trend of saturation of $1/T_1$ around room temperature in Pd.

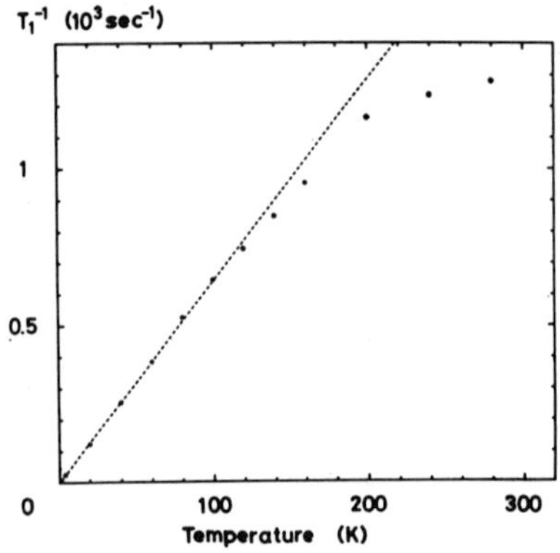

Fig.5 Temperature dependence of the nuclear spin lattice relaxation rate of ^{59}Co in $CoSe_2$ (by YASUOKA et al. [49], whom the author wishes to thank for permission to quote their result here prior to publication)

More generally, the concept of strong temperature variation of S_L^2 is expected to have quite important significance in explaining various properties of magnetic substances. For example, the magnetic and thermal properties of nearly ferro- or antiferromagnetic semiconductor FeSi have been an enigma for more than 10 years and a plausible explanation was given with the use of this concept [50]. In this case γ is expected to be negative at low temperatures and the magnetic susceptibility increases rapidly with temperature owing to the negative coupling between spin fluctuation modes. As temperature goes up γ changes its sign and at high temperatures strong positive coupling between spin fluctuation modes gives the CW behavior for the susceptibility. Negative mode-mode coupling is also expected in $CoSe_2$ and Pd at low temperatures. It should finally be noted that the saturation behavior as discussed here can be described naturally within the present theory by properly calculating $L[x]$, without an explicit use of the condition (4.24). For further details we refer to [21], [22] and [46].

4.6 A Systematic Analysis of Ferromagnetic Metals

It should be pointed out here that the above-explained theory gives a general account of magnetism in metals with the use of relatively small number of physical parameters which can mainly be deduced from a given band structure. Our parameters are $S_L^2(T)$, $\sigma_q(S_L^2)$, I and $\bar{\chi}(0,0;S_L^2)$ or the longitudinal stiffness constant $\gamma(S_L^2)=-4JI\,\partial\bar{\chi}(0,0;S_L^2)/\partial S_L^2$ as a function of S_L^2. When the temperature variation of S_L^2 is small γ may be regarded as constant and

$\{\sigma_q\}$ may be replaced by a temperature-independent value of $\langle\sigma\rangle$ and $P(\sigma)$, the distribution function for σ. $\langle\sigma\rangle$ gives a measure for the degree of delocalization of the spin fluctuation and the degree of short-range order can be deduced from $P(\sigma)$. We can make analyses of the experimental data through the above-discussed expressions with the use of the parameters $S_L{}^2(0)$, $\eta(T_C)=S_L{}^2(T_C)/S_L{}^2(0)$, I, γ and $\langle\sigma\rangle$. Although this type of simple analysis will not be justified when $S_L{}^2(T)$ changes rapidly with temperature, the physical properties are generally described by $S_L{}^2(T)$ as a function of temperature, which is controlled primarily by the longitudinal stiffness constant, and the degree of delocalization of the spin fluctuation as given by $\{\sigma_q(S_L{}^2)\}$.

We here discuss first the temperature dependence of $S_L{}^2(T)$ or $\eta(T)=S_L{}^2(T)/S_L{}^2(0)$. It is obvious that we have $\eta(T)=1$ in the local moment limit. In the weakly ferromagnetic limit a simple calculation shows $\eta(T_C)=3/5$ [46] and $\eta(T)$ increases with T above T_C. These and other typical behaviors of $S_L{}^2(T)$ are sketched in Fig.6. In general $\eta(T)$ is expected to decrease below T_C and to increase above T_C as temperature goes up. The lines a and d represent the local moment limit and the weakly ferromagnetic limit, respectively. The line b corresponds to the case of relatively large longitudinal stiffness constant γ and may be realized for example in Fe. The line c is a possible example for the case of small γ and the invar alloys may belong to this case. The lines e and f represent the cases of CoS_2 and $CoSe_2$, respectively; $S_L{}^2(T)$ is saturated above around a certain temperature T*. For quantitative purposes we have to make numerical calculations as will be discussed in the following section.

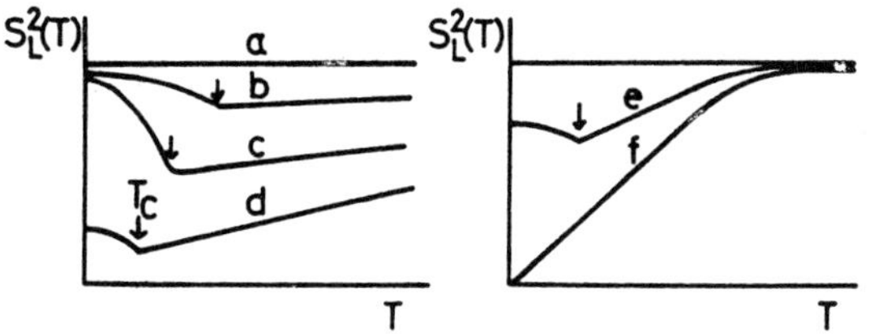

Fig.6 Typical examples for the temperature dependence of $S_L{}^2$, the average squared amplitude of the local spin fluctuation (see text)

Table 1 Physical parameters for various ferromagnetic materials. We used $J=7000\ k_B$ for all the examples and the values $\eta_C=\eta(T_C)=0.6\sim1$ were used for the numerical estimations of J/k_BT_0 and t_C for CoS_2, Fe, Co and Ni.

	T_C	p_s	$S_L^2(0)$	$p_{eff}^2/4$	$\langle 1-\sigma/\sigma\rangle_C^{-1}$	J/k_BT_0	t_C
Sc_3In	6	0.04	0.03	0.12	0.07	0.8	0.3
$ZrZn_2$	25	0.12	0.09	0.44	0.1	0.3	0.25
Ni_3Al	41.5	0.075	0.06	1.4	0.3	0.9	0.15
$Fe_{.5}Co_{.5}Si$	50	0.17	0.13	0.56	0.14	0.4	0.3
MnSi	30	0.36	0.27	1.3	0.04	0.05	0.3
CoS_2	116	0.84	0.63	0.82	$0.04/\eta_C$	$0.2\sim0.3$	$2\sim7$
Fe	1043	2.2	2.35	2.53	$0.1/\eta_C$	$0.2\sim0.8$	$2.5\sim25$
Co	1388	1.7	1.63	2.46	$0.2/\eta_C$	0.3	$1.5\sim4$
Ni	627	0.6	0.45	0.64	$0.3/\eta_C$	2	$1.5\sim5$
$CrBr_3$	33	3	15/4		0.002	∞	∞

As for the values for $\langle\sigma\rangle$, we expect from the theories as briefly mentioned in §2 that $\langle\sigma\rangle$ is relatively large when the occupied or empty fraction of d orbitals is small while $\langle\sigma\rangle$ is relatively small when the d orbitals are nearly quarter-filled. For the half-filled case $\langle\sigma\rangle$ is expected to be negative and large. We show in Table 1 the values for $\langle(1-\sigma)/\sigma\rangle^{-1}$ for various ferromagnets as estimated from the general expression (4.12) for T_C and those for $J/T_0 = 3\gamma/4J$ as estimated from the Curie constant. The latter values are given just to show the idea of the order of magnitude and should not be taken seriously, especially for the materials other than the weakly ferromagnetic metals. The values for t_C were evaluated by assuming the distribution function $P(\sigma)$ given by (4.21).

5. Calculations Based on the Band Structure

The unified theory as discussed in the previous section has turned out to be successful not only in general understanding of itinerant electron magnetism, including the generalization of the Curie-Weiss law, but also in elucidating several hitherto unexplained phenomena. In order to deal with each specific substance, however, we have to make actual calculation for the free energy functional from a given band structure. For this purpose there are two types of mutually complementary approaches. One is the approach from the long-wave limit and the other is the one from the local limit.

5.1 Approaches from the Long-Wave Limit

The simplest but rather qualitative method in this category may be to use the local density of states $\rho_L(\varepsilon)$ [22,50,51]. Calculation of $L(S_L^2)$ is simply made just as in the Stoner theory, replacing ρ and S_0, the density of states and the uniform spin polarization, by ρ_L and S_L. For the calculation of σ we need more information on the band structure. This method is useful for qualitative purposes and was actually employed for the discussions on nearly ferromagnetic semiconductors, etc., by properly assuming $\langle\sigma\rangle$ and $P(\sigma)$ [49,50].

CAPELLMANN [34,52] has previously presented a method of calculating the free energy functional with the use of HFA. He assumes S_L^2=constant and makes use of an expansion which is reasonable in the long-wave limit. KORENMAN and PRANGE [53] have presented a different method in the long-wave approximation again with fixed S_L^2. Although their assumption of fixed S_L^2 naturally limits the validity of their results, good orders of magnitude were reported for T_C of Fe and Ni. This may be related with the fact that our general expression for T_C does not explicitly include the longitudinal stiffness constant.

Tha latter authors treated, in particular, sloppy spin waves above T_C in Fe and Ni. Their model corresponds to the classical Heisenberg model with strong short-range order.

It is desired that these approaches from the long-wave limit are extended further to include the variation of S_L^2.

5.2 Approaches from the Local Limit

Let us now discuss the approaches from the local limit starting from (4.3) and (4.4). The free energy functional Ψ_1 is now expanded in terms of the local potential $\{V_j\}$. If one takes all the terms related with a single site only, and then those with two sites only, and so on, the expansion corresponds to the one discussed in §2.

We may get a better expansion by renormalizing the Green's function taking account of the effect of potentials in average [21,54]. We have a matrix eqation

$$G = (g^{-1} - \Sigma)^{-1} = F + \tilde{G}, \tag{5.1}$$

where Σ is the self-energy or the coherent potential to be determined later and F and $\tilde{G}$ are site-diagonal and site-non-diagonal parts of G, respectively. Now Ψ_1 is expanded as follows:

$$\Psi_1 = T\,\mathrm{Tr}\{\log G - \log g - \log[1 - F(V - \Sigma)] - \log(1 - \tilde{G}t)\}, \tag{5.2}$$

where

$$t = (V - \Sigma)[1 - F(V - \Sigma)]^{-1}, \tag{5.3}$$

is the t matrix which is site diagonal. We now expand Ψ_1 as

$$\Psi_1 = \Psi_1^{(1)} + \Psi_1^{(2)} + \cdots, \tag{5.4}$$

where $\Psi_1{}^{(n)}$ includes all the terms which are related with n sites. The first three terms in (5.2) corresponds to $\Psi_1{}^{(1)}$ and the last term gives $\Psi_1{}^{(n)}$ for n>1. We expect that in many cases it is a good approximation to take terms up to n=2, or those related with atomic pairs. $\Psi_1{}^{(1)}$ corresponds to the local term $\pi N_0 L[\xi_0;x,y,z]$ and $\Psi_1{}^{(2)}$ to the non-local first term of the model functional (4.5).

Now we determine Σ so as to minimize $\Psi_1{}^{(1)}$. Such a variational procedure leads to the well-known coherent potential approximation:

$$\sum_j t_j = \sum_j (V_j - \Sigma)[1 - F(V_j - \Sigma)]^{-1} = 0, \tag{5.5}$$

and at the same time we get

$$\Psi_1^{(1)} = \frac{N_0 T}{\pi}\int d\varepsilon\, \log[1 + e^{-\varepsilon/T}]\mathit{Im}[\sum_\sigma F_\sigma(\varepsilon) - \frac{2}{N_0}\sum_k g_k(\varepsilon)]. \tag{5.6}$$

We do not reproduce here the results for $\Psi_1^{(2)}$ or the expressions for $X^\alpha_{j\ell}[\xi_0;x,y,z]$, $Y_{j\ell}[\xi_0;x,y,z]$, etc. but refer to [21] and [54]. We emphasize that *we have obtained a closed-form expression for the free energy functional encompassing the limits of small and large amplitude of the spin fluctuations or the weak and strong coupling limits.* This expression was shown to lead to the RPA results in the weak coupling limit as it should and in the strong coupling limit with a half-filled band it reduces to the local moment system with the Anderson super-exchange interaction.

Thus we have given a method of actually calculating the thermodynamical properties of magnetic substances from a given band structure. Since the free energy functional obtained here has the form as discussed in the previous section, all the general conclusions in §4 are now well founded. Note that *the effect of short-range order is taken into account in this approximation.* Actual calculation along this line, however, is rather laborious and the case of a tight-binding band in a simple cubic lattice has been studied so far.

Before showing these results we will briefly mention some simpler approximations. The simplest approximation may be the molecular field approximation where the spin fluctuation is assumed to be local and thus the effect of short-range order is totally neglected. HASEGAWA [55] calculated the free energy functional corresponding to the local term $L[\xi_0;x]$ in (4.5) for Fe by using a single site CPA and evaluated the functional integral by a saddle point approximation for the local field variables. He obtained $\langle S_0 \rangle$ and $S_L{}^2$ as functions of temperature and discussed various physical properties. In particular the value of $S_L{}^2$ is shown to be reduced with increasing temperature up to T_C. Although the conclusions of this approximation are qualitatively reasonable in a rough sense, this approximation always gives a constant value for $S_L{}^2$ above T_C which is not reasonable in view of the general result in the previous section or we cannot presuppose in general a fixed value for $S_L{}^2$ above T_C. Also, applicability of this approximation was found to be rather limited because of the necessary requirement for the existence of two saddle points for the local field variable which is taken to be unidirectional. HUBBARD [56] made a calculation for the free energy functional for Fe for a special configuration where the spin direction at one site is rotated relative to the ferromagnetic moment in the ground state and evaluated T_C from the analogy of the functional to the Heisenberg model.

A simple calculation which follows more closely the above mentioned unified theory has been given by USAMI and MORIYA [57]. In place of calculating all the non-local terms in (4.5) or all the σ_q, they calculated the uniform and local irreducible susceptibilities, $\bar{\chi}_0(S_L{}^2)$ and $\bar{\chi}_L(S_L{}^2)$, respectively, with the use of CPA. From the relation:

$$\langle\sigma\rangle = 1 - [\bar{\chi}_L(S_L^2)/\bar{\chi}_0(S_L^2)], \tag{5.7}$$

and the assumed form (4.21) for $P(\sigma)$ they made crude but simple calculations for various physical quantities. We show in Fig.7 the results of calculations for $\langle\sigma\rangle$ vs. electron occupation number for d metals with body-centered and face-centered cubic structures, respectively. The sign of $\langle\sigma\rangle$ is consistent with the general rule mentioned in §2 and indicates that Fe is ferromagnetic in b.c.c. lattice and antiferromagnetic in f.c.c. lattice. Furthermore the values for $\langle\sigma\rangle$ for Fe, Co and Ni are roughly consistent with those in Table 1. If we use (4.21) for $P(\sigma)$ the values for T_C for Fe and Ni are estimated as $\sim$1000 K and $\sim$500 - 1000 K, respectively, while the use of a δ function for $P(\sigma)$ corresponding to the molecular field approximation leads to about twice of these values. As was mentioned already in the previous section we cannot draw any definite conclusion on the degree of short-range order before actually calculating $P(\sigma)$ from the band structure. However, the fact that the use of (4.21) for $P(\sigma)$ rather than a δ function leads to a better agreement with experiment seems to indicate the possible importance of the short-range order effects in Fe and Ni. The magnetic susceptibilities above T_C as calculated by using (4.21) for $P(\sigma)$ also compare fairly well qualitatively with experiment. The contribution of the temperature variation of $S_L{}^2$ to the Curie constant seems to be fairly large in Co and Ni and rather small in Fe.

This type of calculation has also been extended to the case of invar alloys with roughly reasonable results [46]. It has been well known that when we increase the number of holes in the f.c.c. d bands there occurs an instability of ferromagnetism at a certain point near the invar region. It was shown by calculation that the value of $\eta(T_C)$ as defined in the previous section decreased significantly as this instability was approached. For details we refer to [46].

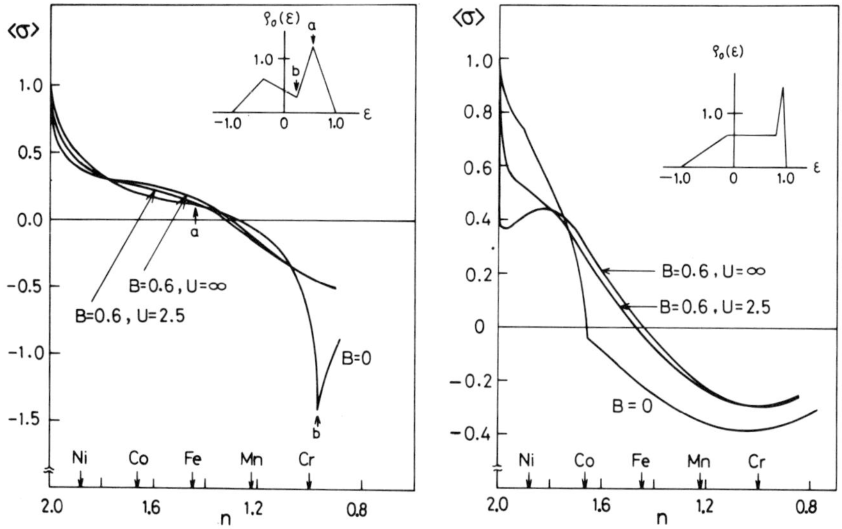

Fig.7 The values of <σ> against electron occupation number for the d metals with (a) b.c.c. and (b) f.c.c. structures. B is the magnitude of the local field giving rise to the local spin fluctuation.

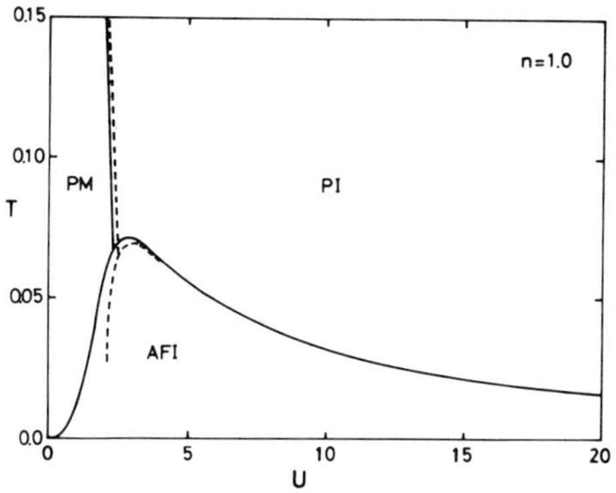

Fig.8 Theoretical phase diagram for the half-filled simple cubic tight-binding band. AFI, PM and PI correspond to antiferromagnetic insulator, paramagnetic metal and paramagnetic insulator phases, respectively. U and T are in units of the transfer integral.

A calculation with the use of the full closed form expression for the free energy functional has been carried out for a simple cubic tight-binding band [54]. When the band is half-filled we get a phase diagram as shown in Fig.8 covering the metallic paramagnetic phase, the paramagnetic and antiferromagnetic insulator phases, indicating the Mott transition at the phase boundary. This seems to be the first calculation of the entire phase diagram of this kind covering between the weak and strong coupling limits continuously. In the both ends the theory reproduces known correct results.

This theory was also applied to the cases of general occupation of electrons. In particular, the coupling between the local spin fluctuations has been calculated for various electron occupation numbers and for various values for the amplitude of the local spin fluctuation. The sign of the coupling obeys the rule derived previously and mentioned in §2 just as expected. The range of coupling decreases with increasing amplitude of the local spin density. It is highly desired that this type of calculation is carried out for actual ferromagnetic metals with the use of realistic band structures.

We would like to mention here some recent work based on the local limit approach. Recently HUBBARD [58] extended his molecular field calculation for Fe by actually evaluating the functional integral numerically. According to this calculation the amplitude of local spin fluctuation is more or less constant and the value for T_C is 1800 K. HASEGAWA [59] also improved his local saddle point approximation for Fe by taking account of the fluctuations around the saddle points and applied it to Ni, too. S_L^2 now increases slowly with temperature above T_C as in the above-mentioned results by Usami and Moriya but T_C for Fe and Ni are evaluated as 2500 K and 1400 K, respectively. These high values for T_C obtained by molecular field approximations seem to agree with the Usami-Moriya's values with the δ function for $P(\sigma)$ and may indicate the importance of the short-range order in Fe and Ni. Hasegawa also applied this molecular field approximation to the case of a half-filled band and obtained qualitatively similar results to the previous ones by Moriya and Hasegawa. Possible merits of these molecular field or local limit approximations may be its simplicity and the results are expected to be good when the effect of short-range order is insignificant.

It should be noted finally that the closed form expression for the free energy functional as mentioned above, though enables us to take account of the effect of short range order, is still an approximation from the local limit for the spin fluctuations. This is because we determine the coherent potential $\Sigma(S_L^2)$ within a single site approximation and then make an expansion in pairwise terms, etc., even though S_L^2 as a function of temperature is determined by taking account of the effect of short-range order. Thus for a more consistent theory we should determine the coherent potential taking account of the short-range order. This type of extension of the theory may be required in quantitative discussions of some real materials. In the case of strong short-range order, the extension of the approach from the local limit in q space is also quite important. These two types of approaches should be extended further in order to get quantitative understanding of magnetism in metals.

6. Concluding Remarks

It seems to be very clear now that the physical properties of magnetic materials in general at finite temperatures are described in terms of exchange-enhanced and mutually interacting modes of spin density fluctuations in a very general sense, ranging between the local moment limit and the weakly ferro- and antiferromagnetic limit as was shown in Fig.1.

The unified picture discussed here is considered to be a sound one covering both metals and insulators. This theory has given several new physical results including the general expression for the Curie temperature, a unified explanation for the Curie-Weiss type susceptibility, temperature-induced local magnetic moment, etc., elucidating some of hitherto unexplained phenomena. However, the use of the static approximation naturally limits its validity especially at low temperatures. Also it seems to be essential to take account of the quantum effect in order to give quantitative explanations of experimental results.

Existing theoretical descriptions of the spin fluctuations including the dynamical properties are more or less satisfactory for the limiting cases in the both ends. However, the theories so far developed for the intermediate range, shown in Fig.1b, including those for Fe and Ni are practically confined to the adiabatic or static approximation. It is one of the most important problems in future to work out a theory for the intermediate situation taking account of the dynamical effects of spin fluctuations including the quantum effect.

References

1. P. Langevin: J. Phys. (France) 4, 678 (1905).
2. P. Weiss: J. Phys. (France) 6, 667 (1907).
3. J. H. Van Vleck: *The Theory of Electric and Magnetic Susceptibilities* (Clarendon Press, Oxford, 1932).
4. W. Heisenberg: Z. Phys. 49, 619 (1928).
5. P. W. Anderson: Phys. Rev. 115, 2 (1959).
6. N. F. Mott: Proc. Phys. Soc. (London) A62, 416 (1949).
7. F. Bloch: Z. Phys. 57, 545 (1929).
8. E. C. Stoner: Proc. Roy. Soc. A154, 656 (1936); A165, 372 (1938).
9. J. C. Slater: Phys. Rev. 49, 537, 931 (1936).
10. J. H. Van Vleck: Rev. Modern Phys. 25, 220 (1953).
11. P. W. Anderson: Phys. Rev. 124, 41 (1961).
12. J. Friedel: Canad. J. Phys. 34, 1190 (1956); Nuovo Cimento, Suppl. 2, 287 (1958).
13. P. A. Wolff: Phys. Rev. 124, 1030 (1961).
14. B. T. Matthias and R. M. Bozorth: Phys. Rev. 100, 604 (1958).
15. B. T. Matthias, A. M. Clogston, H. J. Williams, E. Corenzwit and R. C. Sherwood: Phys. Rev. Lett. 7, 7 (1961).
16. D. M. Edwards and E. P. Wohlfarth: Proc. Roy. Soc. A303, 127 (1968).
17. C. Herring and C. Kittel: Phys. Rev. 81, 869 (1951); C. Herring: Phys. Rev. 87, 60 (1952).
18. T. Izuyama, D. J. Kim and R. Kubo: J. Phys. Soc. Japan 18, 1025 (1963).
19. T. Moriya and A. Kawabata: J. Phys. Soc. Japan 34, 639 (1973); 35, 669 (1973); Proc. Intern. Conf. on Magnetism, Moscow, IV, 5 (1973).
20. K. K. Murata and S. Doniach: Phys. Rev. Lett. 29, 285 (1972).
21. For a recent review see: T. Moriya: J. Magn. Magn. Mat. 14, 1 (1979).
22. T. Moriya and Y. Takahashi: J. Phys. Soc. Japan 45, 397 (1978); J. Phys. (France) 39, C6-1466 (1978).
23. S. Alexander and P. W. Anderson: Phys. Rev. 133, A1594 (1964).
24. T. Moriya: Solid State Commun. 2, 239 (1964); Prog. Theor. Phys. 33, 157 (1965); see also: *Theory of Magnetism in Transition Metals,* ed. by W. Marshall (Academic Press, New York, 1967) p.206.
25. B. Caroli: J. Phys. Chem. Solids 28, 1427 (1967).
26. M. Inoue and T. Moriya: Prog. Theor. Phys. 38, 41 (1967).
27. J. R. Schrieffer, W. E. Evenson and S. Q. Wang: J. Phys. (France) 32, C1 (1971).
28. R. L. Stratonovitch: Soviet Phys. Doklady 2, 416 (1958); J. Hubbard: Phys. Rev. Lett. 3, 77 (1959).
29. W. E. Evenson, S. Q. Wang and J. R. Schrieffer: Phys. Rev. B2, 2604 (1970).
30. J. Friedel, G. Lehman and M. Olszewski: J. Appl. Phys. 32, 325S (1961).
31. P. Lederer and A. Blandin: Phil. Mag. 14, 363 (1966).
32. M. Cyrot: J. Phys. (France) 33, 125 (1972); Phil. Mag. 25, 1031 (1972).
33. P. Lacour-Gayet and M. Cyrot: J. Phys. C7, 400 (1974).
34. H. Capellmann: J. Phys. F4, 1466 (1974).
35. S. Doniach and S. Engelsberg: Phys. Rev. Lett. 17, 750 (1966).

36. N. Berk and J. R. Schrieffer: Phys. Rev. Lett. 17, 433 (1966).
37. T. Moriya: Phys. Rev. Lett. 24, 1433 (1970); 25, 197 (1970).
38. M. T. Beal-Monod, S.-K. Ma and D. R. Fredkin: Phys. Rev. Lett. 20, 929 (1968).
39. J. A. Hertz and M. A. Klenin: Phys. Rev. B10, 1084 (1974).
40. T. Moriya: J. Phys. Soc. Japan 40, 933 (1976); Physica 91B, 235 (1977).
41. J. A. Hertz and M. A. Klenin: Physica 91B, 49 (1977).
42. H. Hasegawa and T. Moriya: J. Phys. Soc. Japan 36, 1542 (1972).
43. W. Weller: Phys. Status Solidi (b) 54, 611 (1972).
44. S. H. Liu: Phys. Rev. B15, 4281 (1977); B17, 3629 (1978).
45. T. Moriya: Prog. Theor. Phys. Suppl. to be published.
46. T. Moriya and K. Usami: Solid State Commun. 34, 95 (1980).
47. T. Moriya: Solid State Commun. 26, 483 (1978).
48. K. Adachi, K. Sato and M. Takeda: J. Phys. Soc. Japan 26, 631 (1961).
49. H. Yasuoka: private communication.
50. Y. Takahashi and T. Moriya: J. Phys. Soc. Japan 46, 1451 (1979).
51. Y. Takahashi: thesis, University of Tokyo 1979 (unpublished).
52. H. Capellmann: Z. Phys. B34, 29 (1979).
53. V. Korenman, J. L. Murray and R. E. Prange: Phys. Rev. B16, 4032, 4048, 4058 (1977); V. Korenman and R. E. Prange: Phys. Rev. B19, 4691, 4698 (1979).
54. T. Moriya and H. Hasegawa: J. Phys. Soc. Japan 48, 1490 (1980).
55. H. Hasegawa: J. Phys. Soc. Japan 46, 1504 (1979).
56. J. Hubbard: Phys. Rev. B19, 2626 (1979).
57. K. Usami and T. Moriya: J. Magn. Magn. Mat. 20, 171 (1980).
58. J. Hubbard: Phys. Rev. B20, 4584 (1979).
59. H. Hasegawa: J. Phys. Soc. Japan 49, 178; 936 (1980).

Calculation of the Magnetic Properties of Iron and Nickel by the Functional Integral Method

John Hubbard*

IBM Research Laboratory, San Jose, CA 95193, USA

1. Introduction

This talk will review the application of the functional integral method to the theory of magnetic systems from a somewhat personal point of view. Some effort is made to discuss the technical difficulties and developments and to compare alternative approaches. The strength of the method is that it provides a mathematical formalism to describe a physical model discussed in section 2. The method is formulated in section 3 and an inherent ambiguity is discussed. The role of Coulomb interactions is described. Section 4 deals briefly with the coherent potential approximation (CPA) approach to the associated problem of electron motion in random fields. Some alternative approaches are mentioned. In section 5 the results of calculations on iron and nickel are reported and in section 6 some conclusions are drawn.

2. The Physical Model

The itinerant model postulates the existence in the ferromagnet of an exchange field which leads to different band structures for the up and down spin electrons, and which is itself generated by the resulting differences in up and down spin electron occupation numbers. The nature of the exchange field postulated by the simple itinerant model is illustrated schematically in Fig.1. The field is the same at all atoms, has its maximum value at the temperature $T = 0$, decreases as T increases, and finally becomes zero at and above the Curie temperature T_c.

The unified theory to be described is similar to the itinerant model, but postulates a more general kind of behaviour for the exchange field. First it recognizes that since the exchange field at an atom depends upon the spin of that atom, which is a vector quantity, the exchange field really has the character of a vector (in spin space). The itinerant model assumes that these vector exchange fields are parallel at all atoms, as illustrated in Fig.1. To obtain the unified theory we abandon this assumption and allow the possibility that the atomic exchange fields might vary in direction and strength from atom to atom as illustrated in Fig.1. This opens up the possibility that above T_c the exchange fields may not vanish, but merely be randomized in direction, producing a net zero magnetization. In fact, the model is sufficiently general to admit either this possibility or the ordinary itinerant picture in which the exchange fields vanish in the paramagnetic regime. On the other hand, we suppose that at $T = 0$ the exchange

---

**John Hubbard* died on November 27, 1980, and this paper has been prepared from his notes by *D.M. Edwards*

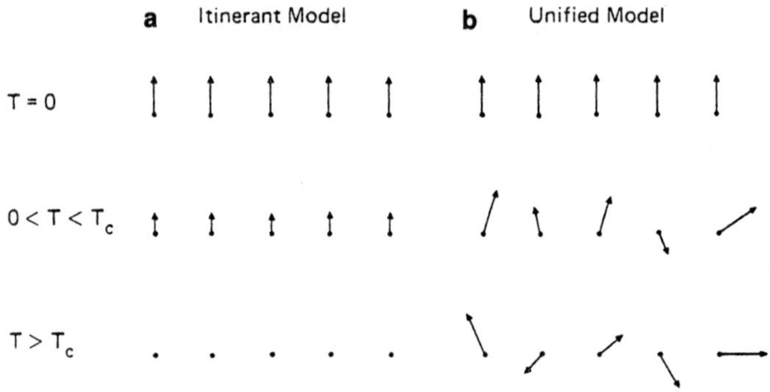

Fig.1 Exchange field configurations in (a) the simple itinerant model and (b) the unified model

field configuration coincides with that of the simple itinerant model (see Fig.1).

The manner in which the model unifies the itinerant and localized theories may now be seen. As in the itinerant model we suppose the electrons to move about from atom to atom, albeit under the influence of a different kind of exchange field. One relates to the localized model by thinking of the different exchange field configurations illustrated in Fig.1 as corresponding to the spin configurations of, say, the Heisenberg model (although, of course, the exchange field configurations are more general than those of the Heisenberg model since the exchange fields may vary in magnitude as well as direction from atom to atom whereas in the Heisenberg model a spin may only vary its direction).

In the detailed calculations reported in section 5 we make use of the single-band Hubbard model

$$H = \sum_{ij\sigma} t_{ij} c^{\dagger}_{i\sigma} c_{j\sigma} + U \sum_{i} n_{i\uparrow} n_{i\downarrow} \quad . \tag{1}$$

In the transition metals iron and nickel this represents one of five d subbands, which are assumed to be independent and equivalent. The density of states corresponding to the one-electron term h_0 in (1) is thus taken to be one fifth of the total d-band density of states of the metal.

3. The Functional Integral Method

The basic mathematical formulation of the above physical picture was provided ten years ago by SCHRIEFFER et al. [1] and CYROT [2] using the functional integral method [3]. The procedure is to write the Hamiltonian (1) in the form

$$H = h_0 - \sum_{\alpha} a_{\alpha} X_{\alpha}^2 \tag{2}$$

where X_α is a one-electron operator (bilinear in the $c^{\dagger}_{i\sigma}$ and $c_{j\sigma}$). The exact grand partition function may then be expressed as

$$Z \propto \int \prod_\alpha [Dv_\alpha(t)] \mathrm{Tr}\, \exp_t \{- \int_0^\beta [\sum_\alpha \frac{v_\alpha^2(t)}{a_\alpha} + H_0(t) + 2\sum_\alpha v_\alpha(t) X_\alpha(t)] dt\} \tag{3}$$

where

$$H_0 = h_0 - \mu N \quad . \tag{4}$$

Here N is the total number operator and μ the chemical potential, $v_\alpha(t)$ is a "field" defined for t on $(0,\beta)$, $Dv_\alpha(t)$ indicates a functional integration over $v_\alpha(t)$ and $\exp_t$ is the time-ordered exponential. We make the static approximation, in which the time fluctuating part of the fields $v_\alpha(t)$ is neglected, and write the functional integral as an ordinary integral:

$$Z \propto \int \prod_\alpha dv_\alpha \exp [-\beta \sum_\alpha \frac{v_\alpha^2}{a_\alpha}] Q(v_1, v_2, \ldots) \tag{5}$$

$$Q(v_1, v_2, \ldots) = \mathrm{Tr} \exp(-\beta H_0 - 2\beta \sum_\alpha v_\alpha X_\alpha) \quad . \tag{6}$$

A fundamental ambiguity of the method arises from the fact that it is possible to write the Hamiltonian (1) in the form (2) in many ways. An exact evaluation of (3) must always give the same result but, owing to the static approximation, this is no longer true of (5). Some alternative ways of writing the interaction term of (1) in the form $-\sum a_\alpha X_\alpha^2$ are as follows:

$$U\sum_i n_{i\uparrow} n_{i\downarrow} = \tfrac{1}{2} U \sum_i n_i - 2U \sum_i (S_i^z)^2 \tag{7a}$$

$$= \tfrac{1}{4} U \sum_i n_i^2 - U \sum_i (S_i^z)^2 \tag{7b}$$

$$= \tfrac{1}{4} U \sum_i n_i^2 - \frac{1}{3} U \sum_i \underline{S}_i^2 \tag{7c}$$

$$= \tfrac{1}{4} U \sum_i n_i^2 - U \sum_i (\underline{e}_i \cdot \underline{S}_i)^2 \quad . \tag{7d}$$

Here $n_i = n_{i\uparrow} + n_{i\downarrow}$, $S_i^z = \frac{1}{2}(n_{i\uparrow} - n_{i\downarrow})$ and $\underline{S}_i$ is the total spin of atom i. In case (a) the first term may be absorbed in h_0. In case (d) $\underline{e}_i$ is an arbitrary unit vector and one can restore rotational symmetry by averaging the partition function over all directions of each $\underline{e}_i$. Of the above cases the only other one with rotational symmetry is (c). Thus in cases (a) and (b) there is only one operator X_α associated with spin for each site so that we have scalar fields v_i. Since the fields associated with spin in the functional integral method are to be considered the exchange fields of the physical model discussed in section 2, they should be vector quantities. Form (a) has been used by SCHRIEFFER et al. [1] and CYROT [2], (b) by HASEGAWA [4], (c) by CYROT [5], (d) by HUBBARD [6-9] and PRANGE and KORENMAN [10]. Of the two rotationally invariant methods (d) has the advantage that at $T = 0$ it leads to the correct Hartree-Fock equations for the Hamiltonian (1) [6]. In section 5 we report results of calculations which use generalisations of both (c) and (d). These generalisations effectively replace (7c) and (7d), the interaction terms in (1), by

$$\tfrac{1}{2}C\sum_i n_i^2 - K\sum_i \underline{S}_i^2 \tag{8c}$$

$$\tfrac{1}{2}C\sum_i n_i^2 - K\sum_i (\underline{e}_i.\underline{S}_i)^2 \tag{8d}$$

respectively. However this is done at the stage of equations (5) and (6) and direct correspondence with any single-band Hamiltonian is abandoned. The introduction of two interaction parameters C and K enables us later, by a suitable choice of C, to enforce charge neutrality at each site. In case (d) equations (5) and (6) become

$$Z \propto \int \prod_i (d\phi_i dv_i) \exp\left[-\frac{\beta}{2C}\sum_i \phi_i^2 - \frac{\beta}{K}\sum_i v_i^2\right] Q \tag{9}$$

$$Q(v_1,v_2,\ldots;\phi_1,\phi_2,\ldots) = \mathrm{Tr}\,\exp\{-\beta H_0 - \beta\sum_i (i\phi_i n_i + 2v_i\underline{e}_i.\underline{S}_i)\} \quad . \tag{10}$$

Here ϕ_i and v_i are the fields associated with the operators $in_i/2$ and $\underline{e}_i.\underline{S}_i$ respectively. We now average over all directions of each $\underline{e}_i$ and find, on writing $v_i\underline{e}_i = \underline{v}_i$,

$$Z \propto \int \prod_i (d\phi_i d^3v_i)\left[\prod_i \frac{1}{v_i^2}\right] \exp\left[-\frac{\beta}{2C}\sum_i \phi_i^2 - \frac{\beta}{K}\sum_i v_i^2\right] Q \tag{11}$$

$$Q(\underline{v}_1,\underline{v}_2,\ldots;\phi_1,\phi_2,\ldots) = \mathrm{Tr}\,\exp\{-\beta H_0 - \beta\sum_i (i\phi_i n_i + 2\underline{v}_i.\underline{S}_i)\} \quad . \tag{12}$$

In case (c), where the vector field $\underline{v}_i$ is introduced directly, we find the same equations but without the factor $\prod v_i^{-2}$ in (11). In previous work on iron [6,7], and in initial work on nickel [8,9], it was stated that method (d) was used. However the factor $\prod v_i^{-2}$ was omitted in the calculations so that the results correspond to case (c). In section 5 results for nickel are reported with method (d) applied correctly.

To evaluate the ϕ_i integrals in (11) approximately we use the method of steepest descent, thus determining the ϕ_i as functions of $\underline{v}_1,\underline{v}_2,\ldots$. We further assume that Coulomb forces are sufficiently powerful to enforce perfect charge neutrality, which amounts to the approximation $C \to \infty$. The fields ϕ_i determined by steepest descent are pure imaginary so it is convenient to introduce related fields w_i which are real. Following appendix A of ref. [7] we find

$$Z \propto \int \prod_i d^3v_i \begin{bmatrix} 1 \\ 1/\prod_i v_i^2 \end{bmatrix}$$

$$\times \exp\{-\frac{\beta}{K}\sum_i v_i^2 + \beta n_d \sum_i w_i - \beta F(\underline{v}_1,\underline{v}_2,\ldots)\} \tag{13}$$

where

$$e^{-\beta F} = \mathrm{Tr}\,\exp\{-\beta[H_0 + \sum_i (w_i n_i + 2\underline{v}_i.\underline{S}_i)]\} \quad . \tag{14}$$

Here n_d is the number of electrons per atom and the w_i are determined as functions of $\underline{v}_1, \underline{v}_2, \ldots$ by the charge neutrality conditions $\langle n_i \rangle = n_d$. The expectation value $\langle n_i \rangle$ is calculated in the fields $\underline{v}_1, \underline{v}_2, \ldots; w_1, w_2, \ldots$ i.e. using the one-electron Hamiltonian contained in square brackets in (14). Clearly F is the thermodynamic potential for this Hamiltonian. In (13) the upper factor in square brackets corresponds to method (c) described above and the lower factor to method (d).

The parameter K is determined to give the correct magnetic moment in the ferromagnetic ground state. In this state w_i and $\underline{v}_i$ are independent of i so the w_i terms in (13) and (14) cancel and $\underline{v}_i = \underline{\mu}$, a constant vector. As $T \to 0$ $(\beta \to \infty)$ the integral (13) is dominated by the configuration $\underline{v}_i = \underline{\mu}$ which minimises the coefficient of $-\beta$ in the exponent. Thus

$$\frac{1}{K}\underline{\mu} + \tfrac{1}{2}\left[\frac{\delta}{\delta \underline{v}_1} F(\underline{v}_1, \underline{v}_2, \ldots)\right]_{\underline{v}_i = \underline{\mu}} = 0 \tag{15}$$

or, from (14),

$$\underline{\mu} = K\langle \underline{S}_i \rangle \quad . \tag{16}$$

$\langle \underline{S}_i \rangle$ is the moment calculated in the uniform exchange field $\underline{\mu}$ so, for a given band structure and K, (16) determines the ground state moment self-consistently. K is adjusted to give the observed value and, using band structures by TAWIL and CALLAWAY [11] and CALLAWAY and WANG [12], we find $K = 0.9$eV for iron and 1.32eV for nickel.

4. The Calculation of F

The problem of calculating F, defined by (14), is that of calculating the thermodynamic potential of non-interacting electrons in random potential fields. The simplest case is the impurity problem in which the field $\underline{v}$ of only one atom deviates from the ground state value of $\underline{\mu}$. Here a precise solution is available [6,7]. A more general approach is CYROT's cluster expansion [2]

$$F(\underline{v}_1, \underline{v}_2, \ldots) = \sum_i F_1(\underline{v}_i) + \sum_{i \neq j}\sum F_2(\underline{v}_i, \underline{v}_j) + \ldots \quad . \tag{17}$$

Yet another method is due to PRANGE and KORENMAN [10]. They assume the only configurations $\underline{v}_1, \underline{v}_2, \ldots$ which contribute importantly are those in which $|\underline{v}_i| = |\underline{\mu}|$ and where the angle between $\underline{v}_i$ and $\underline{v}_j$ is small if i,j are near neighbours. F can then be expanded in terms of these small angles.

A method which is well adapted to the present problem is to adopt a mean-field approach in which correlations between fields at different atoms are neglected. This corresponds to retaining only the first term in (17). It is plausible from (13), and shown in appendix B of ref. [7], that the distribution of exchange fields at each atom is given by

$$f(\underline{v}) \propto \exp[-\beta E(\underline{v})] \quad , \tag{18}$$

where

$$E(\underline{v}) = K^{-1}v^2 - n_d w(\underline{v}) + \langle F \rangle_{\underline{v}} \tag{19}$$

with $w(\underline{v})$ chosen so that

$$\langle\langle n_0 \rangle\rangle_{\underline{v}} = n_d \quad . \tag{20}$$

Here $\langle X \rangle_{\underline{v}}$ means that X is to be calculated for the configuration with fields $\underline{v}_0 = \underline{v}$, $w_0 = w(\underline{v})$ at site 0 and the fields $\{\underline{v}_i\}$, $\{w(\underline{v}_i)\}$ at other sites, and is then to be averaged over the $\underline{v}_1, \underline{v}_2, \ldots$ at sites other than 0 with the distribution function (18). In carrying out these averages over $\underline{v}$ space a weight factor 1 or v^{-2} is to be introduced into the integrals depending on whether method (c) or (d) is used (cf. (13)). One has an involved self-consistent problem; the definitions of $E(\underline{v})$ and $w(\underline{v})$ depend on $f(\underline{v})$ through the $\langle \ldots \rangle_{\underline{v}}$ averaging process and $f(\underline{v})$ depends on $E(\underline{v})$ through (18). The averages involved are conveniently carried out using the CPA and calculations of this type have been performed by HASEGAWA [4] and HUBBARD [7-9]. The results of the author's calculations on iron and nickel are reported in the next section.

5. Results

The only input is the density-of-states curves of refs. [11] and [12] and the values of K , chosen to give the correct ground state moments. The average moment per atom may be calculated according to

$$\underline{M} \propto \int \langle\langle \underline{S} \rangle\rangle_{\underline{v}} f(\underline{v}) g(\underline{v}) d^3v \;/\; \int f(\underline{v}) g(\underline{v}) d^3v \tag{21}$$

where

$$g(\underline{v}) = \begin{cases} 1 \quad , & \text{method (c)} \\ v^{-2} \quad , & \text{method (d)} \quad . \end{cases}$$

The paramagnetic susceptibility $\chi(T)$, above the Curie temperature T_c , may be similarly calculated by introducing an external magnetic field. It is also possible to evaluate at any temperature the distribution function

$$P(S) = \int \delta(S - |\langle\langle \underline{S} \rangle\rangle_{\underline{v}}|) f(\underline{v}) g(\underline{v}) d^3v \;/\; \int f(\underline{v}) g(\underline{v}) d^3v \tag{22}$$

giving the probability that an atom has a spin of magnitude S . The root-mean-square average $[\langle S^2 \rangle]^{\frac{1}{2}}$ of the atomic spin length may then be determined as a function of temperature. We now summarize the numerical results obtained.

Values of T_c obtained are given in the following table:

Method	Metal	Calculated (K)	Observed (K)
(c)	Fe	1840	1044
(c)	Ni	1200	630
(d)	Ni	578	630

The paramagnetic susceptibility $\chi(T)$ was calculated by method (c) only. A very good fit to the Curie-Weiss law

$$\chi = \frac{1}{3} m_{eff}^2/(T-T_c) \tag{23}$$

was obtained and values of m_{eff} were $2.6\mu_B$ and $0.49\mu_B$ for iron and nickel, respectively.

The temperature dependence of the magnetization is shown in Fig.2. There is a significant difference between methods (c) and (d) for nickel, the latter being in better agreement with experiment. This difference between the two methods is also seen in Fig.3 where the root-mean-square atomic spin is plotted as a function of temperature. This quantity was found to be almost exactly constant for iron, using method (c). Several spin distribution functions P(S) are shown in Fig.4 and the difference between methods (c) and (d) for nickel is particularly striking. The quite different behaviour at small S reflects the different behaviour of $g(\underline{v})$ at small v .

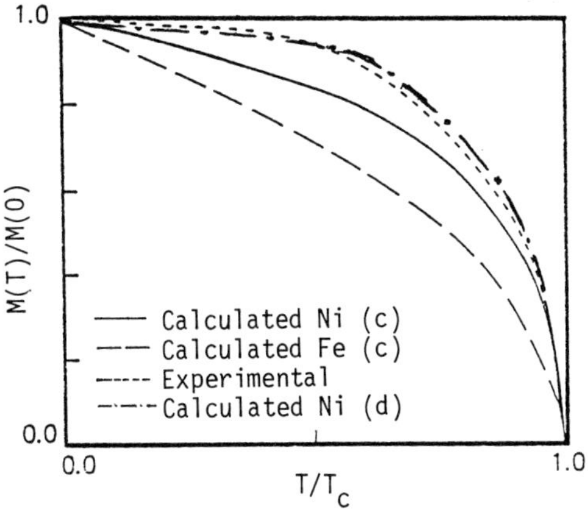

Fig.2 The calculated magnetization curves for nickel and iron compared with the experimental curve for nickel

6. Conclusions

The functional integral method has won us some ground in the long struggle to understand the magnetism of iron and nickel. It combines a correct band-theoretical description of the ground state, yielding the correct ground state moment, with a Curie temperature of the right order of magnitude. It predicts a Curie-Weiss susceptibility and the existence of atomic moments

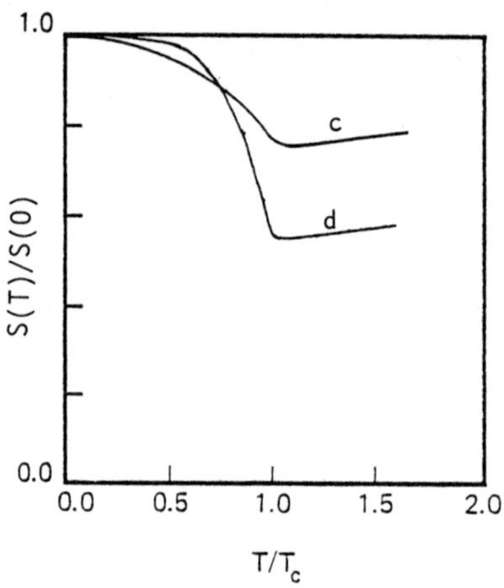

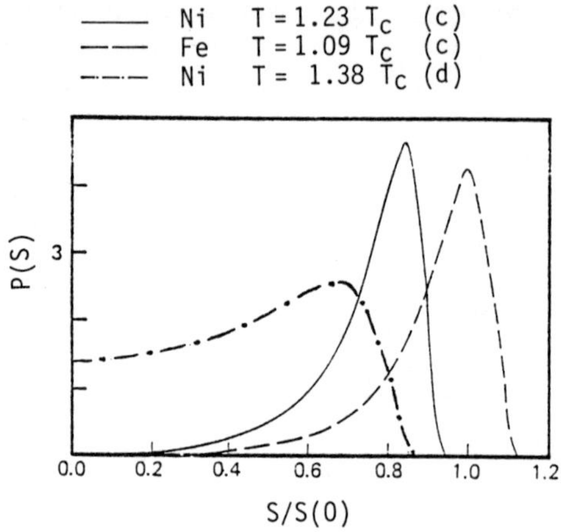

Fig.3 The root-mean-square atomic spin magnitude S(T) calculated for nickel by methods (c) and (d), as a function of temperature; S(0) equals the ground state spin

Fig.4 The distribution functions P(S) for the atomic spin lengths calculated for iron and nickel in the paramagnetic region; S(0) is the ground state atomic spin and curves (c) for nickel and iron show P(S)S(0) rather than P(S).

above T_C of the same order as the ground state moment. However the ambiguity of present theories, which arises from the static approximation, has been stressed. Two different methods of calculation for nickel yield values of T_C differing by a factor 2, significantly different magnetization-temperature curves and quite different atomic spin distributions. There is therefore no precision in these matters. In particular the present model throws no light on the problem of why the experimental reduced magnetization-temperature curves of iron and nickel are so closely similar. Present theories could be improved by a proper treatment of the multi-orbital aspect of the d band. However the fundamental ambiguities would remain and it may be that some new technique is required for further progress.

References

1. S.Q. Wang, W.E. Evanson and J.R. Schrieffer, Phys. Rev. Lett. 23, 92 (1969); W.E. Evanson, J.R. Schrieffer and S.Q. Wang, J. Appl. Phys. 41, 1199 (1970); J.R. Schrieffer, W.E. Evanson and S.Q. Wang, J. Phys. (Paris) 32, C1-1 (1971).

2. M. Cyrot, Phys. Rev. Lett. 25, 871 (1970).

3. J. Hubbard, Phys. Rev. Lett. 3, 77 (1959); R.L. Stratonovitch, Dokl. Akad. Nauk SSSR 115, 1097 (1957).

4. H. Hasegawa, J. Phys. Soc. Japan 46, 1504 (1979); 49, 178 (1980); 49, 963 (1980); Solid State Commun. 31, 597 (1979) and to be published.

5. M. Cyrot, J. Phys. (Paris) 33, 125 (1972); P. Lacour-Gayet and M. Cyrot, J. Phys. C 7, 400 (1974).

6. J. Hubbard, Phys. Rev. B 19, 2626 (1979).

7. J. Hubbard, Phys. Rev. B 20, 4584 (1979).

8. J. Hubbard, Phys. Rev., to be published.

9. J. Hubbard, 26th Annual Conf. on Magnetism and Magnetic Materials, Dallas 1980 (J. Appl. Phys., to be published).

10. R.E. Prange and V. Korenman, Phys. Rev. B 19, 4691 (1979).

11. R.A. Tawil and J. Callaway, Phys. Rev. B 7, 4242 (1973).

12. J. Callaway and C.S. Wang, Phys. Rev. B 7, 1096 (1973).

A Spin Fluctuation Theory of Itinerant-Electron Systems – The Single-Site Approximation

Hideo Hasegawa

Institute for Solid State Physics, University of Tokyo, Roppongi, Tokyo, 106, Japan

1. Introduction

It is well known that the band theory based on the Hartree-Fock approximation has succeeded in describing the ground-state properties such as the magnetic moment, the Fermi surface and the cohesive energy of transition metals. Despite its great success, however, the band theory cannot account for physical phenomena such as the Curie-Weiss susceptibility, the persistence of local moments above the Curie temperature and the large magnetic entropy. In recent years much progress has been made in the theoretical understanding of the finite-temperature properties of transition metals [1-4]. It has been recognized that the effect of spin fluctuations, which is completely neglected in the Hartree-Fock approximation, plays predominant roles in discussing the thermodynamical properties of magnetic metals. Recently, the present author has developed a spin fluctuation theory in which the effect of local spin fluctuations is taken into account by means of the functional-integral method combined with the alloy-analogy approximation [5-8]. We will discuss in this note the finite-temperature properties of ferromagnetic metals (Sec.2), of antiferromagnetic metals (Sec.3) and of concentrated ferromagnetic alloys (Sec.4), calculating the transition temperature, the average magnetic moment, the amplitude of local moments, the spin susceptibility and the specific heat. Supplementary discussions on the validity of the single-site approximation and its extention to a pair theory are given also (Sec.5).

2. Basic Formulation

2.1 Self-Consistent Equations

For a simplicity of calculations, we adopt the single-band Hubbard model. We employ the functional integral method for the calculation of the partition function. Although there are many possible transformations, we use the two-field scheme. The partition function within the static approximation is given by [5,6]

$$Z = e^{-\beta F} = \int \prod_j d\zeta_j \prod_j d\nu_j \exp[-(\beta U/4) \sum_j (\zeta_j^2+\nu_j^2)] \, Z(\zeta,\nu), \qquad (1)$$

where

$$Z(\zeta,\nu) = \mathrm{Tr} \exp[-\beta\{\sum_{\sigma j}\sum \varepsilon_{j\sigma} n_{j\sigma} + \sum_{\sigma j \ell}\sum\sum t_{j\ell} c^{\dagger}_{j\sigma} c_{\ell\sigma}\}],$$

$$\varepsilon_{j\sigma} = \varepsilon_0 - (U/2)(\sigma\zeta_j + i\nu_j).$$

$Z(\zeta,\nu)$ denotes the partition function of the free-electron system in the exchange fields, $-\sigma(U/2)\zeta_j$, and the electric fields, $i(U/2)\nu_j$; this is therefore a one-body problem in random fields. An evaluation of (1) is, however, difficult and it is necessary to use some approximations. By adopting the alloy-analogy concept first initiated by CYROT [9], we regard the system as a random alloy, which is treated by the single-site coherent potential approximation (CPA). Since the effect of electric (charge) field is less important than that of exchange fields in magnetic metals under consideration, we take into account the effect of the former by a simpler steepest descent method. As a result the self-consistent equations for averages of ζ, $\langle\zeta\rangle$ and $\langle\zeta^2\rangle$, by which the system is characterized, are expressed by

$$\langle\zeta\rangle = \int d\zeta_j \zeta_j \, C(\zeta_j), \tag{2}$$

$$\langle\zeta^2\rangle = \int d\zeta_j \zeta_j^2 \, C(\zeta_j), \tag{3}$$

where

$$C(\zeta_j) = e^{-\beta\psi(\zeta_j)} / \int d\zeta_j \, e^{-\beta\psi(\zeta_j)}, \tag{4}$$

$$\psi(\zeta_j) = (U/4)[\zeta_j^2 + \nu_j^2(\zeta_j)] + \int d\varepsilon f(\varepsilon)(1/\pi)\,\mathrm{Im}\sum_\sigma \ln\{1 - [\varepsilon_{j\sigma} - \Sigma_\sigma]F_\sigma\}, \tag{5}$$

$$\tilde{\Sigma}_\sigma = -\sigma(U/2)\langle\zeta\rangle + [(U^2/4)\langle\zeta^2\rangle - \tilde{\Sigma}_\sigma^2]F_\sigma, \tag{6}$$

$$\tilde{\Sigma}_\sigma = \Sigma_\sigma - \varepsilon_0 - (U/2)n, \tag{7}$$

$$F_\sigma(\varepsilon) = F^0(\varepsilon - \Sigma_\sigma) = \int d\omega \rho^0(\omega)/[\varepsilon - \Sigma_\sigma - \omega]. \tag{8}$$

Here Σ_σ and F_σ are the coherent potential and the site diagonal Green's function, respectively, $\nu_j(\zeta_j)$ is the saddle-point value for a given ζ_j, n the number of electrons per atom and ρ^0 the unperturbed density of states. The local free energy, $\psi(\zeta_j)$, denotes the free-energy variation when we vary the amplitude of the exchange field at the j-th atom which is surrounded by the effective medium characterized by Σ_σ, and $\psi(\zeta_j)$ depends on $\langle\zeta\rangle$ and $\langle\zeta^2\rangle$ through Σ_σ. Once $\langle\zeta\rangle$ and $\langle\zeta^2\rangle$ are determined, we easily obtain $\langle m\rangle$ and $\langle m^2\rangle$ as a function of the temperature. It is necessary to note that, if we keep only the single most important contribution with the lowest free energy in (2), we obtain the self-consistent equation in the Hartree-Fock approximation. We intend to take into account the effect of longitudinal local spin fluctuations by including other configuration. We may also realize that self-consistent equations given by (2) and (3) are generalization of the concept of the molecular-field approximation for Heisenberg magnets to the case of the itinerant-electron systems. Indeed, if we calculate (2) in the limit of $U\to\infty$ by using the local saddle-point approximation (LSA), in which the functional integrals are evaluated at two saddle points [5], (2) reduces to the self-consistent equation equivalent to that of the molecular-field theory. Although (3) yields a trivial condition in this limit, it is shown to be generally an indispensable equation because $<\zeta^2>$ or $<m^2>$ has a rather strong temperature dependence in magnetic metals [2].

2.2 Spin Susceptibility

By applying an infinitesimal magnetic field, we calculate the spin susceptibility given by

$$\chi = \Lambda/(T - \Xi), \tag{9}$$

expressions of Λ and Ξ being given in [6].
A) In the limit of $U\to\infty$, we get from (9) the Curie-Weiss susceptibility

$$\chi = C/(T - T_C), \tag{10}$$

with

$$C = <m^2> = (1/3)<\vec{m}^2>.$$

The Curie constant, C, corresponds to that in the molecular-field theory for Heisenberg magnets given by

$$C^{mf} = g_e^2\, S(S+1)/3, \tag{11}$$

where g_e denotes the g factor.
B) In the case of weakly ferromagnetic metals, in which $U\chi^0 \gtrsim 1$ (χ^0: the non-interacting susceptibility), the susceptibility is given by (10) with the Curie constant expressed by

$$C = U\chi^0<\zeta^2>[1 + a\partial<\zeta^2>/\partial T + \cdots]. \tag{12}$$

Eq.(12) shows that the Curie constant of weakly ferromagnetic metals is in general larger that the value expected from the ground-state magnetic moment because of the significant variation of the amplitude of local spin fluctuations [2].
C) In the limit of vanishing interaction, or $T = 0$ K, (9) reduces to

$$\chi = 2\chi^0/(1 - U\chi^0), \tag{13}$$

which is nothing but the susceptibility in the Hartree-Fock approximation. Thus the susceptibility expressed by (9) yields good results in the above-mentioned limits. As numerically shown later, the susceptibility obeyes the Curie-Weiss law approximately in the comparatively wide range of physical parameters.

2.3 Specific Heat

The calculated specific heat consists of two terms, $C = C_1 + C_2$. C_1 arises from the thermal excitation of electrons near the Fermi level and leads to the linear specific heat at $T \simeq 0$. On the other hand, contribution from random magnetic moments, C_2, yields a large peak near the transition temperature. It should be noted that, in the case of $U\to 0$, we get $C_2 = 0$ and the expression of the specific heat reduces to that in the Hertree-Fock theory. On the contrary, in the limit of $U\to\infty$, we get the result equivalent with that in the molecular-field approximation.

2.4 Iron

The assumed density of states function, $\rho^0(\varepsilon)$, is shown in the inset of Fig.1, which is a simplification of the band calculation by WAKOH and YAMASHITA [10]. The energy and the temperature are hereafter measured in reduced units where a half of the total band width of the model band is unity. We assume that the ground-state moment, m_0, is 0.5 and n=1.45, which leads to U=1.317 for the adopted band [6].

Solid lines in Fig.1 show the results of $\langle m \rangle$, $\sqrt{\langle m^2 \rangle}$ and χ^{-1}, while dotted lines those in the LSA. The calculated average magnetic moment well reproduces the overall behavior of experimental data. When we assume that the total band width of d bands is 0.5 Ryd. [10], the calculated Curie temperature is about 800 K. On the contrary, the Hartree-Fock theory yields 4600 K, which is too high compared with the experimental value of 1044 K [11]. When the temperature is raised, the amplitude of local moments decreases slightly and it remains finite even when the long-range order disappears. This is supported by several experimental results such as the neutron diffraction [12] and the photoemission [13,14]. The calculated spin susceptibility shown in Fig.1 obeys the Curie-Weiss law approximately. The ratio of the effective Bohr magneton number, m_{eff}, to the ground-state moment, m_0, is 1.37∿1.79, while experimental value is reported to be 1.44 [15]. The calculated specific heat is shown in Fig.2. For a comparison the result in the LSA is shown by the dotted line [5]. Because of the contribution from disordered moments, the specific heat has a large peak at T_C as experimentally observed [16].

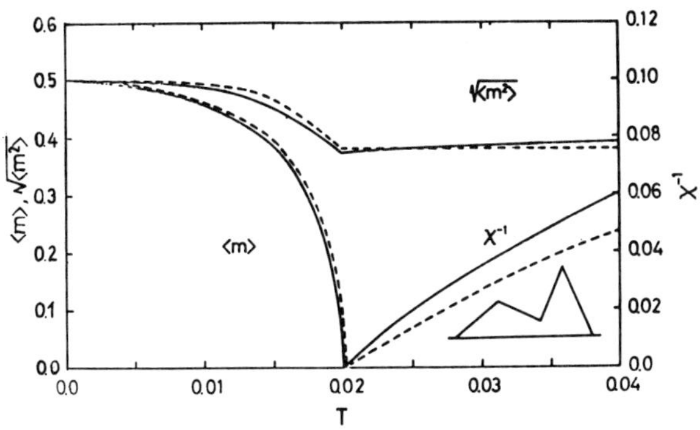

Fig.1 The average moment, $\langle m \rangle$, the rms value of local moment, $\sqrt{\langle m^2 \rangle}$, and the inversed susceptibility of Fe. The dotted lines express those in the LSA. The inset shows the adopted model band.

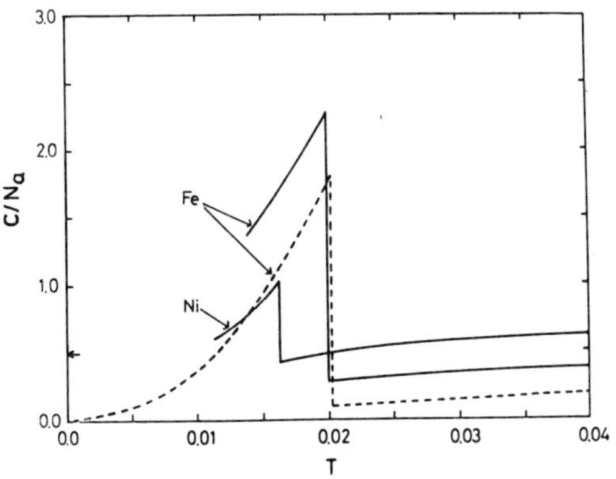

Fig.2 The specific heat of Fe and Ni. The dotted line expresses that of Fe in the LSA. The arrow is the specific heat at T=0 K corresponding to the Dulong Petit term, which arises from the classical treatment of spin fluctuations in the present theory, while the specific heat in the LSA is zero at T=0 K as they should be.

Since the effect of the short-range magnetic order is not taken into account, the specific heat has a discontinuity at T_C.

2.5 Nickel

The inset of Fig.3 expresses the assumed density of state function, $\rho^0(\varepsilon)$, which is a simplification of the density of states of Ni calculated by CONNOLY [17]. We assume that m_0=0.12, n=1.88 and U=1.40 [6].

The temperature dependence of the average magnetic moment shown in Fig.3 follows closely the Brillouin function for spin 1/2. When the total band width is assumed to be 0.35 Ryd. [17], the calculated Curie temperature is 460 K while the Hartree-Fock theory leads to 1400 K and the experimental value is 631 K [11]. The amplitude of the local moments remains finite even in the paramagnetic states as in the case of Fe. This behavior is consistent with the results of the neutron diffraction [18] and the photoemission [19]. The calculated spin susceptibility in Fig.3 shows an approximate Curie-Weiss behavior. The calculated ratio m_{eff}/m_0 is 2.41∿2.95 whereas the experimental value is reported to be about 2.68 [20]. The calculated specific heat of Ni is shown in Fig.2. When we compare the result of the specific heat of Ni with that of Fe, we may see that the specific-heat peak near the Curie point of Ni is considerably smaller than that of Fe, which is qualitatively in agreement with experimental results [16].

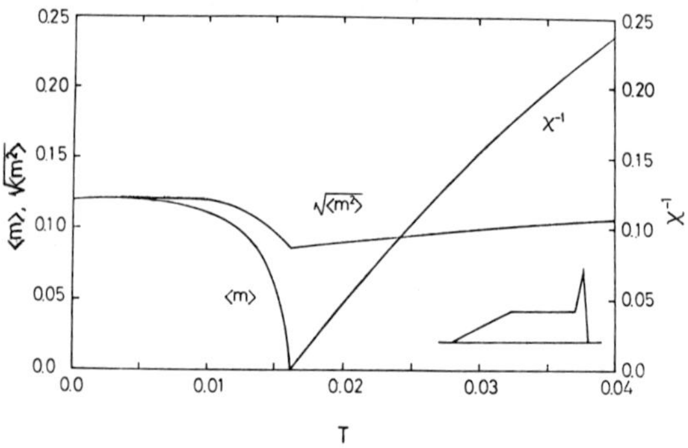

Fig.3 The average moment, <m>, the rms value of local moment, $\sqrt{<m^2>}$, and the inversed susceptibility of Ni. The inset shows the adopted model band.

3. Antiferromagnetic Metals

It is easy that we extend the spin fluctuation theory in Sec.2 to the case of antiferromagnetic metals [6]. We can obtain the self-consistent equations given by (2)-(7) and

$$F_\sigma(\varepsilon) = \{(\varepsilon-\Sigma_{-\sigma})/(\varepsilon-\Sigma_\sigma)\}^{1/2}F^0(\Omega_\sigma), \quad (14)$$

$$\Omega_\sigma = \{(\varepsilon-\Sigma_\sigma)(\varepsilon-\Sigma_{-\sigma})\}^{1/2}, \quad (15)$$

from which we can calculate the temperature dependences of the sublattice magnetization, <m>, and the amplitude of local moments, $<m^2>$.

It is noted that, in the case of the half-filled band (n=1) with $U \to \infty$, the uniform (χ_u) and staggered (χ_s) susceptibilities are given by

$$\chi_u = C_u/(T - \Theta), \quad \chi_s = C_s/(T - T_N), \tag{16}$$

with

$$C_u = C_s = 1, \quad T_N = - \Theta = M_2^o/U,$$

M_2^o being the second moment of ρ^o. Susceptibilities given by (16) agree with the results in the molecular-field theory for Heisenberg antiferromagnets with Anderson's superexchange interaction.

In order to discuss the itinerant-electron antiferromagnetism with this theory, we adopt the half-filled Bell shape band, which may be regarded as a simplification of the density of states of a simple-cubic lattice with the nearest-neighbour hopping [6].

Figure 4 shows the phase boundary line separating the antiferromagnetic insulator (AFI), the paramagnetic insulator (PI) and the paramagnetic metal (PM). The solid lines express the result in the present theory while dashed lines that in the LSA. The calculation well reproduces the general features of the phase diagram observed in real materials like V_2O_3 [21]. The calculated transition between PM and PI is not sharp in contrast with the first-order transition observed experimentally. This may be due to a neglect of the lattice distortion in the theory [22]. The solid lines in Fig.5 show the temperature dependence of the sublattice magnetization and the root-mean-square (rms) value of local magnetic moments for various values of U in the present theory whereas dashed lines those in the LSA. Although the temperature dependence of <m> for U=1.5 closely follows the Brillouin function for

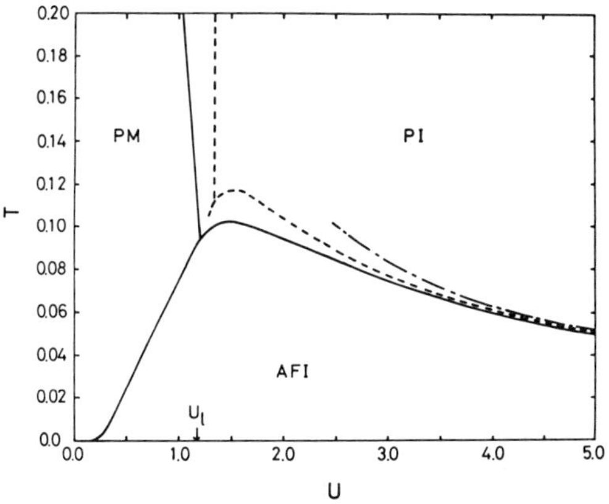

Fig.4 The phase diagram of the half-filled Bell shape band, showing the paramagnetic metal (PM), the paramagnetic insulator (PI) and the antiferromagnetic insulator (AFI) in the present theory (solid lines) and in the LSA (dashed lines). The dot-dashed line is the asymptote; $T_N=1/4U$. The arrow is the critical value of U for formation of local moment, $U_\ell(=1.178)$.

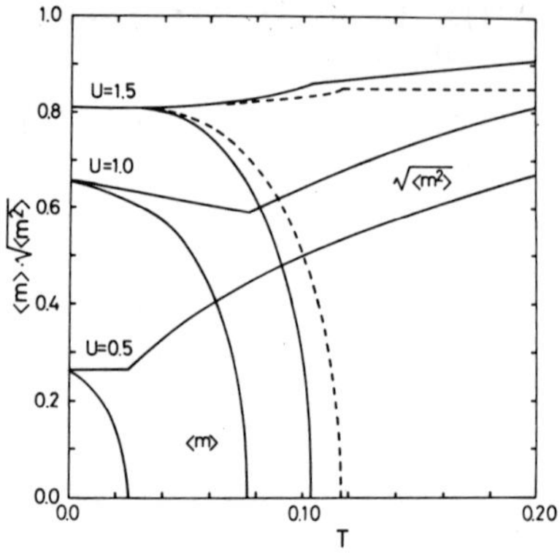

<u>Fig.5</u> The temperature dependence of <m> and √<m²> in the present theory (solid lines) and in the LSA (dashed lines).

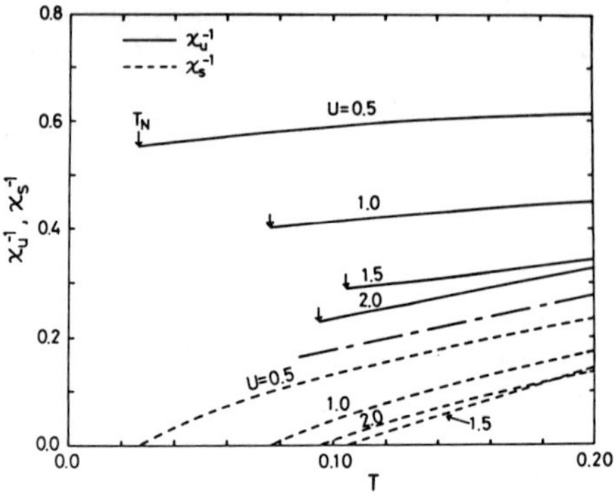

<u>Fig.6</u> The uniform (χ_u) and staggered (χ_s) susceptibilities for various U values. The dot-dashed line shows the gradient of the inversed susceptibility in the atomic limit.

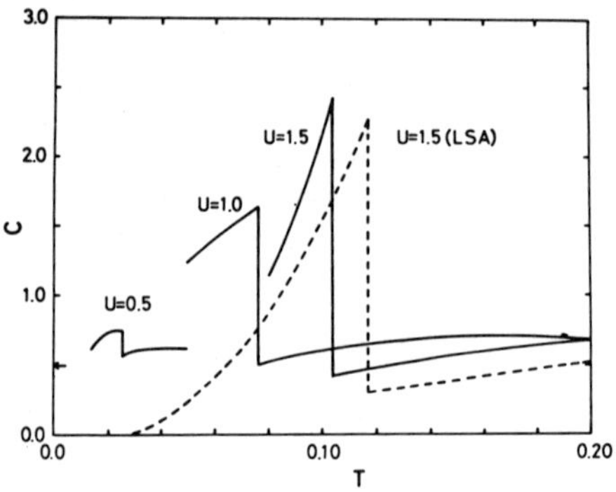

<u>Fig.7</u> The temperature dependence of the specific heat in the present theory (solid lines) and in the LSA (dashed lines). The arrow shows the specific heat at T=0 K in this theory (see also the caption of Fig.2).

S=1/2, those for U=0.5 and 1.0 show an apparent deviation from the Brillouin function. It has been reported that the sublattice magnetization of $\gamma Fe_{1-x}Mn_x$ (T_N=500∿350 K for x=0.4∿0.8) [23] and CrB_2 (T_N=88 K) [24] follows the Brillouin curve whereas that of weakly antiferromagnetic metals such as V_3S_4 (T_N=9 K) and V_3Se_4 (T_N=18 K) [25] decreases faster than the Brillouin curve with raising the temperature. The present calculations are qualitatively in agreement with these results. The solid and dashed lines in Fig.6 express the uniform (χ_u) and staggered (χ_s) susceptibilities, respectively. Both susceptibilities obey the Curie-Weiss law approximately in a comparatively wide range of U values. The gradient of inversed susceptibilities, which is proportional to the Curie constant, varies continuously with varying the U value. The susceptibilities of $FePt_3$ are reported to show the Curie-Weiss behavior; its slope of χ_s^{-1} is nearly half of that of χ_u^{-1} [26]. Fig.7 shows the calculated specific heat for various U values. We may see that, with increasing the strength of the interaction, the specific-heat peak at the Néel temperature increases, yielding large magnetic entropy.

As was shown in numerical calculations, we may observe a gradual transition of the temperature dependences of various physical quantities when the strength of the interaction is varied. These are consistent with experimental results.

4. Concentrated Ferromagnetic Alloys

We consider a random alloy A_xB_{1-x}, where both elements are transition metals. We adopt the single-band model [27] in which the average atomic energy ε_j is assumed to take the value ε_A or ε_B depending on the atom at the j-th site and the electron-electron interaction U_j is assumed to be equal to $U_A(U_B)$ if the A(B) atom is on the j-th site. The generalization of the spin fluctuation theory discussed in the preceding sections to the case of concentrated alloys is straightforward [7]. In this case we have to determine four quantities, $\langle\zeta\rangle_\alpha$ and $\langle\zeta^2\rangle_\alpha$ (α=A and B), self-consistently for a given set of parameters, ε_A, ε_B, U_A, U_B and $\rho^0(\varepsilon)$, where the bracket $\langle\rangle_\alpha$ denotes the conditional average specifying the α atom at a given site. Once they are determined, we obtain the average magnetic moment, $\langle m\rangle_\alpha$, and the amplitude of local magnetic moments, $\langle m^2\rangle_\alpha$, relevant to the α atom. The total magnetic moment, $\langle m\rangle$, is given by

$$\langle m\rangle = x\langle m\rangle_A + (1 - x)\langle m\rangle_B. \tag{17}$$

By applying an infinitesimal magnetic field to the system, we can calculate the partial spin susceptibility relevant to the α atom, χ_α, which is nothing but the Knight shift of the α atom. The equation for χ_α is given by

$$\begin{pmatrix} T - \Xi_{AA} & -\Xi_{AB} \\ -\Xi_{BA} & T - \Xi_{BB} \end{pmatrix} \begin{pmatrix} \chi_A \\ \chi_B \end{pmatrix} = \begin{pmatrix} \Lambda_A \\ \Lambda_B \end{pmatrix}, \tag{18}$$

the expressions of $\Xi_{\alpha\beta}$ and Λ_α being given in [7].

We may show that the expression given by (18) in the vanishing temperature reduces to that obtained by HASEGAWA and KANAMORI [27], who have first discussed magnetism of transition metal alloys by using the CPA combined with the Hartree-Fock approximation.

It is important to investigate the susceptibilities in the dilute limit ($x\to 0$), where they are given by

$$\chi_A = [(T - T_{CB})\Lambda_A + \Xi_{AB}\Lambda_B]/T(T - T_{CB}), \tag{19}$$

$$\chi_B = \Lambda_B/(T - T_{CB}).$$

Both χ_A and χ_B diverge at the Curie temperature of pure B metals, T_{CB}. On the contrary, in the one-electron theory of magnetic alloys, the condition of an instability of the paramagnetic state in the dilute limit is given by [27]

$$(1 - U_B\chi_B^0)(1 - U_A\Gamma_A) = 0. \tag{20}$$

We have either $1-U_B\chi_B=0$ or $1-U_A\Gamma_A=0$ as the instability condition. The former is the Stoner condition of the pure B metals and the latter is nothing but the condition for the appearance of a local moment on the impurity A atom embedded in otherwise pure nonmagnetic B metal. It is unreasonable that the phase transition is induced by a single impurity, which arises from a fallacy inherent in the one-electron theory. This defect is removed in the present theory by including the effect of spin fluctuations.

We apply the theory to bcc $Fe_{1-x}Cr_x$ alloys, which are typical among Fe based alloys. We have adopted the simplified density of states depicted in the inset of Fig.1 and the same parameters for Fe as in the Sec.3.1. Paramaters for Cr are chosen such that we obtain reasonable results of the concentration dependences of the total magnetic moment as well as the average moments of Fe and Cr at T=0 K [28,29].

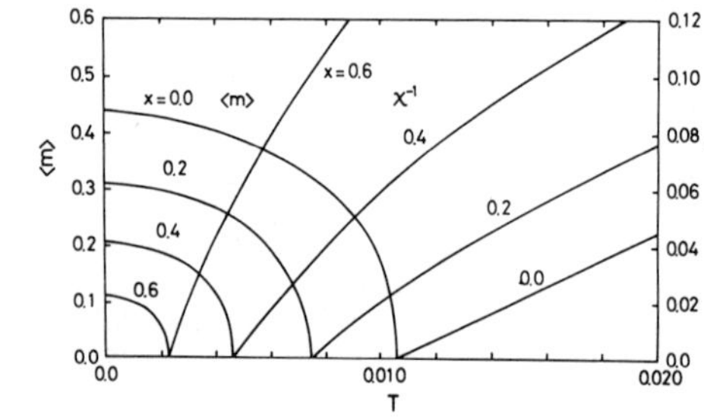

Fig.8 The temperature dependence of the average magnetic moment, ⟨m⟩, and the total spin susceptibility, χ, for various Cr concentration.

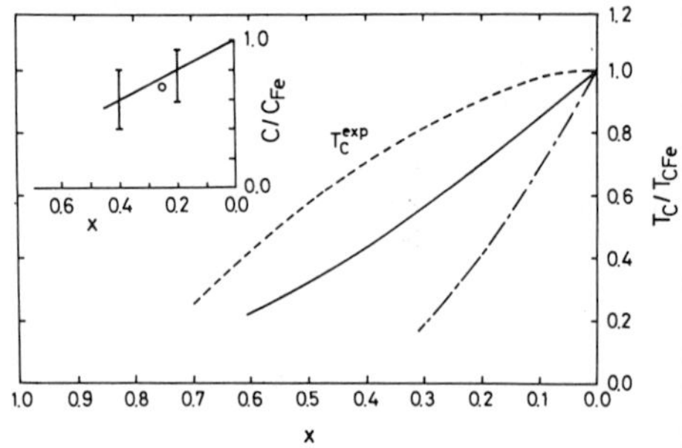

Fig.9 The concentration dependence of the Curie temperature. The solid line expresses the calculated result in the present theory and the dot-dashed line that in the "rigid band" approximation (see text) whereas experimental result is shown by the dashed line. The solid line in the inset is the calculated concentration dependence of the Curie constant and o is the experimental data of $Fe_{0.75}Cr_{0.25}$ [31].

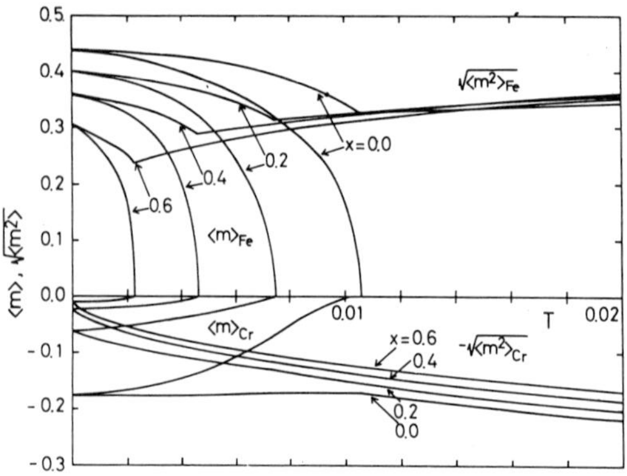

Fig.10 The temperature dependences of the average moments, $\langle m\rangle_{Fe}$ and $\langle m\rangle_{Cr}$, and of the rms values of local moments, $\sqrt{\langle m^2\rangle_{Fe}}$ and $\sqrt{\langle m^2\rangle_{Cr}}$.

The calculated temperature dependence of ⟨m⟩ is shown in Fig.8. with increasing Cr atoms, the Curie temperature decreases as is shown by the solid line in Fig.9. Although experimental data given by the dashed line in Fig.9 shows the maximum near $x \simeq 0.05$ [30], the calculated result decreases monotonously with an increase of the Cr concentration. For a comparison, we show by the dot-dashed line the result with the use of the "rigid band" approximation in which the average electron-electron interaction is chosen such that it reproduces the concentration dependence of the observed average magnetic moment at T=0 K [28]. Although the general trend of experimental result of T_C is explained by the present theory, the agreement between the two is not satisfactory. The calculated total spin susceptibility, χ, is shown in Fig.8. When we increase the Cr concentration, the Curie constant, which is inversely proportional to the gradient of χ^{-1}, decreases as shown in the inset of Fig.9. The experimental result of the susceptibility of $Fe_{0.75}Cr_{0.25}$ [31] is consistent with this calculation. In Fig.10 we show the temperature dependences of the average magnetic moment and the rms value of local moment relevant to each consituent atom. We may note that the local moment, especially that of Fe atom, has almost the same amplitude in the paramagnetic state irrespective of the concentration. Unfortunately, there is little experimental data to be compared with our calculations. Experiments for these quantities of the alloy are highly desired.

5. Discussions

We have discussed the single-site spin fluctuation theory, which is shown to give a good account for finite-temperature properties of itinerant-electron systems. We may show that the present approach has the following advantages [6]: 1) The theory provides us an interpolating description yielding good results in some important limits; in the limit of $U \to 0$ or T=0 K it reduces to the Hartree-Fock theory, and in the limit of $U \to \infty$, it reproduces the results of the molecular-field theory. 2) The theory preserves the variational property; self-consistent equations given by (2)-(8) are obtainable from the condition given by $\partial F/\partial\langle\zeta\rangle = \partial F/\partial\langle\zeta^2\rangle = 0$.

We now examine the adopted approximations and the validity of the present theory. Since our theory is quite similar in spirit to that of the molecular-field theory for the Heisenberg model, it has the same disadvantages as the molecular-field theory.

We have used the two-field functional-integral method within the static approximation, which means a neglect of the effects of dynamical spin fluctuations and of the vectorial nature of magnetic moment. It is the most difficult to justify the static approximation and to remove it though it is widely used. Quite recently, Hubbard [4] has developed a spin fluctuation theory respecting the spin rotational invariance by using the four-field static functional integral method. The present theory can be regarded as a simplified scheme of Hubbard's, though it preserves the essential features of calculated results with simpler calculations. The advantage of the present approach is that the calculated Curie constant of the half-filled band with $U \to \infty$ agrees with the exact result [see (16)] and that the theory can be easily applied to the cases of antiferromagnetic metals [6] and of concentrated alloys [7].

Since we have employed the single-site approximation, we have completely neglected the effect of the short-range magnetic order (SRMO), whose existence in the paramagnetic states of transition metals is confirmed by several experiments [12,18]. The simplest way including the effect of SRMO is to extend the present single-site approximation to the two-site one [8]. We consider a configuration such that a pair of neighbouring two atoms are embedded in the effective medium characterized by the coherent potential, which is determined by a self-consistency condition. The local free energy relevant to neighbouring sites, j and ℓ, is expressed by

$$\psi(\zeta_j,\zeta_\ell) = (U/4)(\zeta_j^2+\zeta_\ell^2) + \int d\varepsilon\, f(\varepsilon) \sum_\sigma \mathrm{Im}\ln D_\sigma, \tag{21}$$

with

$$D_\sigma = \det\,[1 - V_\sigma G_\sigma], \tag{22}$$

$$(V_\sigma)_{j\ell} = -\,\sigma(U/2)\zeta_j\delta_{j\ell} - (\Sigma_\sigma)_{j\ell}, \tag{23}$$

$$(G_\sigma)_{j\ell} = G_{j\ell\sigma}, \tag{24}$$

where we use an approximation for charge fields to set $-i\nu_j=n$. The coherent potential Σ_σ, which is determined by the generalized CPA condition, can be expressed in terms of the averages of ζ_j as

$$\Sigma_\sigma = \Sigma_\sigma\,(\langle\zeta\rangle, \langle\zeta^2\rangle, \tau), \tag{25}$$

where

$$\langle\zeta\rangle = \int d\zeta_j d\zeta_\ell (1/2)(\zeta_j+\zeta_\ell)C(\zeta_j,\zeta_\ell), \tag{26}$$

$$\langle\zeta^2\rangle = \int d\zeta_j d\zeta_\ell (1/2)(\zeta_j^2+\zeta_\ell^2)C(\zeta_j,\zeta_\ell), \tag{27}$$

$$\tau \equiv \langle\zeta_j\zeta_\ell\rangle = \int d\zeta_j d\zeta_\ell\ \zeta_j\zeta_\ell\ C(\zeta_j,\zeta_\ell), \tag{28}$$

$$C(\zeta_j,\zeta_\ell) = e^{-\beta\psi(\zeta_j,\zeta_\ell)}\Big/\int d\zeta_j d\zeta_\ell\ e^{-\beta\psi(\zeta_j,\zeta_\ell)}. \tag{29}$$

Eqs.(21)-(29) form the self-consistency conditions for $\langle\zeta\rangle$, $\langle\zeta^2\rangle$ and τ. As for a preliminary calculation, we adopt the simpler single-site coherent potential given by (6), and determine $\langle\zeta\rangle$ and $\langle\zeta^2\rangle$ self-consistently from (6), (21)-(24), (26), (27) and (29) in order to study the SRMO of bcc iron [8].

Calculated results are shown in Fig.11, where solid lines express the results in the pair theory and dashed lines those in the single-site theory. We may realize that the Curie temperature is lowered to some extent because of the effect of SRMO. The amplitude of local moments in the pair theory

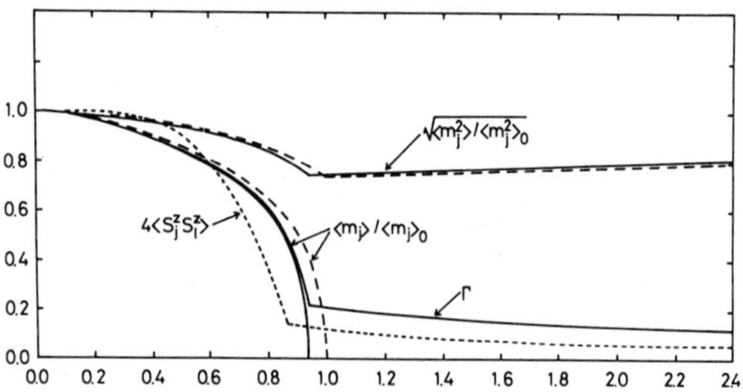

Fig.11 The solid lines express the long-range order, $\langle m_j\rangle$, the rms value of local moments, $\sqrt{\langle m_j^2\rangle}$, and the short-range order, $\Gamma\equiv\langle m_j m_\ell\rangle/\langle m_j^2\rangle$, in the two-site approximation against the reduced temperature, $t=T/T_C$ (in the single-site approx.). The dashed lines denote results in the single-site theory. The dotted line expresses the short-range order of the Ising model (spin 1/2) with the bcc structure having the nearest-neighbour exchange interaction in the constant-coupling approximation [32] against the reduced temperature, $t=T/T_C$ (in the molecular-field approx.).

is not so different from that in the single-site theory. The short-range order for neighbouring sites, j and ℓ, $\Gamma \equiv \langle m_j m_\ell \rangle / \langle m_j^2 \rangle$, is about 0.21 at T_C, and its curve has a long tail above T_C.

KORENMAN et al. [1] have developed a fluctuating band theory in which they assume the existence of a large amount of SRMO in the paramagnetic state though no model exhibiting a large SRMO is proposed. They have claimed that, from the analysis of sloppy spin waves of Fe [12], the short-range magnetic order, Γ, is about 0.8, which is considerably larger than the present result.

Quite recently, SHASTRY et al. [33] have discussed sloppy spin waves of Fe by using the spherical Heisenberg model with exchange interactions extending to far nearest neighbours and obtained $\Gamma \simeq 0.12 \sim 0.18$, which coincides with our result. This small value of SRMO is also consistent with the observed large magnetic entropy of Fe in the paramagnetic state [34] although it is controversial whether sloppy spin waves can exist in the paramagnetic phase with such a small SRMO.

The Curie temperatures of Fe and Ni calculated by the present theory are in the right order of magnitude compared with experimental data [11] as was shown in Sec.2. However it is necessary to include the effects of dynamical spin fluctuations, the SRMO, the orbital degeneracy and the Fermi-distribution function for a quantitative comparison of T_C with experimental data. It seems rather difficult to take into account the dynamical effect of spin fluctuations. When we consider the effect of the orbital degeneracy on the Curie point, the Curie temperature for the d-fold band, $T_C(d)$, is roughly estimated to be [6]

$$T_C(d)/T_C(d=1) = d \ln 2/\ln(2S + 1). \qquad (30)$$

If we assume that S=1 and d=5 for Fe, and S=1/2 and d=3 for Ni, the theoretical Curie temperatures become 2500 K for Fe and 1400 K for Ni. Recently KAKEHASHI [35], making calculations similar to the present one, has shown that the Curie temperatures of Fe calculated with and without taking into account the effect of Fermi-distribution function are 1180 K and 1530 K, respectively. Since the Hartree-Fock theory for Ni yields T_C^{HF}=1400K, which coincides with the value including the effect of the orbital degeneracy, it is indispensable to take account of the Fermi-distribution function for a quantitative discussion on the Curie temperature of Ni. Detailed calculations including the orbital degeneracy as well as the Fermi-distribution function is under consideration and will be reported elsewhere.

To summarize, the present theory, having clear physical meanings, can well account for various finite-temperature properties of itinerant-electron systems and it is considered to be a fruitfull method for the study of magnetism of ferromagnetic and antiferromagnetic metals and alloys. The application and extension of the theory are left for future investigations.

Acknowledgement

The author would like to express his sincere thanks to Prof. T. Moriya for valuable discussions.

References

1. V. Korenman, J. L. Murray and R. E. Prange, Phys. Rev. B 16, 4032, 4048 (1977).
2. T. Moriya and Y. Takahashi, J. Phys. Soc. Jpn. 45, 397 (1978).
3. H. Capellmann, Solid State Commun. 30, 7 (1979); Z. Phys. B 38, 29 (1979).
4. J. Hubbard, Phys. Rev. B 19, 2626 (1979); B 20, 4584 (1979).
5. H. Hasegawa, J. Phys. Soc. Jpn. 46, 1504 (1979); Solid State Commun. 31, 597 (1979); J. Magn. Magn. Mat. 15-18, 272 (1980).
6. H. Hasegawa, J. Phys. Soc. Jpn. 49, 178 (1980); 49, 963 (1980); Proc. Intern. Conf. on Transition Metals, Leeds 1980.
7. H. Hasegawa, Tech. Rep. ISSP A 1086, (October 1980).
8. H. Hasegawa, Solid State Commun. (to be published).
9. M. Cyrot, Phys. Lett. 37, 189 (1971).
10. S. Wakoh and J. Yamashita, J. Phys, Soc. Jpn. 21, 1712 (1966).
11. J. Crangle and G. M. Goodman, Proc. Roy. Soc. (London) A321, 477 (1971).
12. J. W. Lynn, Phys. Rev. B 11, 2624 (1975).
13. L. G. Petersson, R. Melander, D. P. Spears and S. B. M. Hagstron, Phys. Rev. B 14, 4177 (1976).
14. D. E. Eastman, F. J. Himpsel and J. A. Knapp, Phys. Rev. Lett. 44, 95 (1980).
15. W. Sucksmith and R. P. Pearce, Proc. Roy. Soc. (London) A167, 189 (1938).
16. M. Braun and R. Kohlhaas, Phys. Stat. Sol. 12, 429 (1965).
17. J. W. D. Connolly, Phys. Rev. 159, 415 (1967).
18. H. A. Mook, J. W. Lynn and R. M. Nicklow, Phys. Rev. Lett. 30, 556 (1973).
19. D. E. Eastman, F. J. Himpsel and J. A. Knapp, Phys. Rev. Lett. 40, 1514 (1978).
20. S. Arajs and R. V. Colvin, J. Phys. Chem. Solids 24, 1233 (1963).
21. D. B. McWhan and J. P. Remeika, Phys. Rev. B 2, 3734 (1971).
22. M. Cyrot and P. Lacour-Gayet, Solid State Commun. 11, 1767 (1972).
23. Y. Endoh and Y. Ishikawa, J. Phys. Soc. Jpn. 30, 1614 (1971).
24. Y. Kitaoka, H. Yasuoka, T. Tanaka and Y. Ishizawa, Solid State Commun. 26, 87 (1978).
25. Y. Kitaoka, H. Yasuoka, K. Oka, K. Kosuge and S. Kachi, J. Phys. Soc. Jpn. 46, 1381 (1979).
26. K. Kohgi and Y. Ishikawa, J. Phys. Soc. Jpn. 49, 994 (1980).
27. H. Hasegawa and J. Kanamori, J. Phys. Soc. Jpn. 31, 382 (1971); 32, 586 (1972); 33, 1599 (1972); 33, 1607 (1972).
28. A. T. Aldred, Phys. Rev. B 14, 219 (1976).
29. A. T. Aldred, B. D. Rainford, J. S. Kouvel and T. J. Hicks, Phys. Rev. B 14, 228 (1976).
30. M. Hansen, Constitution of Binary Alloys (McGraw-Hill, New York, 1958).
31. Y. Nakagawa, J. Phys. Soc. Jpn. 11, 855 (1956).
32. P. W. Kasteleijn and J. Van Kranendonk, Physica 22, 317 (1956).
33. B. S. Shastry, D. M. Edwards and A. P. Young, preprint.
34. D. M. Edwards, J. Magn. Magn. Mat. 15-18, 262 (1980).
35. Y. Kakehashi, preprint.

Comments on Weakly and Nearly Ferromagnetic Metals

Michel Cyrot

Laboratoire Louis Néel, C.N.R.S., 166X,
F-38042 Grenoble Cédex, France

The Stoner theory is known to have serious drawbacks. For instance, the most often quoted are the too high calculated Curie temperature and the inability to produce a Curie-Weiss susceptibility above T_C. For weak ferromagnetic metals such as $ZrZn_2$ Murata and Doniach (M.D.) |1| and Moriya and Kanabata (M.K.) |2| presented theories which explain this behaviour. The more fundamental last one reduces to the first one above T_C. However, the basic physics behind such a theory has never been very clear as the mathematics is rather involved. On the other hand, detailed theories have appeared for nearly ferromagnetic materials |3-7|. These theories, generally valid at low temperature, calculated the contribution of spin fluctuations which do not appear in the usual Stoner theory. In both cases, we are faced with a real improvement of the Stoner theory and these authors go one step further than the mean field theory.

The purpose of our comment is to place these various improvements of the mean field theory in a general context and to show that they can be obtained in a very simple way (practically an handwaving way) from an old concept introduced by Onsager |8| "the reaction field". In liquid dielectric, Onsager was faced with a complete break-down of the mean field theory. He explained the situation by the fact that a dipole was changing its own Weiss field when changing its direction. The medium is not rigid enough in order that the spin feels the same field when it is up or down. He notices that the orientating part of the field for the dipole is not the Weiss field (or mean field) but the Weiss field minus that part of the Weiss field which follows the dipole. One named it the reaction field as it is due to the fact that the dipole polarized the medium around it and this polarization reacts back on the Weiss field. If the dipole turns its direction the reaction field follows it. This method is an handwaving way to take into account correlations between dipoles which are not introduced in the Weiss theory. Our purpose is now to show that introduction in magnetism of this concept permits to recover previous results in the theory of nearly and weakly ferromagnetic metals and even in some cases to extend them.

In magnetism the Weiss field at point r is given by

$$H_w(r) = \int \lambda(r-r') \langle S(r')\rangle \, d_3r' \qquad (1)$$

where $\langle S(r')\rangle$ is the magnetization at point r' and $\lambda(r-r')$ is related to the molecular field coefficient. In the Heisenberg model, $\lambda(r-r')$ is the interaction between two spins at sites r and r' and is usually written J_{ij}. In the Stoner theory, $\lambda(r-r')$ is a delta function with weight $U/2\mu_B^2$ where U is the intraatomic Coulomb interaction.

The reaction field is that part of the field seen by the spin at r which follows instantaneously the spin. This part can be written in the non-magnetic case or above T_c

$$H_R = \int \lambda(r-r') <S(r,t)\ S(r't)> d_3r' <S(r,t)> . \qquad (2)$$

If there is no correlation between spins at site r and r', this reaction field is zero. But in general one cannot neglect it and we have to subtract it from the Weiss field. For instance in an Ising model with nearest-neighbour approximation at the Curie temperature

$$H_W \sim zJ ,$$

z is the number of nearest neighbours and J the interaction

$$H_R \sim z \frac{J^2}{kT_c} = J.$$

The reaction field is negligible if z is large, which is the criterium for the validity of the mean field approximation. In a spin glass system, we have shown |9| that both fields are of the same order of magnitude. In general one cannot neglect it when the medium can easily react. Of course, this is the case of itinerant electrons.

Using the fluctuation dissipation theorem |10|, we can write

$$\delta \equiv \int \lambda(r-r') <S(r,t)\ S(r't)> d_3r' = \int d_3q \int d\omega\ \lambda(q) \frac{1}{1-e^{-\omega/T}} \operatorname{Im} \chi(q,\omega+i\varepsilon) \qquad (3)$$

where $\chi(q,\omega)$ is the dynamical susceptibility of the system.

The field which orientates the spins is then

$$h(\omega,q) = \lambda(q)\ S(q,\omega) - \delta S(q,\omega) , \qquad (4)$$

the dynamical susceptibility is given by

$$\chi(q,\omega) = \frac{1}{\chi_0^{-1}(q,\omega) - \lambda(q) + \delta} \qquad (5)$$

which is the usual R.P.A. result except for the term δ which is introduced and which has to be determined self-consistently by equation 3.

First, let us consider the weak ferromagnetic metals above T_C. $\lambda(q) = U$ (we take $(2\mu_B^2 = 1)$). $\chi_0(q,0)$ is the susceptibility of the non-interacting electron gas. In equation 3, we take the high-temperature limit or classical limit

$$\begin{aligned}\delta(T) &= 3U \int d_3q \int d\omega \frac{T}{\omega} \operatorname{Im} \chi(q,\omega+i\varepsilon)\\ &= 3UT \int d_3q\ \chi(q,0) \qquad (6)\end{aligned}$$

and by (5) we have

$$\chi(q,0) = \frac{1}{\chi_0^{-1}(q,0) - U + \delta(T)} . \qquad (7)$$

Except for coefficients we have the same kind of theory as M.D.

Writing $\chi_0(q) = n(E_F)(1 - Aq^2)$,

T_c is given by $1 - Un(E_F) + \delta(T_c)n(E_F) = 0$

i.e.

$$T_c = \frac{|Un(E_F) - 1|A}{4\pi q_c 3U\, n^2(E_F)},$$

q_c is the cut-off wave vector. The Curie-Weiss susceptibility is given by the behaviour of $\delta(T)$ which tends to grow linearly as a function of temperature. This is the result of M.D. and Moriya |11,12|. However, our derivation does not lie on an expansion in power of the magnetization and relies on more general assumptions. Our introduction of Onsager reaction field not only takes into account parts of the correlations but guarantees the fluctuation theorem of statistical mechanics |13|.

We now apply the theory at nearly ferromagnetic metals. Equations (3) and (5) are rewritten

$$\chi(q,0) = \frac{1}{\chi_0^{-1}(q,0) - U + \delta(T)} \qquad (8)$$

$$\delta(T) = 3U\, T \int d_3q \sum_{\omega_n} \chi(q,\omega_n) \qquad (9)$$

where ω_n are the Matsubara frequencies. We are interested here in a low-temperature expansion of δ.

We can write $\delta(T) = \delta(0) + KT^2 + \ldots$

The quantity $\delta(0)$ can be eliminated by defining an effective value of U. If we are concerned only by the coefficient of T^2 in an expansion of the static susceptibility, we can calculate K in neglecting δ in the denominator of equation (8), i.e. taking the R.P.A. result for the susceptibility and calculating δ by equation (9) . We obtain

$$\chi = \frac{n(E_F)}{1 - Un(E_F)} \left| 1 + \frac{\delta\chi^0(T)+n(E_F)KT^2}{1 - Un(E_F)} \right|$$

where $\delta\chi^0(T)$ is the T^2 term in the expansion of the static susceptibility of a non-interacting gas. This is exactly the result obtained in the paramagnons theory. We point out however that our result (equations (8) and (9)) are not restricted to the low-temperature regime. If apply to a band structure which gives an increase of the susceptibility at low temperature, as we recover an approximate Curie-Weiss law at high temperature, we obtain a maximum of the susceptibility near the spin fluctuation temperature. There has been some controversy about the possibility of obtaining such a maximum in the paramagnons theory. This seems definitively set up by this more general theory where such a behaviour appears naturally. Other physical quantities have been calculated in the literature , specific heat, resistivity, etc. We would obtain the same results as for instance the ground state energy can be obtained from the transverse susceptibility.

Our comment is the following : with a very simple method, one can correct the mean field approximation. This gives very sensible results in all cases where this has been done. We remind also that we have shown |9| that the spin glass theory of Thouless Anderson and Palmer |14| can be obtained in the same way.

Acknowledgments

I like to thank Professor T. Moriya who gives us the opportunity to try to sum up the recent progresses in itinerant magnetism.

References

1. K.K. Murata and S. Doniach, Phys. Rev. Lett. 29, 285 (1972).
2. T. Moriya and A. Kawabata, J. Phys. Soc. Japan 34, 639 (1973) ; 35, 669 (1973).
3. S. Doniach and S. Engelsberg, Phys. Rev. Lett. 17, 750 (1966).
4. N. Berk and J.R. Schrieffer, Phys. Rev. Lett. 17, 433 (1966).
5. M.T. Beal-Monod, S.K. Ma and D.R. Fredkin, Phys. Rev. Lett. 20, 929 (1968) ; Phys. Rev. 174, 227 (1968).
6. T. Izuyama and R. Kubo, J. Appl. Phys. 39, 642 (1968).
7. K. Makoski and T. Moriya, J. Phys. Soc. Japan 38, 10 (1975).
8. L. Onsager, J. Am. Chem. Soc. 58, 1486 (1936).
9. M. Cyrot, Phys. Rev. Lett. 43, 173 (1979).
10. R. Kubo, J. Phys. Soc. Japan 12, 893 (1957).
11. T. Moriya, Prog. Theor. Physics (supplement).
12. T. Moriya, J. Magn. Magn. Mat. 14, 1 (1979).
13. R. Brout and H. Thomas, Physics 1, 313 (1967).
14. D. Thouless, P.W. Anderson and R.G. Palmer, Phil. Mag. 35, 593 (1977).

Ferromagnetism in Iron and Nickel

R.E. Prange*

Department of Physics and Astronomy and Center for Theoretical Physics,
University of Maryland,
College Park, MD 20742, USA

1. Introduction

This talk summarizes the Fluctuating Band Picture of ferromagnetism in pure iron and nickel--that picture developed by KORENMAN, MURRAY and PRANGE, as well as by CAPELLMANN [1], and based on the concept of short-range magnetic order (SRMO). Most of the material in this talk has been published in a series of papers and conference reports [2-13]. Some of the unpublished work will be discussed in shorter talks.

Of all magnetic materials, iron and nickel are by far the most extensively studied. Good single crystals exist, as elemental materials with cubic symmetry extensive numerical calculation is feasible, and there is great motivation, both basic and applied, to study these materials. The basic interest, of course, is that there is no widely accepted picture of the fundamental physics of iron and nickel.

As compared with other magnetic materials, therefore, much is demanded of a theory of iron and nickel. A theory as good as that of normal or superconducting metals, i.e. one whose parameters can be estimated from first principles to, say, 20% accuracy, can be hoped for. This goal is perhaps too ambitious at present, but it should not yet be abandoned. We summarise the experimental and band-theoretical situation as follows:

1.1 The ground state and elementary excitations are given by band theory and the RPA. For the ground state, the above goal is almost met: the density functional method provides an understanding of the ground state and, to a lesser extent, of the elementary excitations of iron and nickel (and of almost everything else). This theory tells when magnetism does and does not exist [14], gives the magnetisation and form factors [15], the cohesive energy, lattice constant and bulk modulus [14] and the Fermi surface [15]. The main part of the cohesion of the 3d metals is from the banding of d electrons and the s-d hybridisation [16]. These energies evidently remain nearly intact at experimental temperatures. There is a giant internal magnetic pressure which expands the lattice considerably in the magnetic state [14,17]. This expansion gains by reducing the kinetic energy needed to exchange split the bands. It is characteristic of itinerant exchange. Experimentally, the magnetic pressure remains above T_c which argues that the local electron structure in the paramagnetic state remains very close to that in the ground state. This indicates that such notions as the Fermi surface continue to have a meaning not too unlike their ground state counterparts. The fact that the magnetic energy per atom is considerably larger than T_c [14] leads to the same conclusions.

---

* Supported in part by NSF Grant No. DMR-79-08819

By using an effective (Hubbard) Hamiltonian, meant to be treated in the Hartree-Fock and extensions thereof, (e.g., the RPA), the spin wave states, including their intersection with the Stoner continuum are found [18] in agreement with experiment [19]. Self-energy corrections to Koopman's theorem are interesting and important in the case of nickel [20]. The two elementary excitations, magnons and 'Stoner', suffice to give the thermal properties of iron and nickel below about $T_c/2$. (Doubt of this existed for nickel, but LONZARICH [21] has resolved the discrepancy by taking the entire experimental spinwave spectrum into account.)

1.2 Standard finite temperature band (Stoner) theory is on too high an energy scale to agree with the observed thermodynamics, and if the Stoner transition temperature T_s is arbitrarily scaled down to agree with the observed T_c, the theory would predict far greater changes in single particle states and spin wave states than are actually observed. In fact, all those quantities that can be calculated from underlying band wavefunctions and energies show little temperature dependence even above T_c. A list would include the magnetic pressure, the density of states (photoemission, specific heat, resistivity), the work function and core level chemical shifts, and the Landau-Lifshitz-Ornstein-Zernike magnetic stiffness parameter [4]. Stoner excitations describe longitudinal spin fluctuations coupled in an important way to density fluctuations. Estimates by HUBBARD [22], HASEGAWA [23] and others, reported at this meeting, which include possible collective effects, do not change the conclusion that the magnetisation magnitude fluctuations are high in energy.

1.3 Standard (RPA) spin wave theory gives a somewhat lower scale of energy, but has no good way of including the needed self-consistent softening of the spin waves with temperature. In the past decade, several interesting but crude methods to account for this nonlinearity have been devised which get T_c right to within a factor of two or so. Methods presented at this conference include versions of the alloy analogy (CPA) method, [HUBBARD, HASEGAWA] and versions of the effective Heisenberg Hamiltonian method, [EDWARDS, PRANGE] followed by a mean field approximation. The remarkable thing is that ostensibly quite different pictures give very similar results to this degree of accuracy.

1.4 The thermodynamics are roughly described by a Heisenberg model (HM). More precisely, M(T) is quite well given by a spin 1/2 HM in mean field theory for iron nickel and cobalt. The magnetic specific heat, entropy, and (thermodynamic) magnetic energy, however, are considerably smaller than expected from a HM for both iron [24] and nickel [25] (contrary to EDWARDS [26]). In fact, this small entropy is consistent with the scale of short-range magnetic order deduced from the measurements of the dynamic susceptibility [10]. The static susceptibility has a Curie-Weiss form with Curie constant somewhat greater than the Heisenberg S(S+1)/3 (S=1,1/2 for Fe,Ni) but not by so much that a great deal of short-range order is easily understood.

1.5 The temperature dependence of the dynamic susceptibility and the single particle Green's function is important, in part because this dependence is much less than the Stoner model predicts. Spin waves, intersecting the Stoner continuum, exist above T_c much as at low temperatures [19], and the broadening, shift, line skewing, intersection, and cross over to diffusive behavior at long wavelengths can be understood on the basis of short-range magnetic order [5]. The temperature dependence of the single electron Green's function for nickel, as measured by angle-resolved photoemission [27,28] has attracted considerable attention. However, the resolution is such that the experiments are consistent with both SRMO [13] and the alloy analogy model [28].

1.6 Pure iron and nickel are thus remarkably similar magnetically in almost all directly observable ways. This is in spite of the fact that the exchange splitting of Fe (2eV) is an order greater than that of Ni (.3eV). In good part, this is due to the importance of intraatomic (Hund's rule) exchange in Fe, which has two or three d holes per atom, whereas Ni has none or one. This means that the existence of an atomic moment on iron is much less dependent on environment than is the case in nickel, a fact which is very important in distinguishing between dilute alloys of these materials, for example. Some approaches make much of this difference, and require two distinct pictures, both of which must be able to explain the very similar phenomena described above. A major virtue of the fluctuating band picture is that it encompasses iron and nickel in a single theory. This can be done because the interatomic (itinerant) exchange is comparable in the two cases, and it is just this part which gives rise to the temperature dependence. Another difference is that, unlike iron, the majority d band in nickel is full, but the primary effect of this is on the self-energy corrections in the single-electron Green's function. Finally, because of the smaller exchange splitting, Fermi surface smearing effects probably have some observable effect in nickel [29].

2. Formulation of the Theory

An outline of all the current spin fluctuation theories is as follows: An effective Hamiltonian is chosen, to be treated in Hartree-Fock theory and its extensions, so as to avoid the necessity of carrying along high-energy correlations (e.g. screening) in the wave function. Next, the partition function is represented (by means of a Hubbard-Stratonovitch transformation) as a sum over an ensemble of generalised Hartree-Fock states. The sum takes the form of a functional integral over spin (or exchange field) configurations, weighted by the Boltzmann factor for that configuration. To do this sum, it is decided which are the important kinds of configurations, and their energies are estimated. Finally, the configuration sum is approximately carried out and the observable quantities are calculated. The theories differ in their choice of dominant configuration and how to estimate its energy, on how important the time dependence of the configurations are, on how to count them, and on what further ad hoc corrections to include.

The fluctuating band theory starts from a generalised effective Hubbard Hamiltonian, with nine s-p-d bands. (It is usual to use an even simpler Hamiltonian but if better than factor-of-two accuracy is desired, it is not clear how much further simplification is allowed [18,26].) In terms of the operator $c_{i\nu\sigma}$ which annihilates an electron in atomic orbital ν of spin σ on site i one such Hamiltonian is

$$H = \sum t_{i\mu,j\nu} c^{\dagger}_{i\mu\sigma} c_{j\nu\sigma} + \sum U_{\nu} n_{i\nu\uparrow} n_{i\nu\downarrow} . \qquad (1)$$

Here $n_{i\nu\sigma}$ is the number operator. The hopping matrix t is short range between d orbitals, but is longer range when s-p orbitals are involved. The intraatomic Coulomb interaction is taken to be diagonal in the atomic basis, an approximation which is justified because it is adequate for the ground state. U_{ν} thus takes on two values for t_{2g} and e_g symmetry respectively, and is zero for s or p symmetry.

States far removed from the ground state cannot be reliably found on the basis of (1). In particular, the intraatomic configurations must not be too different from the ground state configuration. For example, the mean charge on an atomic site must remain nearly zero, and if there are two or more d holes on a site, their spin must be parallel.

The parameters of (1) are found indirectly by fitting to the band structure derived from the density functional method, or they can be fit directly to experiment. (There is some uncertainty in this fit for nickel.)

The next step is the transformation to magnetisation variables. (A separate discussion will be provided of some of the technical problems encountered.) Although the method in principle is exact, an exact solution is not desired since we start with an effective Hamiltonian. Rather, we want a generalization of the Hartree-Fock approximation.

We choose a formulation which maintains spin rotational invariance throughout and is capable of dealing with spin waves accurately. Our philosophy is not to ignore the experimental evidence that the nonlinear effects of the longwavelength spin fluctuations are of importance. This contrasts with the alloy analogy approaches using the CPA which considers only short-wavelength fluctuations. Using the HUBBARD-KORENMAN-PRANGE representation [22,7] and simplifying its writing somewhat, the partition function is

$$Z = \int D\vec{\mu}_{i\nu\tau} \exp(-\sum \beta U_\nu \int \vec{\mu}_{i\nu\tau}^{\,2} d\tau) \mathrm{Tr}\{e^{-\beta H_o} \exp(\sum 2\beta U_\nu \int \vec{\mu}_{i\nu\tau} \cdot \vec{M}_{i\nu}(\tau) d\tau)\} \tag{2}$$

For large β, a saddle point approximation is reasonable, and this gives the Hartree-Fock approximation to the problem. We remark that β may be 'large' even above T_c since the energies that appear in the underlying single-particle problem are band energies in the electron volt range. The condition for the saddle point is $\vec{\mu}_{i\nu\tau} = \langle\vec{M}_{i\nu}(\tau)\rangle_\mu$. The expectation value is in the single-particle system (defined by the above trace) for a given configuration $\mu_{j\mu\tau}$. The minimizing solution is $\vec{\mu}_{i\nu\tau} = M^S{}_\nu \hat{e}_z$, a constant independent of site i and (imaginary) time τ, dependent only on ν, and equal to the corresponding component of the Stoner theory magnetization $M^S{}_\nu$ at temperature $1/\beta$. Thus the Stoner free energy, $F_S(T)$, (which gives the linear specific heat) is obtained.

Next the exponent is expanded about the minimizing path which leads to a Gaussian form

$$Z = \exp(-\beta F_S(T)) \int D\vec{\mu}_{i\nu\tau} \exp[-\sum \delta\vec{\mu}_{i\mu\tau} \cdot \overleftrightarrow{K}_{\mu,\nu}(i-j,\tau-\tau') \cdot \delta\vec{\mu}_{j\nu\tau'}] \;. \tag{3}$$

Here $\overleftrightarrow{K}$ is the inverse RPA susceptibility. Eq.(3) provides the standard spin wave corrections to the free energy.

Finally, we expand about locally minimizing paths, but paths which globally can have any spin direction. At this stage we introduce a static approximation and neglect the time dependence of the direction of μ completely. The static approximation amounts to the neglect of certain quantum effects, which will be discussed later. Also, because of the importance of Hund's rule energetics, μ is taken to have the same direction for all ν. The results are numerically not bad and can be applied even above T_c. They do lead to a certain lack of self-consistency however.

The expansion parameters in this case are $\delta\mu$, the fluctuations in the magnitude of the vectors $\vec{\mu}$, and $|\hat{\mu}_i - \hat{\mu}_j| \equiv a_{ij}$, the difference in the directions on neighboring sites. The smallness of a_{ij} is what is meant by SRMO in this context. The fluctuations in $|\mu|$ are again handled by expansion and Gaussian integration. The direction dependence gives an free energy expression which has the form of a classical Heisenberg energy with exchange (Fourier transformed)

$$J(q) = -(1/4)\sum |\tilde{\Delta}|^2 [f_{n,p+q,\sigma} - f_{m,p,-\sigma}]/[E_{n,p+q,\sigma} - E_{m,p,-\sigma}] \;. \tag{4}$$

Here the E's are Stoner band energies, the f's are Fermi functions, and $\tilde{\Delta}$ is

$\sum_\nu A_{n,\nu}(p+q,\sigma)U_\nu M^S{}_\nu A^\dagger{}_{m,\nu}(p,-\sigma)$, an expression which takes into account the matrix elements A which transform from the atomic-like basis to band states. (We do not agree with CAPELLMANN [1] and HUBBARD [22] that the matrix elements drop out, except in the one-band case.) It is precisely these matrix elements that are needed to get accurate RPA spin wave energies.

The resulting J(q) has been computed by C.S. WANG [30], using energy bands fit to ones she obtained from first principles [15]. (Some of the details are reported in a separate discussion.) Using J in mean field HM or in a spherical model formula, a T_c about 300K for nickel, and about 1000K for iron results. (There are some band corrections to be made that likely will bring the value for iron down somewhat.) Using the spherical model formulation and finding the susceptibility as a function of T/T_c gives a Curie law with the observed Curie constant. WANG's calculation seems to be the best, in terms of its freedom from unnecessary approximations of several [1,22,31] that exist for what amounts to J(q) averaged over q, i.e. T_c in mean field theory. These calculations disagree with each other by as much as a factor of four, so expressions like (4) seem to be quite sensitive to approximations made in their evaluation. WANG's is the only calculation of the full J(q) however.

The inconsistency that arises is that the mean angle between neighboring exchange fields as predicted from this classical HM is too large to agree with that found from neutron scattering, as well as being too large to justify the expansion in a_{ij}.

The quantum corrections, i.e., the corrections arising from the time dependence of the exchange fields may account for this discrepancy. These corrections in a quantum Heisenberg model raise T_c by about a factor of $1+1/S$ as compared with a classical Heisenberg model of the same exchange, as much as a factor of three in nickel. Already at low temperatures, the classical Heisenberg model considerably overestimates the decrease in magnetization, because the Bose factor $n_q = 1/(e^{\beta\omega(q)}-1)$ which suppresses the thermal excitation of short-wavelength spin waves is replaced by $1/\beta\omega(q)$ in the classical case. Thus, quantum effects are more important for short-wave-lengths, and have the consequence that short-wavelength fluctuations are less highly excited than is predicted classically.

Therefore quantum effects tend to raise T_c by inhibiting short-wavelength fluctuations and thus increase the short-range order. The magnitude of these effects may be fairly large, judging by the size of the factor $1+1/S$. There is some reason to feel that the energies calculated classically may not be so sensitive to the 'time' dependence, but more study is needed on this question.

3. Short-Range Magnetic Order

The concept of SRMO has come under rather fierce attack from other participants at this symposium. Nevertheless, by accepting the experimental evidence on its behalf, one is able to explain many things, which no other presently extant theory is able to do.

There are two arguments against SRMO. One is based on the observed static susceptibility, the other on classical (static approximation) spin fluctuation theory. (The results just reported as well as those of HUBBARD and HASEGAWA fall into this class.) The second argument is compelling only insofar as there is reason to believe that the quantum corrections are insignificant.

Mean field theory in a HM gives something like the observed susceptibility, so a Heisenberg model with lots of SRMO tends to give too great a Curie constant. The susceptibility χ is $\sum_j\langle\vec{S}_i\cdot\vec{S}_j\rangle/T$. With local, temperature-independent spins, each term in the sum tends to be positive and SRMO makes χ large. However, the corresponding calculation for paramagnons gives both positive and negative contributions to the sum and a much smaller susceptibility.

The states which on the classical calculation destroy SRMO are short wavelength since the 'moment' flips direction at random from one neighbor to the next. It is known from experiment that such high-energy, high-wavenumber excitations decay rapidly into Stoner states: we call these states paramagnons rather than magnons, for lack of a better label. Theories up to now, (except for the RPA) neglect this because of the static approximation. (The velocity of a high-wavenumber spin fluctuation is not zero, and indeed it can match the velocity of a particle-hole pair leading to rapid decay and paramagnon effects.) Thus, the local 'spins' of iron and nickel come apart quite unlike a Heisenberg model when they are excited at short wavelength. It is not implausible that the spin correlation function has both positive and negative parts, as in other paramagnon theories.

SRMO explains a great deal. The arguments against it are not conclusive at this stage, neither is it conclusive that quantum effects play the role just envisaged. The fact is that the states of quantum energy T_c and above are complicated in Fe and Ni. A crucial stumbling block which is preventing the completion of the theories of iron and nickel is the difficulty of dealing convincingly with these states.

4. Conclusions

There is a choice between believing in instantaneous short-range order on the scale indicated by the neutron data or not. Believers can easily explain the neutron data itself, including its intersection with the Stoner continuum, the small magnetic entropy, and many other experimental results which indicate that the band energies do not change much at experimental temperatures. They can naturally understand why iron and nickel are so similar in their thermodynamics and their excitation spectrum. There is no difficulty in seeing how a nickel atom can remember its spin. They must, however, believe that something, maybe quantum effects, can increase the short-range order above that predicted by the static models at the same time giving T_c its observed value. They also must believe that paramagnon effects or something else leads to a Curie constant as small as is observed.

Those who do not believe in instantaneous SRMO have two alternatives. The first is that there is nevertheless a dynamic SRMO. This could happen if the 'paramagnon' fluctuations did not appreciably interact with themselves or with the single electron states even though they are excited enough to reduce the scale of instantaneous SRMO. For example, maybe the quantum time dependence of these excitations is so important that their action on other excitations is reduced by 'motional narrowing'. To choose between this and the previous scenario means dealing with the 'paramagnon' states convincingly.

Finally, some deny the existence of SRMO altogether, (at least in iron,) although this may exacerbate the problem of dealing with the paramagnons. In this case the length scale coming from the neutron data is regarded as simply an artifact of the analysis. One may then treat iron as a system of classical (or quantum) spins with no SRMO and by putting in some quantum corrections, achieve a reasonable picture of the phase transition and paramagnetic state.

It is plausible to argue that adding in the long wavelength fluctuations will fix up the critical and low-temperature regions while not perhaps having much other effect. One then must believe that it will be possible to show that the giant internal magnetic pressure continues to be exerted in the disordered state, and all those things such as the Landau-Lifshitz parameter, will come out to be independent of temperature. 'Propagating' spin waves will exist in the theory and some substitute for a Stoner continuum will be at just the right energy to cut them off. And there will be two quite different theories, neither of which now exists, which accomplish all this.

Acknowledgement. The work reported has been done with V. Korenman.

5. References

1. H. Capellmann:J. Phys. F4, 1966 (1974); Z. Phys. B34, 29 (1979)
2. R. E. Prange and V. Korenman:AIP Conf. Proc. 24, 325 (1975)
3. V. Korenman, J. L. Murray and R. E. Prange:AIP Conf. Proc. 29, 321 (1975)
4. V. Korenman, J. L. Murray and R. E. Prange:Phys. Rev. B16, 4032 (1977)
5. V. Korenman, J. L. Murray and R. E. Prange:Phys. Rev. B16, 4048 (1977)
6. V. Korenman, J. L. Murray and R. E. Prange:Phys. Rev. B16, 4058 (1977)
7. V. Korenman and R. E. Prange:Phys. Rev. B19, 4691-97 (1979)
8. V. Korenman and R. E. Prange:Phys. Rev. B19, 4698-702 (1979)
9. V. Korenman and R. E. Prange:Int. Phys. Conf. Ser. 39, 562,567 (1978)
10. V. Korenman and R. E. Prange:Solid State Comm. 31,909-912 (1979)
11. V. Korenman and R. E. Prange:J. Appl. PHysics 50, 1779-81 (1979)
12. V. Korenman and R. E. Prange:J. Mag. and Mag. Mat. 15-18, 355-356 (1980)
13. V. Korenman and R. E. Prange:Phys. Rev. Lett. 44, 1291-1294 (1980)
14. J. F. Janak and A. R. Williams:Phys. Rev. B14, 4199 (1976)
15. C. S. Wang and J. Callaway:Phys. Rev. B15, 298 (1977); B16, 2095 (1977)
16. C. D. Gellatt,Jr.,H. Ehrenreich and R.E. Watson:Phys.Rev.B15,1613 (1977)
17. M. Shiga:Proc.Conf.Phys. Transition Metals (Leeds,August 1980)
18. J. F. Cooke:Phys.Rev.B7,1108 (1973);J.Mag. and Mag.Materials 14,112-(1980); J.F.Cooke,J.W.Lynn and L.C.Davis:Phys. Rev. B21,4118 (1980)
19. H. A. Mook, J. W. Lynn and R. M. Nicklow:Phys. Rev. Lett. 30, 556 (1973); J. W. Lynn:Phys. Rev. B11, 2624 (1975) and (to be published)
20. J. Hertz and D. M. Edwards:Phys. Rev. Lett. 28, 1334 (1972); L.A.Feldkamp and L.C. Davis:Phys. Rev. B22,3644 (1980); A. Liebsch:(to be published)
21. G. Lonzarich:(to be published)
22. J. Hubbard:Phys. Rev. B19, 2626 (1979); B20, 4584 (1980)
23. H. Hasegawa:J. Phys. Soc. Japan, 46, 5 (1979)
24. G. Grimvall:Physica Scripta 13, 59 (1976)
25. R. E. Pawel and E. E. Stansbury:J. Phys. Chem. Solids, 26, 757-65 (1965)
26. D. M. Edwards:J. Mag. Mag. Mat. 15-18, 262 (1980)
27. D. E. Eastman, F. Himpsel, and J. A. Knapp:Phys.Rev.Lett. 40,1514,(1978); U. Gerhardt, C. Maetz, A. Schutz, and E. Dietz:(to be published)
28. K. Usami and T. Moriya:(to be published)
29. J. F. Cooke:Bull.Am.Phys.Soc. 25, 183 (1980)
30. C.S. Wang and R. E. Prange:(to be published)
31. P. J. Lin-Chung and A. J.Holden:(to be published)
32. B. S. Shastry, D. M. Edwards and A. P. Young:Proc.Conf.Phys. Transition Metals (Leeds,August 1980)
33. J. W. Lynn:Proc.Conf.Phys. Transition Metals (Leeds,August 1980)
34. J. B. Sokoloff,et.al:J. Appl. Phys. 50(3)(1979)
35. M. V. You, V. Heine, A. J. Holden and P.J.Lin-Chung:Phys.Rev.Letters 44, 1282 (1980)
36. W.E.Evanson, S.O.Wang and J.R.Schrieffer:J. Appl. Phys. 41,1199 (1970)
37. M. Cyrot:Phys. Rev. Letters 25, 487 (1970)

38. K. K. Murata and S. Doniach:Phys. Rev. Letters 29, 285 (1972)
39. J. A. Hertz and M. A. Klenin:Physica(Utrecht)B91,49 (1977)
40. T. Moriya and Y. Takahashi:J. Phys. Soc. Japan, 45, 397 (1978);
T. Moriya: J. Magnet. and Mag. Materials 14, 1-46 (1979).

6. Appendix

I have been told that about a million haiku and half that many tanka are published each year in Japan, some in the Japanese equivalent of the Wall Street Journal. This has emboldened me to submit the following poem, which like most haiku, attempts to evoke the thoughts one had at a special time and place.

A Haiku Composed in Honor of the Taniguchi Symposium

First snow on Fuji--
Where do spinning sparks wander?
How do they come home?

An Imaginary Tanka Contest to Express the Results of the Symposium.

(In a tanka contest, the given haiku must be completed with a fourteen syllable couplet.)

Swimming and swirling
the 'lectrons are in chaos
well arrang'd. But how?

....

Orderly, orderly, dance
the close neighbors together.

(Prange)

Iron spins are tightly bound
nickel spins go round and round.

(Edwards)

Up down down up up up down
Coherently at random.

(Hasegawa)

Not knowing which way is up
Like drunkards who remember.

(Hubbard)

Spinning weakly, more or less,
Is good for every one.

(Moriya, Hertz)

....

Short-Range Magnetic Order in Iron and Nickel

R.E. Prange[1]

Department of Physics and Astronomy and Center for Theoretical Physics, University of Maryland, College Park, MD 20742, USA

The concept of short-range order is important and widespread in physics. That short-range order exists and explains many phenomena is accepted as commonplace in the theories of liquids, alloys, and glasses. The reason for short-range order at a given temperature is the same as the reason for any order, namely that in the competition to raise the entropy, while lowering the energy, the amount of energy gained by ordering wins out over the entropy lost. Clearly, an understanding of the energetics of ordering is important, and so is an understanding of the entropy. It is the latter which is usually the most difficult to come to grips with.

The Fluctuating Band Picture of itinerant electron ferromagnetism is based on the concept of short-range magnetic order (SRMO). As is the case with short-range positional order, it is difficult to calculate theoretically whether and to what extent it exists. It is also rather difficult to measure directly. Evidence in favor of SRMO is as always: The outcome of experiment is predicted successfully, on the assumption of SRMO. One consequence of SRMO is that, in a definite sense, the band structure of an itinerant ferromagnetic does not depend very much on long-range order as long as there is an appropriate degree of SRMO. Such quantities as the total density of states, the work function, the Ornstein-Zernike magnetic stiffness parameter, and several others depend only on this band structure and therefore should be weakly dependent on temperature, even above T_c, as is observed [4][2]. Other quantities, such as the dynamical susceptibility and the Green's function are much less temperature dependent than in the absence of SRMO.

The specific heat and thermodynamic energy involve the SRMO. It is not easy to extract the magnetic part of these quantities. The thermodynamics of iron especially, with its α-γ-δ phase transitions in the range above T_c is not well understood. EDWARDS [26] claims that the specific heat for iron can support little SRMO. We have been reluctant to attempt a reanalysis of the data because the axe we are grinding is too obvious. However, we have recently realised that GRIMVALL [24] has provided an independent current estimate. He finds that the total magnetic entropy of iron is about half that predicted by the formula Nkln 3. The total magnetic energy is also small, 5.68 kJ/mole. Setting this equal to the Ginzberg-Landau energy, $\int d^3xA|\nabla M|^2$, which the fluctuating band theory indicates is a reasonable approximation, gives for $a\equiv\sqrt{<|\nabla M|^2>/2}$ the value .29 A^{-1}, in agreement with the neutron data. We [4] have previously used an analysis [25] of the nickel data to arrive at a similar estimate for it. GRIMVALL's energy actually includes contributions from single-particle excitations above and beyond that given by the linear specific heat. Thus the scale of SRMO might be greater than given by this analysis.

---

[1] Supported in part by NSF Grant No. DMR-79-08819

[2] References in this paper are to those cited in the summary paper of *R.E. Prange*, this symposium (pp. 55-62)

SRMO is most directly involved in the magnetic susceptibility, $\chi(q,\omega)$ or, equivalently, the magnetic structure function $S(q,\omega)$. Indeed, one might be inclined to say that a measurement of S is ipso facto a measurement of the magnetic order. Unfortunately, although various techniques exist for making this measurement, some of the more interesting features of the susceptibility are experimentally unknown.

Among the more interesting data is the dynamic susceptibility [19], which for a range of q values from about .3 -.8($A^{\circ -1}$) and for all temperatures shows a propagating mode which is very similar to the spin wave mode to which it reduces at low temperature. Recent suggestions [32] that this mode is diffusive rather than propagating are experimentally unfounded [33]. Even without background subtractions, there is a peak in $S(q,E)$ at finite energy for fixed q. Suggestions that the background subtractions made by the experimenters are incorrect are also dubious. The background consists of a <u>temperature-dependent, q-independent</u> part, coming from multi-phonon scattering, plus a <u>temperature-independent, q-dependent</u> part, decreasing with increasing q, coming from the fact that the steep energetic dispersion requires detection of neutrons at very small angles and thus the detector is very close to the incident beam. The supposition made by EDWARDS [33] that the slope (with q) of the background is temperature dependent is thus ruled out by the experiment.

The details of this mode agree very satisfactorily with the predictions made on the basis of SRMO. Other theories of this propagating mode of a more qualitative nature, also rely on the idea of SRMO, directly or indirectly. The dynamic susceptibility in the range of q values less than about .3 $A^{\circ -1}$ changes from propagating to diffusive at T_c, and the susceptibility for large q values is very broad in energy which supports the idea of a Stoner continuum of states. The former feature is quite general to any reasonable theory, but the latter is representative only of itinerant magnets and depends on the detailed band structure. So, the observed dynamic susceptibility supports the idea of SRMO.

In the fluctuating band theory, the short-wavelength excitation rides on a long-wavelength thermal fluctuation with typical wavenumber a. One can apply the RPA in a reference frame which straightens out the thermal fluctuation. The biggest effect (of order a/q) is the inhomogeneous broadening of the line. The neutron can excite waves of momentum $q \pm a$ while giving up only momentum q. In the case of inhomogeneous broadening it can well be simpler to look at the line shape for fixed E, and that is the case here. It is found, [19] and theory predicts [10] that the line should be approximately Gaussian in the magnitude of the momentum. Its shape for fixed q is quite shifted and skewed, and of course the intersection with the Stoner continuum is conspicuous. The excitation is propagating in the sense that $S(q,\omega)$ has a peak at finite ω for fixed $q \gg a$. For $q \ll a$, the peak is at $\omega = 0$ and the mode is diffusive. Theories without SRMO have difficulty in explaining propagating modes [32]. The magnitude of the SRMO determined from the spin wave data is that the rms gradient of the magnetisation direction is about .5 $A^{\circ -1}$ in both iron and nickel.

The theory also predicts some interesting polarisation dependence below T_c. There should be three separate lines, in which the neutron's spin is flipped up, down, or not at all. Only the first is allowed at low temperature, and the second has already been observed [34]. Each of these lines should have somewhat different widths and shifts. This will eventually provide a nice test of the theory.

Unfortunately, the neutron data cannot be integrated over frequency to give the instantaneous structure function, $S(q) \equiv \int d\omega S(q,\omega)/2\pi$. What can be measured is the static, long-wavelength susceptibility, $\chi(0,0) \equiv \chi_o$, which shows a Curie-Weiss behaviour, with a Curie constant only somewhat larger than $S(S+1)/3$ (with $S=1,1/2$ for Fe, Ni respectively). χ_o is directly related to $S(0)$. This experimental result has been used [32] to argue against SRMO on the basis of a Heisenberg model but the evidence is not completely compelling. Further discussion is given in the summary talk[2].

Before leaving the dynamic measurements, it should be mentioned that the amplitude of the observed lines appears to decrease with increasing temperature above T_c, although the line shape and position remain more or less constant. This is of course expected from the effects of the Debye-Waller factor, i.e. it is less and less probable that a single magnon will be emitted or absorbed with no accompanying phonon (or second magnon). It is estimated that the amplitude decreases even faster than does the phonon Debye-Waller factor. This decrease may be important, if true.

Effective Spin Fluctuation Energies: How Should They Be Calculated? What Do They Mean?

R.E. Prange[1]

Department of Physics and Astronomy and Center for Theoretical Physics,
University of Maryland,
College Park, MD 20742, USA

In the spin fluctuation theory of itinerant ferromagnetism the concept of the energy of a configuration of spins or spin exchange fields is central. If the energy is high, the probability of the configuration appearing thermally is small, so finding the energy is the first step in understanding which are the most important configurations. Some of the questions which arise in the course of this computation are numerical in nature, and some conceptual. The numerical questions are very important at this stage of theoretical development, since the answer to many questions hinge upon them.

The energy of a configuration is usually meant to be calculated for a system of independent electrons. In principle this reduction of the problem allows one to bring to bear much of the calculational firepower that has been developed over the past decade. There is, however, still the problem that the electrons are in a random potential. Therefore, additional approximations, such as the coherent potential approximation (CPA), or the short-range order approximation, are necessary. So far, the random potential has been taken to be static, so the problem is, at this level, well defined.

There is left a definite calculational problem which in a sense is purely numerical. From its solution, the basic input exists to find the magnitude of observable quantities such as T_c and the Curie constant, (after further theory and approximation). Unfortunately, this problem is one of band theory, and is quite difficult and time consuming. Such calculations are mostly left to the specialists. In this case, only a few specialists have so far enlisted their services. As a result, some of the rest of us have tried our hand, and have made approximations that are not technically necessary in the interest of saving time. The first thing is to choose a single-particle effective Hamiltonian. Since the calculations involved are difficult even for the specialist, a simplified form for the Hamiltonian is used rather than proceeding directly from first principles by means of spin density functional theory. The many-band 'Hubbard' Hamiltonian, with sp-states taken into account, is usually considered to be adequate [18,30,35][2]. One must be a little careful, because certain important energies, such as the exchange between different orbitals on the same atom, are often left out in favor of a renormalisation of some other parameter. This works only for certain states sufficiently resembling the ground state, however.

It is dangerous to simplify the Hamiltonian beyond this point, because in some cases at least, really bad numerical results are obtained, and no rule is known with any exactitude telling which cases are sensitive to further approximation and which not [26]. Sometimes the extreme simplification of using just one band is made, but there is no good way to know how important this ap-

---

[1] Supported in part by NSF Grant No. DMR-79-08819

[2] References in this paper are to those cited in the summary paper of *R.E. Prange*, this symposium (pp. 55-62)

proximation is without comparing it with a better one. Or, one would like to leave out the s-p bands. This, for magnetic problems is intuitively much better, but it could make the difference between ferromagnetism and some kind of spin density wave state! Our results below indicate that these approximations can be quite important in certain problems.

Many of the important configurational energies are of order T_c, i.e. of order 100 meV in iron and nickel, and of course much less in the weak magnets. Band energies are uncertain to this order of magnitude, and the main features of the d-band structure have characteristic energies 10 times higher. Thus one must obtain, by a complicated computational procedure based on a complex of energies ranging over a few volts, another complex of energies ranging over a few tenths of a volt. It is not clear just what causes this reduction in scale. Getting small numbers out of large ones, however, is a task which is often sensitive to small differences in the method.

To illustrate, I report some computations done together with C. S. WANG, where I hasten to add that she did the hard part. The results did not conpletely conform to my preconceptions, so they are quite believable. From these calculations a single number can be derived, which has the interpretation of being T_c (for iron or nickel), computed in a certain mean field approximation. It turns out that CAPELLMANN[1], HUBBARD[22], and LIN-CHUNG and HOLDEN [31] (only for iron) have also computed the identical number, or more precisely, this number can be inferred from their published results. (It is <u>not</u> the T_c they quote, however.) The computational schemes are quite distinct. CAPELLMANN and HUBBARD use one-band methods in their evaluation, but HUBBARD's result is a factor of two larger than CAPELLMANN's. WANG, and LIN-CHUNG and HOLDEN differ in that the latter neglect the s-p states. Since the important states in this calculation are shortwavelength, close to the zone boundary, it can be argued that this is not serious. These two calculations are presumably better than the other two, which is only fitting as they are the specialists. However, WANG's result is a factor of <u>two</u> smaller than that of the Cambridge group. The method used by WANG is more tried and true, and she makes the fewest subsidiary approximations, so her result is a priori the most believable. (LIN-CHUNG and HOLDEN made their calculation to test a new method.)

WANG also 'repeated' CAPELLMANN's and HUBBARD's work in the following sense. These two authors used what amounts to a one-band approximation. Once that approximation is made, only the density of states appears in the formula. HUBBARD takes for this density of states, one-fifth the total d-band density of states, and repeated five times. CAPELLMANN repeats 1/2 the e_g state density twice and adds 1/3 the t_{2g} density three times. However, WANG did not use the same state density as the other authors but rather used the d density that she found for her fully evaluated, nine-band formula. These results are listed as WC and WH respectively in Table 1. The actual formula evaluated by WANG is that for the effective classical Heisenberg exchange, $J(q)$, which is Eq. (4) of the main summary paper by the present author. T_c in mean field theory is then $T_c = (2/3kN)\sum[J(0)-J(q)]$ (where k is Boltzmann's constant, and the sum is over the N points in the first Brillouin zone). CAPELLMANN evaluated precisely this same formula in the one-band case.

The other authors found the cost in energy to change the exchange field at a single site, while keeping the fields at other sites fixed at their zero temperature value. A <u>special case</u> is if the exchange field does not change its length, but makes a small angle ϕ with all other fields. The formula for J is calculated on the assumption that all spin exchange fields have their zero-temperature magnitude, and that the angle made by any given exchange field with its nearest neighbors is small. This formula also covers the special case. An easy calculation shows that for the special case, the energy is $(kT_c/3)(1-\cos\phi)$, so from the energy of the single rotated spin exchange field, provided by the workers mentioned, the value of T_c can be extracted. These results are listed in Table 1.

Table 1 The Curie temperature from a classical mean field Heisenberg model, described in the text, as evaluated by several authors. W [30], WC,WH [see text], C[1], H [22], LC-H [31]

		W	WC	WH	C	H	LC-H
T_c[K]	Fe	1200	1300	1400	550	1200	1900
	Ni	320	290	280	200		

These results are difficult to understand. Leaving out sp-states and making a one band approximation seem to compensate at the 20% level, but apparently the choice of density of states can have a big effect. (CPA methods are not sensitive to the model density of states, according to HASEGAWA.)

We add a few comments on WANG's results. At small q, $J(0)-J(q)\equiv L(q)$ is known to have the form $(V/N)Aq^2$, with V/N the atomic volume and A the Bloch wall stiffness parameter, known from experiment. In the case of Ni, the value of A found by WANG agrees roughly with experiment. In the case of iron, however, WANG's value is almost a factor of two too high. The underlying bands for iron were taken to be those of WANG and CALLAWAY [15], about which there is relatively little controversy. (However, a direct unpublished calculation by CALLAWAY using his bands gives too great a value for A too.) Unless this indicates that the present calculation is in error over most of the Brillouin zone, the above value for T_c should not be much affected. On the other hand, within the framework of the Heisenberg model, a better formula for T_c is given by the spherical model approximation, i.e. $T_c = (2/3)[(1/N)\sum\{1/L(q)\}]^{-1}$, which is sensitive to the long-wavelength behaviour of L(q). Thus if a way is found to correct WANG's calculation to bring it into agreement with the experimental stiffness, it is likely that the best resulting T_c, within the framework of the classical Heisenberg model, would be somewhat lower than the observed value, but not by a factor of two, as in the case of nickel. One can also estimate that the classical calculation in the case of iron predicts somewhat more short-range order than it does in the case of nickel, but in both cases the scale is insufficient to fit the numbers derived from neutron data.

Many of the questions remaining in the theory of pure iron and nickel hinge on the outcome of numerical calculations such as those above. Are there important corrections to the above theory which raise the transition temperature, or must we seek effects which lower T_c? Are the T_c values predicted by the short-range order theories (those just reported), systematically lower than those found in the alloy analogy theories? Or are we at the limit of what can be calculated just now and must we conclude that there is not much chance of distinguishing between rival approaches? I hope that the latter is not the case, but it will take further hard work to tell.

Functional Integral Representations in Spin Fluctuation Theory: Is There a Right One?

R.E. Prange[1]

Department of Physics and Astronomy and Center for Theoretical Physics,
University of Maryland,
College Park, MD 20742, USA

1. Introduction

Most discussions of spin fluctuation theory can be formulated in the language of functional integration. A list of the practitioners of this method, just in the present application, would include EVENSON, WANG, and SCHRIEFFER [36][2], CYROT [37], MURATA and DONIACH [38], HERTZ and KLENIN [39], MORIYA and coworkers [40], HASEGAWA [23], HUBBARD [22], and KORENMAN and PRANGE [7]. There are, however, important mathematical problems which are encountered when the method is used. The most serious problem is that there are many possible different representations. Although these are equally exact before approximations are made, they can give very different results after formally identical initial approximations.

In this discussion, I shall review some representations that have been employed, and some of the criteria that can be used in choosing among them. The chief criteria I find to be of interest are, a) Is the lowest approximation a sensible one? Is it consistent with the known successes and failures of the Hartree approximation? b) Are spin waves, and therefore spin rotational invariance a manifest part of the theory? c) Is the theory free of Keiter's disease? d) Does the theory incorporate the major quantum effects?

There does not exist a representation for which the answers to the above questions are favorable. For the HUBBARD-PRANGE-KORENMAN [22,7] representation (I) questions a),b), and c) are answered in the affirmative, but the answer to d) is only partially satisfactory. The two-field Ising representation (II), probably introduced first by HAMANN in his study of the Kondo problem, has satisfactory answers to questions a) and c), and question d) is given a partially satisfactory answer. Moreover, it is answered in a sense complementary to the manner in which it is answered by I. However, b) is unsatisfactorly answered, in this case. It is possible to hybridize the two representations, and have the best of both worlds, but at considerable expense.

One other representation, the four-field Heisenberg representation (III), has been used regularly. It is essentially a competitor with I and both of them cannot be right. Here the issue is whether question a) is satisfactorily answered. The consequences of choosing between I and III are not trivial. Some of these consequences will be stressed.

There may be other criteria and justifications which exist, but which have not been published, since a number of well-regarded papers have simply gone ahead and used representations not meeting the above criteria. It is hoped that this discussion will serve to bring these arguments out into the open.

[1] Supported in part by NSF Grant No. DMR-79-08819

[2] References in this paper are to those cited in the summary paper of *R.E. Prange*, this symposium (pp. 55-62)

2. HUBBARD-STRATONOVITCH Transformation

The HUBBARD-STRATONOVITCH (HS) transformation is well known, and will not be repeated in detail here. The purpose is to transform the interaction term in a Hamiltonian so the the result is more amenable to approximation. When applied to the partition function, Z, the result is a functional integral whose integrand is simpler in the sense that it is the 'partition function' of a noninteracting particle problem, albeit a problem with a random space-time-dependent field. The transformation will have a virtue insofar as the important configurations of this field have significant simplifying features. Since the single most important configuration,(i.e. the one which maximises the integrand) corresponds to a particular Hartree approximation of the original Hamiltonian, there is hope that this program can have some success.

The general form for Z is

$$Z = \int Dx(r,\tau,\nu)\exp(-\beta x\circ x)\ \mathrm{Tr}\ \exp_+\{-\beta(H_0 - M\circ x)\}$$

$$\equiv \int Dx\ \exp\{-\beta F[x]\} \equiv \int Dx\ \exp\{-\beta x\circ x - F_0[x]\}\ . \tag{1}$$

Here r is position (often discrete), τ the imaginary time introduced along with the time ordered exponential $\exp_+$. There are generally a number of field variables x introduced and these are labelled by ν. The inner product $x\circ M$ indicates an integral and sum over the full range of these arguments. M is some single-particle operator function of the same arguments, and H_0 is the single-particle Hamiltonian. The exact meaning of the differential Dx can vary from case to case and that can have a very important effect, of course.

The form (1) differs from that encountered in field theory, where F[x] is replaced by L[x], with L the Langrangian and β replaced by $1/\hbar$. The difference is that L has terms in the space and time derivatives of x so that very rapidly varying configurations have large L and are thus suppressed. (In the case of one-dimensional Wiener measure, $L \propto (dx/d\tau)^2$ and discontinuous paths have probability zero, although the only paths with nonvanishing probability are those which have square root singularities almost everywhere.) The remaining paths can be relatively close to the extremal path, so in the limit $\hbar \to 0$, the extremal, classical, path completely dominates.

In the HS case, however, that is not true, and the extremal path is only part of the story for $\beta \to \infty$. The reason is that the independent particle system cannot respond to very rapidly varying paths of finite amplitude, so F_0 for such paths becomes small. More subtly, by starting from discrete τ and taking the continuum limit, it is possible to see that paths of infinite amplitude and infinite frequency make finite contributions which are not suppressed at large β. These paths are inconvenient and without heuristic value in the functional integral context, and so they are integrated out in some approximate way, leaving only the continuous paths. We then suppose that these remaining paths are governed by an effective Hamiltonian. Since, at low temperature, for the continuous paths the saddle point method becomes exact, the effective Hamiltonian must be one which in Hartree approximation adequately represents the ground state of the original true Hamiltonian.

The representation chosen determines the space of continuous paths over which the remaining integral is to be carried out. If this space is inappropriate to the problem, then an effective Hamiltonian may not exist, or it may be very unphysical. On the other hand, if it is known from other considerations that an appropriate effective Hamiltonian exists, then this program has a good chance of success. One knows in fact, what the effective Hamiltonian is in the context of itinerant ferromagnetism.

The tradition is to introduce the effective Hamiltonian in generalised Hubbard form from the start, make the HS transformation as if it were the real Hamiltonian, and then make the saddle point approximation. If this point of view is adopted, then the choice of representation is narrowed to one class, which is also free from Keiter's disease, and which will now be described. (Keiter's disease results from using the operator identity $n^2=n$, and then treating the rhs exactly, and the lhs approximately.) This class results from the decomposition

$$n_{\uparrow}n_{\downarrow} = n^2/4+(\underline{M}\cdot\underline{\zeta})^2 . \tag{2}$$

Here n_σ, n, and $\underline{M}$ are number and vector magnetisation operators, and $\underline{\zeta}$ is a unit vector. These quantities depend on site, time, and orbit. The two choices for $\underline{\zeta}$ which have been made are I, $\underline{\zeta}$ arbitrary, followed by an average over it, and II, $\underline{\zeta}$ constant. The latter approximation has the drawback that the space of continuous paths is not large enough to include spinwave-type fluctuations, which correspond to fluctuations in the direction of $\underline{\zeta}$. This space is Ising-like, which of course has a resemblance to a spin 1/2 Heisenberg system in its single-site or short-wave behavior, namely, each site has two states. It gets the longwave behavior completely wrong, but for certain purposes e.g. calculating T_c in mean field theory, these short-wave properties dominate and the approximation is not unreasonable.

The other obvious choice, I, is to average over all directions of $\underline{\zeta}$ independently, which makes manifest spin rotational invariance at all stages of approximation. PRANGE and KORENMAN [8] have shown that the saddle point approximation, including the small fluctuations about the minimizing path, then leads to the standard spin wave contributions at low temperature, which of course dominate the magnetic part of the thermodynamics. This is also clear from the functional integral method, because the classical configurations corresponding to long-wavelength spin waves cost little energy. The static approximation overestimates greatly the contribution of the shorter wavelength configurations, at low temperature, replacing the Bose function $n(\omega)$ by $1/\beta\omega$, and this is a case where one can see explicitly the mechanism by which the time dependence of the configuration leads to a well-known quantum effect.

An average over directions $\underline{\zeta}$ combining the virtues of I and II can in principle be chosen. A constraint that the angles between nearest neighbor $\underline{\zeta}$ not be allowed to become too large, gives rotational invariance <u>and</u> Ising behavior at short wavelengths. Imposing this nonlinear constraint in a felicitous way is not easy however, and a principle by which to choose the cutoff angle would have to be found.

The third interesting choice, III, is

$$n_{\uparrow}n_{\downarrow} = n^2/4+(\underline{M})^2/3 . \tag{3}$$

This choice has the spin rotational virtues of I, plus some technically happy points to be mentioned, but it gives a incorrect Hartree approximation, that is, one which is not equal to the Hartree-Fock. The form is the same, but there is a peculiar factor of three. The only way out of this contretemps seems to have been first proposed by Hubbard at this meeting. Namely, the starting Hamiltonian after all should not have been the final one, because eliminating the infinite frequency paths will renormalise the starting Hamiltonian into the effective Hamiltonian finally used, and who knows, maybe the mysterious factor of three will disappear in this process.

Perhaps one should have an open mind on this issue, as certainly nothing is known with any rigor. There is, however, an argument against this attitude. A great deal is known about the physical reasons for the renormalisations which carry the original Hamiltonian into the effective one. The most important are best expressed in Feynman graph language and include screening, core state relaxation, KANAMORI renormalisation, and corrections to take into account the omission of higher and lower bands. At zero temperature, the only continuous path left is the constant 'Stoner' path. All the other paths may be treated by formal perturbation theory, which is identical to ordinary perturbation theory, and thus the renormalisations may be carried out. The fact that there is no sign of funny factors of three in this procedure provides an argument against choice III.

The important technical difference between methods I and III is in the meaning of the measure Dx. In the latter case, there is a Gaussian field introduced for each component of M, thus forming a vector field, say $\underline{\mu}$, whose differential is $d^3\mu = \mu^2 d\mu\, d^2\mu$. In method I, only one Gaussian field is introduced, so the differential is $d\mu\, d^2\zeta$. The former differential is much simpler, of course, since it not only rotationally but translationally invariant. (The factor μ^2 cannot be exponentiated and included as part of $F[\mu]$ because the limit of this quantity as the time variable becomes continuous would not then exist.) It is only in method III that approximating $F[x]$ of Eq.(4) (of the summary paper of the author) by a quadratic form will result in a quadrature. This provides a motivation for using method III, if not a justification. Unfortunately, it may be that certain treatments in the literature may stand or fall depending on whether method III can be justified or not.

Finally, it should be remarked that the effects of the neglect of the time dependence are doubtless different between cases I and II. For case I, the time dependence is almost certainly not important for slow variations of direction $\underline{\zeta}$ (even of large amplitude,) whose typical $\nabla\zeta$ satisfies $D|\nabla\zeta|^2 \leqslant kT$, and is certainly important for spatially rapid variations of small amplitude. Little is known about spatially rapid variations of large amplitude, but they may very well be suppressed by these quantum effects. The time dependence is also important for fluctuations in the local magnitude of the decoupling fields, and inhibits large fluctuations. Little is known about case II, since by construction, only the difficult case is left.

The Magnetism of bcc Iron

D.M. Edwards

Department of Mathematics, Imperial College,
London SW7 2BZ, England

1. Introduction

In ferromagnetic insulators, and in rare-earth metals, the magnetic properties are associated with definite atomic magnetic moments which exist above the Curie temperature T_c and whose directions become ordered below T_c. Such systems are described quite well by a simple Heisenberg model, and far above T_c, at $1.5T_c$ say, there is very little correlation between the directions of neighbouring spins. In this lecture we shall discuss the extent to which this picture applies to ferromagnetic transition metals and to iron in particular. There is one essential difference between the transition metals and the insulators and rare earths. In the latter case the atomic moments are associated with electrons localized on the atoms whereas in transition metals the magnetic moments arise from itinerant 3d electrons. We must therefore consider the following points:

1. The concept of well-defined atomic moments above T_c in itinerant-electron systems.

2. The degree of short-range order above T_c.

An atomic moment may be regarded as well defined if

(a) it has a non-zero mean value, when averaged over a time interval long compared with that spent by a d electron on a given atom but shorter than some characteristic spin-fluctuation time,

(b) the magnitude of this mean value is essentially constant although its direction changes on the scale of the spin-fluctuation time.

The "hopping frequency" for d electrons is determined by the d band-width (~5eV for iron) while the spin-fluctuation frequency, of the order of a typical spin-wave frequency, corresponds to an energy k_BT_c (~0.1eV for iron). In the absence of strong short-range order it follows from (b) that the magnitude of the moment should be fairly independent of its local environment, that is of the directions of neighbouring atomic moments.

It is instructive to consider the conditions under which well-defined atomic moments can exist in the disordered state from two points of view. First consider the detailed motion of holes in the d band as they hop from atom to atom. In the case of iron there are about 2.8 holes and owing to strong correlation each atom will normally have either 2 or 3 holes. Hund's rule exchange tends to align the spins of these holes but, in the ferromagnetic ground state of bcc iron, the competing tendency to lower the one-electron band energy leads to incomplete alignment with a mean moment of about

$2.2\mu_B$. Thus in the ground state we expect each atom, at a given instant, to have either 3$\uparrow$ spin holes, 2$\uparrow$ spin holes or 2$\uparrow$ spin and 1$\downarrow$ spin hole. This is shown schematically for a line of atoms in Fig.1(a).

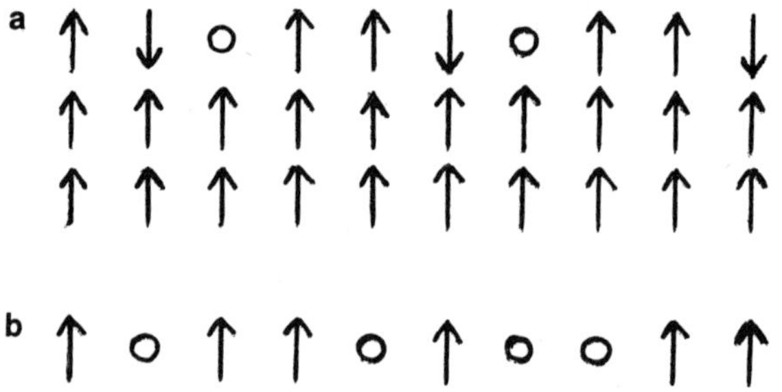

Fig.1(a) Hole occupation of a line of atoms in (a) iron (b) nickel

As holes hop on and off each atom retains at least one pair of holes with spins coupled by Hund's rule, thus defining a persistent spin direction. Even in the disordered state spin directions on atoms can clearly persist during many hopping processes. It is clear that atomic moments cannot persist in this way when there is less than one hole per atom, as in nickel. The ferromagnetic ground state for this case is shown schematically in Fig. 1(b), and many atoms have no hole. In the disordered state an $\uparrow$ spin or $\downarrow$ spin hole is equally likely to hop onto such an atom so there can be no persistence of atomic spin direction. Cobalt with about 1.7 holes is an intermediate case. From this point of view we expect that well-defined atomic moments can never exist with magnitudes less than $1\mu_B$. The second point of view is essentially based on an adiabatic approximation in which the spin direction on each atom is "frozen". For simplicity we consider the Ising-like situation where these directions are either up or down. We then ask whether an approximate self-consistent solution of the many electron problem exists with non-zero magnetic moments $\pm m_p$ on the atoms. If short-range order is unimportant the resulting binary "alloy" problem may be treated within the single-site coherent potential approximation (CPA). This type of procedure emerges from the Schrieffer-Cyrot functional integral formalism [1,2] when the static and local saddle-point approximations are made [3-5]. Numerical calculations show that non-zero moments exist in the disordered state in iron [4] but not in nickel [6]. Hubbard [7,8] and Hasegawa [9], going beyond the local saddle-point approximation, consider a continuous distribution of magnetic moments analogous to a multicomponent "alloy". The method may then be applied to nickel although the physical justification, in terms of atomic moments which are long-lived on the scale of the electron hopping frequency seems to be absent in that case. Possibly the Hasegawa-Hubbard method is only valid when well-defined atomic moments exist, as indicated by the appearance of non-zero moments in the local saddle-point approximation. In this case Hasegawa [9] finds little difference between the results of the saddle-point approximation and those of the more general procedure.

The two points of view discussed above both indicate that well-defined atomic moments may exist in the disordered state in iron but not in nickel. Since Hasegawa's work is based on a one-band model it may seem surprising that he ever finds well-defined atomic moments, in view of the stress laid on orbital degeneracy and Hund's rule in the other description. Presumably the one-band Hubbard model succeeds in so far as it simulates a real d band by a suitable choice of density of states, electron density and interaction parameter. For any genuine s band, with the "alloy" treated self-consistently using spin density functional theory so that there are no adjustable parameters, it seems unlikely that one would find disordered atomic moments in the metallic state. Of course atomic moments will exist in the Mott insulator regime, these being associated with localized electrons.

In section 2 we show how the Hasegawa saddle-point theory is related to a previous phenomenological model. We discuss the extent to which a purely Heisenberg-like model of iron is valid. In section 3 we consider further the nature of the paramagnetic state of iron. In particular we argue that the degree of short-range order above T_c is not unusually large and recent work on the problem of spin waves above T_c is described. Some conclusions are summarized in section 4.

2. The Hasegawa "alloy" theory and an earlier phenomenological model

Ten years ago the author [10,11] introduced a phenomenological model to describe the thermodynamic properties of metals such as iron where Stoner theory clearly fails. The model Hamiltonian may be written

$$H = H_0 - 2J \sum_i \underline{\sigma}(i) . \underline{S}_i - 2 \sum_{i<j} g_{ij} \underline{S}_i . \underline{S}_j . \tag{1}$$

Here H_0 represents an "itinerant" system characterised by an enhanced paramagnetic susceptibility χ_i, $\underline{\sigma}(i)$ is the "itinerant" contribution to the spin density at site i and $\underline{S}_i$ is a "localized" spin at each site. J and g_{ij} are exchange parameters. Although this Hamiltonian is formally of the s-d type its meaning is quite different. The "itinerant" and "localized" contributions to the magnetic moment do not refer to electrons occupying two distinct sets of orbitals, as in the s-d model, but to two different aspects of the same itinerant electron system. The precise meaning of (1) has only recently become clear [12] through establishing its relationship to Hasegawa's "alloy" theory. The model is interesting for the following reasons.

1. The molecular field theory of the model is in one-to-one correspondence with the results of Hasegawa's saddle-point approximation [4] when this approximation yields non-zero moments $\pm m_p$ in the paramagnetic state. This establishes a condition for the validity of the model and provides a way of calculating the parameters from first principles [12]. The molecular field theory works well for the thermodynamic properties of iron [11].

2. The generalization of the model to alloys has been found useful for discussing the temperature dependence of hyperfine fields in systems such as FeMn [13].

3. In iron the last term of (1) is found to be the most important [11,12] which suggests that a purely Heisenberg-like model of iron might not be too unreasonable. Such a model is used to discuss the spin dynamics in section 3.

To develop point 1 we consider Hasegawa's "alloy" near the paramagnetic state with $0.5+\delta c$ ↑ spin sites and $0.5-\delta c$ ↓ spin sites, where δc is small. The moments on the two types of site are then $\pm m_p + \delta m$ where in zero external field δm is proportional to δc. The magnetization per atom (in units of the Bohr magneton μ_B) is given by

$$M = 2m_p \delta c + \delta_m = M_\ell + M_i . \tag{2}$$

The first term is associated with the reversal of local moments and the uniform induced moment $\delta m = M_i$ is an essentially itinerant contribution associated with shifts in the local density of states. The energy per atom, relative to the paramagnetic state, may be written in the form

$$E = \tfrac{1}{2}\chi_i^{-1}M_i^2 - \gamma M_i M_\ell - \tfrac{1}{2}\beta M_\ell^2 \tag{3}$$

and the coefficients are directly related to quantities which can be calculated within the CPA. Thus χ_i, the susceptibility of the system with $\delta c = 0$, is given by

$$\chi_i = d^{-1}(L^-\rho_p^+ + L^+\rho_p^-) \tag{4}$$

where

$$L^\sigma = 1-K^{\sigma\sigma}+K^{\sigma-\sigma},\quad d = (1-K^{++})(1-K^{--})-K^{+-}K^{-+}. \tag{5}$$

Here we have used Hasegawa's notation [4] in which ρ_p^+ and ρ_p^- are the local densities of states at the Fermi level for majority and minority spin electrons at a given site. The quantities $K^{\sigma\sigma'}$ are just the unenhanced partial susceptibilities familiar from alloy theory [14] multiplied by the interaction parameter I, so that d^{-1} is an enhancement factor. For given M_ℓ, that is given δc, E is minimized when $M_i = \gamma\chi_i M_\ell$. We deduce from Hasegawa's work that the quantity $2m_p\gamma\chi_i$ is given by the expression on the right of (4) with ρ_p^σ replaced by Hasegawa's quantity M^σ. This determines γ, and β is determined by the expression

$$\Delta F = 4m_p^2(\gamma^2\chi_i+\beta)\delta c \tag{6}$$

for the energy to reverse a local moment. This energy may be calculated within CPA for a given δc. The expressions given here, which determine the parameters χ_i,β,γ, correspond to Hasegawa's one-band Hubbard model with the one-electron potentials determined by Hartree-Fock theory. In a more realistic calculation the parameters could, in principle, be determined by a full KKR-CPA calculation [15] with the potentials determined by spin density functional theory. We assume, with Hasegawa, that it is sufficient to replace the Fermi function by a step function in calculating χ_i,γ,β. The weak temperature dependence of these quantities can be included so that E becomes a free energy. Sakoh and Edwards [11] included a temperature dependence of χ_i phenomenologically but it played a negligible role. The molecular field acting on a local moment is given by

$$h = -\partial E/\partial M_\ell = \gamma M_i + \beta M_\ell + H \tag{7}$$

in the presence of an external field H. Then in mean field theory

$$M_\ell = m_p B_S(m_p h/k_B T) \tag{8}$$

where B_S is the Brillouin function and in Hasegawa's theory the spin S associated with the local moment is $\tfrac{1}{2}$. The itinerant contribution to the magnetization per atom is given by

$$M_i = \chi_i(\gamma M_\ell + H). \tag{9}$$

Equations (7), (8), and (9) determine $M_i(T)$ and $M_\ell(T)$ and the total magnetization is given by (2). Although the basic equation (3) is strictly only valid for small M_i and M_ℓ, that is δc small, Hasegawa's numerical calculations for Fe show that ΔF and δm are almost linear functions of δc for $\delta c < 0.3$.

With Hasegawa's theory written in the above form we see the precise correspondence with the molecular field theory of the Hamiltonian (1). Comparing (1) and (3) we see that γ and β are related to the exchange parameters J, g_{ij} by the equations $\gamma = \frac{1}{2}J\lambda$, $\beta = \frac{1}{2}\lambda^2\Sigma_j g_{ij}$ where $\lambda = 2S/m_p$. χ_i is the susceptibility of the "itinerant" system H^o. In this paper γ corresponds to $\gamma\mu_B^2N$ in reference [11] and similarly for β. From equations (7)-(9) and (2) we find [11] that the total susceptibility above T_c is given by

$$\chi = (\chi_i^{-1} + \chi_\ell^{-1} + 2\gamma)(\chi_i^{-1}\chi_\ell^{-1} - \gamma^2)^{-1} \tag{10}$$

where

$$\chi_\ell^{-1} = 3k_BT[m_p^2(S+1)/S]^{-1} - \beta. \tag{11}$$

This form makes explicit the interaction between "localized" and "itinerant" contributions. For a real alloy, of fixed composition, the susceptibility would be just the enhanced susceptibility χ_i. However an applied field turns round some local moments so that the composition of the "alloy" changes. The Curie temperature is given by

$$k_BT_c = \frac{1}{3}m_p^2[(S+1)/S]\ (\gamma^2\chi_i + \beta). \tag{12}$$

Clearly the specific heat, also discussed by Hasegawa [5], is a sum of two contributions, one itinerant and one from the localized spins. In the paramagnetic state the entropy is given by $\gamma_o T + k_B \ln(2S+1)$ where γ is related to the density of states $(\rho_p^+ + \rho_p^-)/2$ in the usual way. This agrees well with that observed in Fe [16] when we take S = 1. Sakoh and Edwards [11], taking S = 1 and $m_p = 2$, found that an excellent fit to the magnetic data of Fe was given by J = 0.063eV, $\Sigma_j g_{ij}$ = 0.058eV, $\chi_i/\mu_B^2 = 4.8\text{eV}^{-1}\text{atom}^{-1}$. The theoretical curve for χ^{-1} is shown in Fig. 2 with the experimental data for the bcc phases. With these parameters it is found from (12) that 85% of the contribution of T_c comes from the 'direct' interaction g_{ij} between local moments. The last term in (1) therefore seems to be the dominant one and a purely Heisenberg-like model of iron may not be too unreasonable. However the local moments do not correspond to localized electrons as in straightforward Heisenberg systems.

3. The Nature of the Paramagnetic State in Iron

In section 2 we have talked in terms of single-site approximations and molecular field theory. Thus short-range order above T_c is neglected. Hasegawa

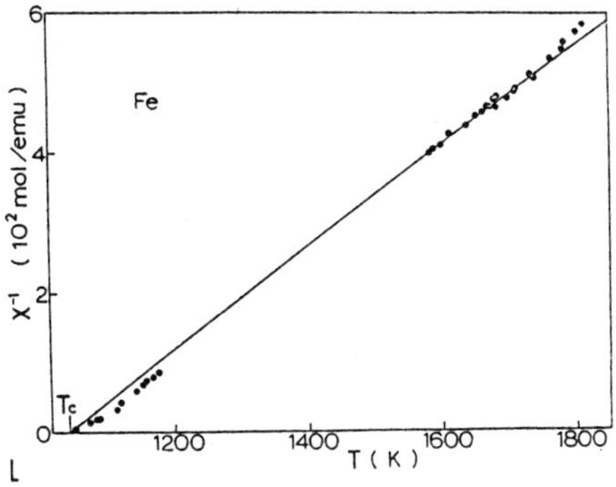

Fig.2 The fitted theoretical curve for χ^{-1} in the bcc iron compared with experimental points for the bcc phases [11]

[17] has recently reported the results of a two-site theory for iron. He finds that the correlation function $<m_i m_j>/<m_i^2>$, where m_i and m_j are moments on nearest neighbour sites i,j, decreases from 0.22 to 0.14 as T/T_c increases from 1 to 2. Values of this magnitude correspond to the small degree of short-range order met with in simple Heisenberg ferromagnets. On the other hand Prange and Korenman [18] propose a high degree of temperature-independent short-range order corresponding to a value of cos 36° = 0.81 for this correlation function [12]. The object of this proposal is to provide a simple explanation of the existence of rather well-defined spin waves far above T_c [19]. There are three arguments against the existence of this giant short-range order in iron:

1. It is incompatible with specific-heat data [12].

2. The resultant paramagnetic susceptibility χ is almost inevitably much larger than observed [20].

3. The behaviour of the magnetic correlation length in iron is indistinguishable from that in EuO and EuS over the range of reduced temperature ($T/T_c \lesssim 1.1$) covered by neutron critical scattering experiments [21]. EuO and EuS are simple Heisenberg ferromagnets with no indication of unusual short-range order.

With regard to point 2 we may consider the exact model-independent formula

$$\chi = (k_B T)^{-1} \int d^3r \int d^3r' < m(\underline{r}) m(\underline{r}')> , \tag{14}$$

where $m(\underline{r})$ is the z component of the magnetization at position $\underline{r}$. If the correlation function falls off rather slowly with $\underline{r}-\underline{r}'$, χ will be very large unless the correlation function becomes negative at large distance, leading to a subtle cancellation. We may summarize by saying that the thermodynamic properties of iron resemble quite closely those of a simple S = 1 Heisenberg model with little short-range order above T_c. This is not too surprising in view of the conclusions of section 2. Shastry et al. [20] therefore decided to investigate the static and dynamic properties of paramagnetic iron by means of an effective Heisenberg model with long-range exchange interactions. You et al. [22] show that the interaction extends to fourth or fifth nearest neighbours and that the more distant neighbours interact antiferromagnetically. We show the following [20]:

(a) Strongly competing ferromagnetic and antiferromagnetic interactions can lead to giant short-range order but as expected the paramagnetic susceptibility is much larger than that observed in Fe.

(b) Exchange parameters chosen to fit the low-temperature spin-wave dispersion curve of Fe lead to satisfactory static properties with little short-range order and to a neutron scattering function $S(\underline{q},\omega)$ compatible with Lynn's "constant ω" plots [19]. In fact we use parameters appropriate to $Fe_{88}Si_{12}$ on which alloy Lynn's detailed measurements were made.

To evaluate the paramagnetic static properties we use the spherical model with S = 1, which is a good approximation to the Heisenberg model except near T_c. Thus writing the Heisenberg model in Fourier form

$$H = - \tfrac{1}{2} \sum_{\underline{q}} J(\underline{q}) \underline{S}(\underline{q}) . \underline{S}(-\underline{q}) \tag{15}$$

we have

$$C(\underline{q}) \equiv \langle S^z(\underline{q})S^z(-\underline{q}) \rangle = k_BT[J(0)-J(\underline{q}) + \chi^{-1}]^{-1} \tag{16}$$

with the inverse susceptibility χ^{-1} determined by the sum rule

$$\frac{1}{3} S(S+1) = N^{-1} \sum_{\underline{q}} C(\underline{q}) \tag{17}$$

where N is the number of atoms. The correlation between a spin at the origin and its nth nearest neighbour at $\underline{R}_n$ is given by

$$\langle S_n^z S_0^z \rangle = N^{-1} \sum_{\underline{q}} \exp(-i\underline{q}.\underline{R}_n)\ C(\underline{q}). \tag{18}$$

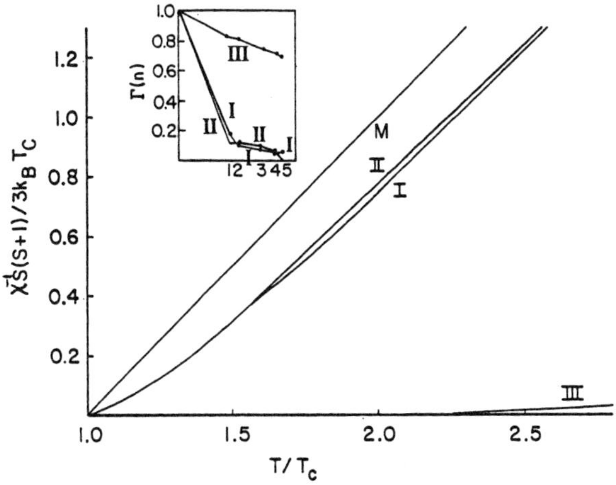

Fig. 3 The reduced inverse susceptibility $\chi^{-1}S(S+1)/3k_BT_C$ as a function of T/T_c for (I) the nearest-neighbour case $\alpha = \beta = \gamma = 0$ (II) $\alpha = 1.412$, $\beta = 1.265$, $\gamma = -1.152$ (III) $\alpha = 1.412$, $\beta = -0.542$, $\gamma = -0.060$. Curve M corresponds to mean field theory. The inset shows the reduced correlation functions $\Gamma(n)$ for n = 1,2,...,5 in cases I, II and III with horizontal scale linear in distance.

In Fig. 3 results are given for the reduced inverse susceptibility $\chi^{-1}S(S+1)/3k_BT_c$ as a function of T/T_c and for the reduced correlation function $\Gamma(n) = \langle S_n^z S_0^z \rangle / [\frac{1}{3}S(S+1)]$ at $T = 1.28T_c$ for n^{th} nearest neighbours. The exchange parameters are specified in the caption by α,β,γ which are, respectively, J_2/J_1, J_3/J_1, J_5/J_1 where J_n is the exchange parameter for n^{th} nearest neighbours. We set $J_4 = 0$ for convenience since the difficult geometry of the 4th nearest neighbours leads to algebraic complexity in our subsequent evaluation of the dynamic properties. Case II corresponds to a least-squares fit of Lynn's [19] room-temperature spin-wave dispersion curve and case III is a highly compensated situation leading to $\Gamma(1) = 0.82 = \cos 35^o$. The giant short-range order postulated by Prange and Korenman [18] corresponds to $\Gamma(1) = \cos 36^o$ [12]. The observed susceptibility χ of bcc Fe follows a Curie-Weiss law quite closely with a Curie constant only about 30% larger than the mean field value for the S = 1 Heisenberg model. The huge susceptibility of case III is thus unrelated to that of Fe. The susceptibility in case II is very close to that for the nearest-neighbour case I and agrees qualitatively with the observed χ. We conclude again that the Prange-Korenman picture does not apply to Fe and therefore does not provide a viable explanation of Lynn's neutron data [19].

To calculate the neutron scattering functions $S(\underline{q},\omega)$ we use the three-pole approximation of Lovesey and Meserve [23] for the relaxation shape function $F(\underline{q},\omega)$. The relation between S and F is

$$S(\underline{q},\omega) = \hbar\omega(1-e^{-\hbar\omega/k_BT})^{-1}\ \chi(\underline{q})F(\underline{q},\omega) \tag{19}$$

where $\chi(\underline{q})$ is the static wave-vector-dependent susceptibility. In the three-pole approximation F is obtained for a given $\vec{q}$ from $<\omega^2>$ and $<\omega^4>$, the second and fourth moments of F which may be expressed in terms of static correlation functions. Following Lovesey and Meserve, we approximate the four-spin correlation functions which appear in $<\omega^4>$ by products of two-spin correlation functions given by (4). For the spherical model this procedure is exact and $\chi(\underline{q})$ is given by $(k_BT)^{-1}C(\underline{q})$ determined by (16). We have calculated $F(\underline{q},\omega)$ as a function of ω, at various temperatures in the range $1.28T_c < T < \infty$, for values of $\underline{q}$ along various symmetry axes in both sc and bcc lattices. Many different sets of exchange parameters extending to 4th (sc) or 5th (bcc) nearest neighbours were considered. In general the shape function has either one maximum at $\omega = 0$ or two maxima, one at a non-zero frequency. The latter situation implies a peak in $S(\underline{q},\omega)$ at a non-zero frequency and hence a damped propagating mode. In no case did we find a propagating mode for $q/q_{max} < 0.4$, where q_{max} corresponds to the zone boundary in the direction considered. The minimum value of q at which a propagating mode exists tends to decrease with increasing range of interaction. We have not considered the case of strong compensation because in this regime, which is anyway not applicable to Fe, factorization of the four-spin correlation functions may be inaccurate. Of greatest interest are the results for case II considered above, with exchange parameters α, β, γ for the bcc lattice chosen to fit Lynn's room-temperature spin-wave dispersion curve for $Fe_{88}Si_{12}$. In fact in the (110) direction, the only direction for which the dispersion curve in this alloy is available, the spin-wave energies depend on J_1 and J_2 only through the sum $J_1 + J_2$. We chose $\alpha = J_1/J_2 = 1.412$ in accordance with results of You et al.[22] and the values of β and γ then follow by curve-fitting. The absolute energy scale is fixed by choosing the nearest-neighbour exchange parameter $J_1 = 9.89$meV so that the observed room-temperature dispersion curve lies 15% lower than the calculated dispersion curve at T = 0, this renormalisation being of the magnitude expected [19]. The root-mean-square error in the fit to the dispersion curve is less than 3meV. The calculated value of T_c is then 1290 K compared with the observed 970 K. To compare with Lynn's data we take $T = 1.28T_c$ and $\underline{q}$ in the (110) direction. We find that $F(\underline{q},\omega)$ shows peaks in "constant q" plots as a function of ω for $q/q_{max} \geq 0.6$. but for $q/q_{max} < 0.6$ is monotonic. However in "constant ω" plots, which is how the experimental data is recorded, the calculated $S(\underline{q},\omega)$ always shows a pronounced peak as a function of q, as shown for three values of ω in Fig. 4. It is the position of such peaks in the experimental data which Lynn plots as a dispersion curve and in Fig.5 we compare this with our similarly plotted theoretical curve for $T = 1.28T_c$. The experimental curve is apparently independent of temperature between T_c and $1.4T_c$ and we find that, even for the largest value of ω plotted, the theoretical value of q increases by less than 2% on raising the temperature from $1.28T_c$ to $1.4T_c$. The observed region corresponds to $q/q_{max} < 0.5$ and the dispersion curve does not correspond to propagating modes in our theory. Lynn fits his "constant ω"

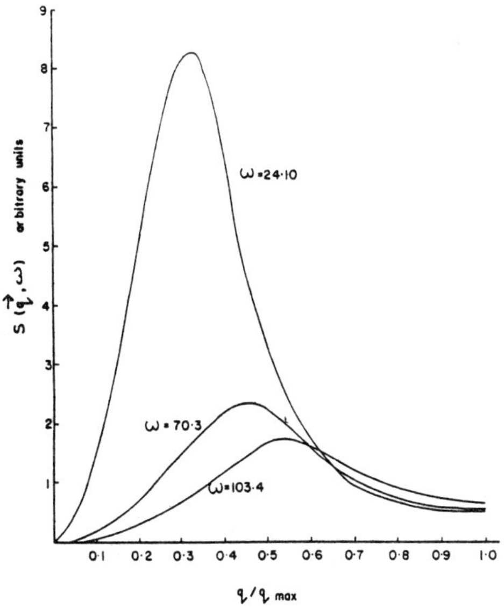

Fig. 4 The scattering function $S(\underline{q},\omega)$ as a function of q in the (110) direction for parameters appropriate to $Fe_{88}Si_{12}$ at $T = 1.28T_c$. The curves correspond to three different values of ω specified in meV.

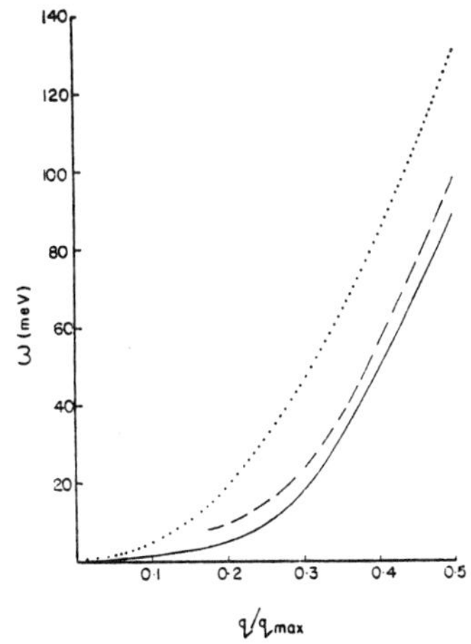

Fig. 5 Experimental (dashed line) and theoretical (full line) dispersion curves for $Fe_{88}Si_{12}$ at $T = 1.28T_c$ in the (110) direction. The dotted curve is the experimental spin-wave dispersion curve at room temperature which was used to determine the exchange parameters of the model.

plots to a Gaussian "spin-wave peak" plus sloping background. If the background lines are taken to slope downwards more steeply asymmetric curves of the type shown in Fig. 4 are obtained.

We conclude that a Heisenberg model with exchange parameters giving reasonable static properties for bcc iron does not lead to spin-wave peaks in constant q scans for $q/q_{max} < 0.6$. This conclusion is particularly convincing since we have shown, by comparing with numerical integration of the equations of motion for a one-dimensional model with long-range interactions, that the method of moments tends to favour propagating modes when there is little short-range order. Thus, although we can understand Lynn's constant ω scans and the renormalization of the dispersion curve we are unable to explain the existence of peaks in his constant q plots [19,24]. If further experimental work confirms that such peaks occur at quite small q in itinerant-electron systems such as iron and nickel, but not in the Heisenberg systems such as EuO, we must reconsider the problem from a fundamental itinerant-electron viewpoint.

4. Summary

The magnetic properties of iron are discussed with special reference to the existence of local moments above the Curie temperature T_c. It is shown how recent work by Hasegawa is related to an older phenomenological model.

The most important interactions in this model are Heisenberg-like ones between local moments. An account is given of recent work by Shastry, Edwards and Young on an effective Heisenberg model of iron, with exchange interactions extending to fifth nearest neighbour. It is shown that strongly competing ferromagnetic and antiferromagnetic interactions can lead to giant short-range order, of the type postulated by Prange and Korenman, above the Curie temperature T_c. However the static properties, in particular the susceptibility and specific heat anomaly, are then incompatible with those observed in Fe. Exchange parameters chosen to fit the spin-wave dispersion curve at room temperature lead to little short-range order above T_c and to a neutron scattering function $S(\underline{q},\omega)$ which is compatible with Lynn's "constant ω" plots. Furthermore the dispersion curve obtained by plotting the positions of peaks in "constant ω" plots is in excellent agreement with the observed one above T_c. However the theory does not predict the observed peaks in "constant q" plots at quite small q.

The work described in section 3 was carried out with B. Sriram Shastry and A.P. Young [20].

References

1. W.E. Evanson, J.R. Schrieffer and S.Q. Wang., J. Appl. Phys. 41, 1199 (1970).

2. M. Cyrot, Phys. Rev. Lett. 25, 871 (1970).

3. L.M. Roth, Inst. Phys. Conf. Ser. No. 39, 473 (1978).

4. H. Hasegawa, J. Phys. Soc. Japan 46, 1504 (1979).

5. H. Hasegawa, Solid State Commun., 31, 597 (1979).

6. D.G. Pettifor, J. Magn. Magn. Met. 15-18, 847 (1980).

7. J. Hubbard, Phys. Rev. B 20, 4584 (1979).

8. J. Hubbard, to be published.

9. H. Hasegawa, J. Phys. Soc. Japan, 49, 178 (1980); 49, 963 (1980).

10. D.M. Edwards, Phys. Lett. A33, 183 (1970).

11. M. Sakoh and D.M. Edwards, Phys. Stat. Sol. (b) 70, 611 (1975).

12. D.M. Edwards, J. Magn. Magn. Mat. 15-18, 262 (1980).

13. P.J. Schurer, K.W. Maring, F. Van der Woude and G.A. Sawatzky, Int. J. Magnetism 4, 291 and 297 (1973).

14. R. Riedinger and M. Nauciel-Bloch, J. Phys. F. 5, 732 (1975).

15. B.L. Györffy and G.M. Stocks, in Electrons in Disordered Metals and at Metallic Surfaces, edited by P. Phariseau, B.L. Györffy and L. Shiere (Plenum, New York 1979).

16. J.A. Hofmann, A. Paskin, K.J. Tauer, and R.J. Weiss, J. Phys. Chem. Solids 1, 45 (1956).

17. H. Hasegawa, Solid State Commun., to be published.

18. R.E. Prange and V. Korenman, Phys. Rev. B 19, 4691 (1979) and references contained therein; Inst. Phys. Conf. Ser. No. 39 567 (1978).

V. Korenman and R.E. Prange, Solid State Commun. 31, 909 (1979).

19. J.W. Lynn, Phys. Rev. B. 11, 2624 (1975).

20. B.S. Shastry, D.M. Edwards and A.P. Young, to be published.

21. Als-Neilsen, J., in Phase Transitions and Critical Phenomena , Vol.5A, edited by C. Domb and M.S. Green pp.139-145.

22. M.V. You, V. Heine, A.J. Holden and P.J. Lin-Chung, Phys. Rev. Lett. 44, 1282 (1980).

23. S.W. Lovesey and R.A. Meserve, J. Phys. C. 6, 79 (1973).

24. Report on Panel Discussion on Itinerant Electron Magnetism, Conference on Physics of Transition Metals, Leeds, 1980 (Inst. Phys. Conf. Ser. to be published).

Thermal Expansion Due to Magneto-Volume Effect in Ferromagnetic Metals

Kan Usami and Tôru Moriya

Institute for Solid State Physics, University of Tokyo, Roppongi, 106 Tokyo, Japan

One of the most important consequences of the recent theory of itinerant electron magnetism is that the average squared amplitude of the local spin density $S_L{}^2$ varies generally with temperature depending on the band structure, the electron occupation and the relative strength of the exchange interaction [1]. Below the Curie temperature T_C, $S_L{}^2$ usually decreases with increasing temperature while it increases with temperature above T_C. This temperature variation can be quite significant in many cases and reflects itself in the thermal expansion.

1. Introduction

The problem of thermal expansion in ferromagnetic metals and alloys, in particular the invar phenomenon, has been the subject of continued interests. There have been two types of theories proposed for this problem [2]. One is the theory based on the localized model with the Heisenberg exchange which varies with the interatomic distance. In this model the temperature variation of the average relative orientation of the neighboring spin pairs gives rise to the volume change of the crystal. The other is based on the itinerant electron model and the Stoner-Wohlfarth theory has been employed in the previous studies [3]. In the Wohlfarth theory [3], the magneto-volume coupling is taken as

$$- D_0 \, S_0{}^2 \omega, \tag{1}$$

where S_0 is the uniform magnetization, ω the volume strain and D_0 the coupling constant. The magnetic contribution to the thermal expansion is given by

$$\omega_{SW} = (D_0/B)[S_0{}^2(T) - S_0{}^2(0)] = - (D_0/B)\Delta S_0{}^2(T), \tag{2}$$

where B is the bulk modulus and $S_0(T)$ the uniform magnetization as a function of temperature. According to this theory the magnetic contribution to the thermal expansion vanishes above the Curie temperature T_C. Below T_C this theory has been of qualitaitve success in explaining the negative thermal expansion at low temperatures in weakly ferromagnetic metals. However, the quantitative estimates of ω_{SW} from the measured values for D_0/B and S_0 do not seem to give any convincing agreement with experiment. On the other hand in the case of Fe-Ni invar alloys, SHIGA [4] proposed on empirical grounds that the volume change is caused by the decrease with temperature of the size of the local magnetic moments rather than the uniform magnetization. This assumption seems to be a plausible one in view of the recent theory of general spin fluctuations [1,5] where the temperature variation of the amplitude of the local spin fluctuation as well as the non-local character of the spin fluctuations are treated.

A general consideration of thermal expansion due to the magneto-volume coupling based on the unified theory of spin fluctuations has been presented recently by the present authors [6] together with some comparisons with experimental results. We here summarize this theory very briefly and give further discussions on this problem, also answering to the recent comments by WOHLFARTH [7] on our previous paper [6] as to the comparisons with experimental results.

2. General Considerations

The magneto-volume coupling may be generally given by

$$F_{m.v.} = - \omega \sum_q D_q S_q^2, \tag{3}$$

where S_q is the Fourier q component of the spin density. We take for brevity only the quadratic terms with respect to the spin density. When necessary we can easily take account of the higher order terms. The magnetic contribution to the thermal expansion is given by

$$\omega_m = \sum_q (D_q/B)(\langle S_q^2\rangle_T - \langle S_q^2\rangle_0), \tag{4}$$

where $\langle\ \rangle_T$ means the statistical average at temperature T.

For ferromagnetic metals we may rewrite (4) as

$$\omega_m = (D_0/B)[(\langle D_q\rangle/D_0)S_L^2(T) - S_0^2(0)], \tag{5}$$

with

$$S_L^2(T) = \sum_q \langle S_q^2\rangle_T,$$

$$\langle D_q\rangle = \sum_q D_q \langle S_q^2\rangle/S_L^2. \tag{6}$$

Now we consider the two limiting cases in the general spin fluctuations [1].
(1) Local moment case: We have S_L^2=const. and the thermal expansion arises solely from the temperature variation of $\langle D_q\rangle$. In order to study this case D_q must be calculated from the electronic band structure.
(2) Weakly ferromagnetic limit: Since only the long-wave components of the spin fluctuations are important in this case we may put $\langle D_q\rangle \simeq D_0$ and get

$$\omega_m = (D_0/B)[S_L^2(T) - S_0^2(0)] = - (D_0/B)\Delta S_L^2(T). \tag{7}$$

Thus the temperature variation of S_L^2 is the only magnetic mechanism for thermal expansion in this case.

In general we have to consider both of the contributions related respectively with $\langle D_q \rangle$ and $S_L^2(T)$. The calculation of the former is rather complicated since details of the band structure is required for this purpose. While $S_L^2(T)$ can be calculated more easily with various approximations. When the temperature variation of $S_L^2(T)$ is strong we can always expect a large magnetic contribution to the thermal expansion. Various typical examples for the temperature variation of $S_L^2(T)$ are shown in Fig.6 of [1].

3. Weakly and Nearly Ferromagnetic Metals

We have previously shown the following relation:

$$S_L^2(T_C)/S_L^2(0) = 3/5, \tag{8}$$

in the weakly ferromagnetic limit [6]. This means that we have $\Delta S_L^2(T_C) = (2/5)\Delta S_0^2(T_C)$ and we approximately generalize this relation to all temperatures below T_C. We have

$$\omega_m(T) \simeq -(2D_0/5B)\,\Delta S_0^2(T), \qquad (T < T_C). \tag{9}$$

Above T_C we get

$$\omega_m(T) - \omega_m(T_C) = \frac{D_0}{B}\,[S_L^2(T) - S_L^2(T_C)] = \frac{3D_0}{5B\gamma}\frac{1}{\chi}, \qquad (T > T_C), \tag{10}$$

where χ is the uniform susceptibility and γ is the coefficient of the M^4 term in the Landau expansion of the free energy. The last equality in (10) is the consequence of the self-consistent renormalization (SCR) theory [5]. Since $S_L^2(T)$ increases almost linearly with temperature above T_C, giving rise to the Curie-Weiss susceptibility, we expect positive magnetic contributions to the thermal expansion in weakly ferromagnetic metals. This is in strong contrast with the Stoner-Wohlfarth theory which gives vanishing magnetic contribution above T_C.

This positive contribution is expected to be important also in nearly ferromagnetic metals at low temperatures, where we have [6]

$$\omega_m(T) = \frac{3D_0}{5B\gamma}\left(\frac{1}{\chi(T)} - \frac{1}{\chi(0)}\right). \tag{11}$$

Since D_0/B can be deduced from the forced magneto-striction and high-field susceptibility data in the low-temperature limit through

$$D_0/B = (\partial\omega/\partial H)/2S_0\chi_{h.f.}, \tag{12}$$

and γ from the Arrott plot, $\omega_m(T)$ can be evaluated from the additional use of the experimental data for $S_0(T)$ and $\chi(T)$. Thus the comparison between the observed $\omega_m(T)$ and the one evaluated from the above procedure through (9)-(11) should constitute a crucial test of the SCR theory without any adjustable parameter. In actuality, however, there is some ambiguity in separating the magnetic contribution $\omega_m(T)$ from the measured value for $\omega(T)$ which includes the nonmagnetic contributions ω_{nm} of both the lattice and the electronic origins.

$$\omega(T) = \omega_m(T) + \omega_{nm}(T). \tag{13}$$

Previously we have evaluated $\omega_m(T)$ for Sc_3In, $ZrZn_2$, Ni_3Al, and Ni-Pt alloys below T_C from the above arguments and found fairly good agreement with the experimental values for $\omega(T)$. This is considered to be reasonable in view of relatively small expected values for $\omega_{nm}(T)$ at low temperatures. However, Wohlfarth has recently made a comment suggesting that the electronic contribution to $\omega_{nm}(T)$ can be comparable with $\omega_m(T)$ even at low temperatures.

We make here more detailed analyses, by using (9)-(11), of the experimental data on $ZrZn_2$ [8] and Ni_3Al [9] where Wohlfarth claims discrepancies between our theory and experimental results. In Fig.1 we plot by a dashed line the values for the magnetic contribution to the linear thermal expansion coefficient $\alpha_m(T)$ as estimated from (9) and (10) by using the experimental data by

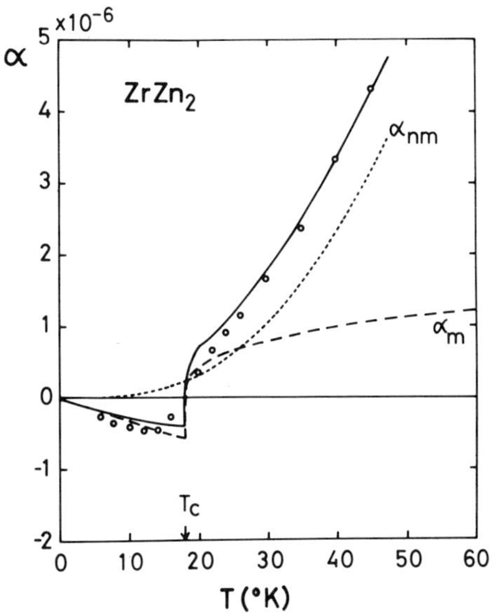

Fig.1 The linear thermal expansion coefficient of $ZrZn_2$. The solid line is a calculated one and α_m (α_{nm}) denotes the magnetic (non-magnetic) contribution (see text). The small circles indicate the experimental values.

OGAWA [10] and OGAWA and WAKI [11]. The experimental values are shown by small circles. The non-magnetic contribution as shown by a dotted line is evaluated by extrapolating $\alpha_{exp}(T) - \alpha_m(T)$ from the high-temperature side. As may easily be seen from the figure one cannot expect any large electronic contribution to α_{nm} at low temperatures and the above extrapolation has been made by assuming α_{nm} to be mainly of lattice origin. The solid line in the figure shows the values for $\alpha(T)$ thus evaluated. This compares fairly well with experiment. α_{nm} at low temperatures is small as expected, in contrast with the analysis by Wohlfarth which neglects $\alpha_m(T)$ above T_C.

As for Ni_3Al, or Ni_xAl_{1-x} with $x \simeq 0.75$, there are not enough low-temperature experimental data to make really convincing evaluations of $\omega_m(T)$. Fig.2 shows our best estimates for $\omega_m(T)$ from (9)-(11). The values for D_0/B are taken from Kortekaas and Franse's data [12] and M(T) and χ from DE BOER et al. [13]. The solid and dashed lines stand for the results of estimations by using the

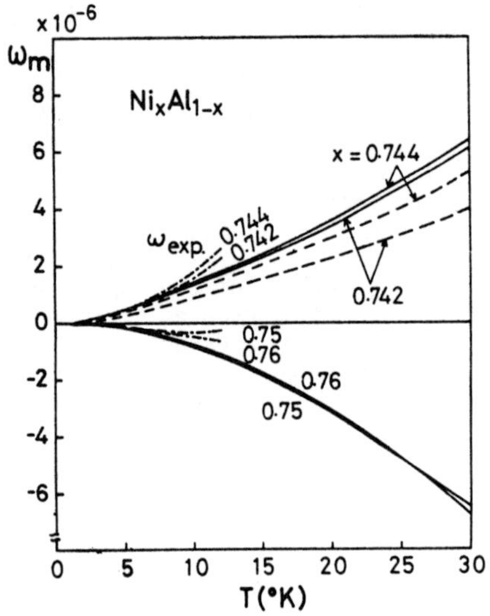

Fig.2 The magnetic contribution to the volume strain in Ni_xAl_{1-x} alloys. The solid and dashed lines are estimated by usind different experimental values for γ (see text). The dot-dashed lines show the exerpimental values for total ω.

values for γ obtained by De Boer et al. and by Kortekaas and Franse, respectively. The dot-dashed lines show the experimental values for the volume strain [9]. This figure indicates that the magnetic contribution to the thermal expansion is dominating at low temperatures and the non-magnetic contributions are almost independent of the Ni concentration. This fact seems to favor the present theory. However, for the purpose of making more convincing quantitative comparisons, more detailed experiments on this and other systems at low temperatures are highly desired.

4. Fe-Ni Invar Alloys

In more general cases, such as Fe-Ni invar alloys, we have to calculate both $\langle D_q \rangle$ and $S_L{}^2(T)$ from a given band structure.

In the previous paper [6] we have made calculations for $\eta(T_C)=S_L{}^2(T_C)/S_0{}^2(0)$ for two model densities of states which simulate those of the d bands for f.c.c. and b.c.c. lattices, respectively. In evaluating $S_L{}^2(T_C)$ we have made use of a simple method by USAMI and MORIYA [14] which is a kind of approach from the local limit with the use of the coherent potential approximation (CPA).

It is well known that as we increase the number of holes in the f.c.c. d-band ferromagnets, the ferromagnetism suddenly disappears at a certain critical hole number and the invar phenomenon takes place in the ferromagnetic region near this instability. The calculated values for $\eta(T_C)$ tend to decrease substantially as the hole number increases toward the instability. This significant reduction of $\eta(T_C)$ near the instability naturally leads to a large negative value for ω_m, which compensates the usual thermal expansion ω_{nm}. This is a possible explanation for the invar phenomena. The value of ω_m for an Fe-Ni invar alloy with 34.6%Ni is estimated from (5), assuming $\langle D_q \rangle/D_0 \simeq 1$ and using the following values: $D_0/B \simeq 10^{-6}$ $(\mathrm{emu/g})^{-2}$, $\eta(T_C) \simeq 0.5$ and $M_0(0) \simeq 170$ emu/g. The estimated value: $\omega_m(T_C) \simeq -0.8\times10^{-2}$ compares reasonably well with experiment.

On the other hand, $\eta(T_C)$ for the b.c.c. metals depends little on the hole number, although its value is significantly reduced from 1. The calculated value of $\omega_m(T_C)$ for Fe is about -0.4×10^{-2}. Since T_C of Fe is considerably higher than that of the invar alloys, the magnetic contribution is quite small compared with the non-magnetic one at any temperature.

In the above arguments we made use of a rigid band approximation. For a more realistic discussion of Fe-Ni invar alloys, however, the effect of random atomic potentials should be taken into account. Recently we have extended the previous theory to concentrated ferromagnetic binary alloys. This theory in the ground state reduces to the CPA theory of ferromagnetic alloys by HASEGAWA and KANAMORI [15] and at finite temperatures the values for $S_L^2(T)$ for two component atomic species are calculated separately.

We consider the case where the Coulomb interaction is much larger than the exchange. Further we assume the following approximate relation:

$$\omega_m(T) \simeq \sum_\alpha P_\alpha \, (D_0/B)_\alpha \, S_{0\alpha}^2 \, (\eta_\alpha(T) - 1), \tag{14}$$

where the suffix α denotes the component atomic species and P_α is its concentration. As the squared amplitude of the magnetic moment of Fe is much larger than that of Ni, $\omega_m(T)$ is considered to be determined mainly by $\eta(T)$ for Fe-site or $\eta_{Fe}(T)$. The calculated value of $\eta_{Fe}(T_C)$ as a function of the Ni concentration is about 0.5 at vanishing Fe concentration and gradually decreases with increasing concentration of Fe. Above the Curie temperature $\eta_{Fe}(T)$ shows a gradual but substantial increase with increasing temperature. The significant reduction of the calculated $\eta_{Fe}(T)$ at small Fe concentrations results in a gradual increase of $|\omega_m(T)|$ with increasing Fe concentration in a wider range of concentration than is actually observed. Since the present approach is expected to be better in the range of higher Fe concentration, where the spin fluctuation should be more localized, our conclusion of significant reduction of $\eta_{Fe}(T_C)$ is expected to be more reliable in the invar region. Experimentally the paramagnetic neutron scattering by COLLINS [16] has shown a significant reduction of $\eta_{Fe}(T)$ above T_C in an Fe-Ni alloy with 35%Ni. The values for $\omega_m(T_C)$ thus estimated in the invar region are of the same order of magnitude as the observed ones.

After all the invar phenomena seem to be explained by the strong temperature variation of the amplitude of the spin density fluctuations, although more detailed calculations are needed for really quantitative explanations for the Fe-Ni and other invar alloys.

References

1. T. Moriya: in this volume.
2. See for example: Proc. International Symposium on The Invar Problem, J. Magn. Magn. Mat. 10, (1979).
3. E. P. Wohlfarth: Physica 91B+C, 305 (1977), and references therein.
4. M. Shiga: AIP Conf. Proc. No. 18, 463 (1974).
5. See for a review: T. Moriya: J. Magn. Magn. Mat. 14, 1 (1979).
6. T. Moriya and K. Usami: Solid State Commun. 34, 95 (1980).
7. E. P. Wohlfarth: Solid State Commun. 35, 797 (1980).
8. S. Ogawa and N. Kasai: J. Phys. Soc. Japan 27, 789 (1969).
9. T. F. M. Kortekaas and J. J. M. Franse: Phys. Stat. Sol. (a) 40, 479 (1977).

10. S. Ogawa: Researches of Electrotechnical Laboratory No. 735 (1972).
11. S. Ogawa and S. Waki: J. Phys. Soc. Japan 22, 1514 (1967).
12. T. F. M. Kortekaas and J. J. M. Franse: J. Phys. F: Metal Phys. 6, 1161 (1976).
13. F. R. de Boer, C. J. Schinkel, J. Biesterbos and S. Proost: J. Appl. Phys. 40, 1049 (1969).
14. K. Usami and T. Moriya: J. Magn. Magn. Mat. 20, 171 (1980).
15. H. Hasegawa and J. Kanamori: J. Phys. Soc. Japan 31, 382 (1971).
16. F. M. Collins: Proc. Phys. Soc. 86, 973 (1965).

Functional Integral Approach to the Low-Temperature Properties of Itinerant Ferromagnets

Yoshinori Takahashi

Institute for Solid State Physics, University of Tokyo, Roppongi, 106 Tokyo, Japan

1. Introduction

Although there have been a lot of investigations of the itinerant electron magnetism based on the functional integral formalism [1], almost all the treatments employ static or classical approximation. From the sense of this approximation, applicability of these theories ought to be restricted to the high-temperature properties. Therefore in order to give more adequate descriptions, in particular, when we discuss the ordered phase, we must be faced with the problem of dealing with the dynamical effects. Even in the paramagnetic state in the local moment limit, we have to take into account the dynamical effects in order to obtain the Curie constant proportional to S(S+1) rather than S^2 obtained in the classical approximation. This has been regarded as one of the main difficulties of the existing treatments. Extension of the theory to take into account dynamical effects is therefore highly desirable.

Up to the present, the functional integral method seems to be indispensable in treating the finite spin density fluctuations at finite temperatures. It is worth while to realize what the dynamical effects mean and to know how to take them into account in the formalism. We present here an approach to deal with the dynamical effects in the functional integral scheme, although it is very crude.

In our treatment, we confine our discussions to the case of long-wavelength spin density fluctuations. In this limit, the static functional has already been obtained by CAPELLMANN [2] and by KORENMAN et al.[3]. They considered, however, only the transverse degrees of freedom of the spin densities. Here we generalize their theories by taking into account the effects of the longitudinal spin density fluctuations and the mode-mode coupling between the transverse and the longitudinal modes.

Another well-known shortcoming of the functional integral theory comes from the ambiguity in representing the interaction part of the Hamiltonian. For example, in the Hubbard model, intra-atomic Coulomb term is rewritten as

$$Un_{\uparrow}n_{\downarrow} = \frac{U}{4}\{(n_{\uparrow} + n_{\downarrow})^2 - (n_{\uparrow} - n_{\downarrow})^2\},$$

in the two-field method, or in a generalized form as recently pointed out by PRANGE and KORENMAN [4] and by HUBBARD [5]:

$$Un_{\uparrow}n_{\downarrow} = \frac{U}{4}(n_{\uparrow} + n_{\downarrow})^2 - U(\vec{S}\cdot\hat{e})^2,$$

with $\hat{e}$, an arbitrary unit vector. Obviously, there is no problem in the weak interaction limit, where perturbation expansion can be justified. But when there is a finite local exchange splitting, the approximate results of the problem depend on the choice of the representation. It is natural to impose the following requirement. That is, in the saddle point approximation, we must have the Hartree-Fock approximation. The second comes from the rotational symmetry requirement. We also briefly discuss this problem.

In the next section, we rewrite the conventional functional integral scheme by introducing the new variables, which have meanings of the spin density and the charge density. The third section is devoted to the evaluation of the functional, coming from the adiabatic transverse motion of spin density. The result of this section is the same as that of CAPELLMANN [2] and KORENMAN et al.[3] except the inclusion of the explicit time dependence. It is shown that the classical phenomenological equation of motion is derived as the saddle point of our functional. In section 4, we discuss the functional from the longitudinal charge and spin density fluctuations. Also, the effects of the spatial gradient of the classical transverse modes on the longitudinal modes are evaluated, which give rise to the mode-mode coupling between these modes. In the final section, a rough thermodynamic arguments are given with some discussions.

2. Functional Integral Formalism

Like the spin Hamiltonian theory of insulator magnets, the evaluation of the free energy is split into two parts. One is to evaluate the functional in terms of the space-time-dependent fictitious charge and spin density variables. Then the free energy is obtained by the functional integration of these variables.

We start with the single-band Hubbard Hamiltonian for simplicity.

$$H = H_0 + H_I,$$

$$H_0 = \sum_{k,\sigma} \varepsilon_k \hat{n}_{k\sigma}, \quad H_I = U \sum_i \hat{n}_{i\uparrow}\hat{n}_{i\downarrow}, \tag{2.1}$$

where H_0 and H_I represent non-interacting part and the intra-atomic Coulomb interaction of the system, respectively, with the standard notations. The free energy ΔF which comes from H_I is given by

$$e^{-\beta\Delta F} = \langle T_\tau \exp(-\int_0^\beta d\tau\, H_I(\tau))\rangle, \tag{2.2}$$

with $H_I(\tau)=e^{\tau H_0}H_I e^{-\tau H_0}$ and $\langle\cdots\rangle$ being defined by $\mathrm{Tr}[e^{-\beta H_0}\cdots]/\mathrm{Tr}\,e^{-\beta H_0}$.

The functional integral theory first rewrite (2.2) in the form:

$$e^{-\beta\Delta F} = \lim_{N\to\infty} \langle \mathrm{Tr}\,(1 - \frac{\beta}{N} H_I(T_N)) \cdots (1 - \frac{\beta}{N} H_I(\tau_1))\rangle, \tag{2.3}$$

and at each instant of time $\tau_k=k(\beta/N)$, each factor $(1-(\beta/N)H_I(\tau_k))$ is expressed as follows:

$$1-\frac{\beta}{N}H_I(\tau_k) = \langle \exp[\sum_i \frac{\beta}{N}(i\xi_i^o(\tau_k)\rho_i(\tau_k)+\vec{\xi}_i(\tau_k)\cdot\vec{S}_i(\tau_k))]\rangle_{G.A.}, \tag{2.4}$$

where $\langle\cdots\rangle_{G.A.}$ means the Gaussian average with the weight:

$$\exp[-\frac{\beta}{4NU}\sum_i (\xi_i^o(\tau)^2+\vec{\xi}_i(\tau)\hat{A}_i(\tau)\vec{\xi}_i(\tau))]. \tag{2.5}$$

In (2.5), we have introduced an arbitrary (3×3) matrix with one constraint coming from the fermion commutation relations

$$\sum_{\alpha=x,y,z} \hat{A}^{-1}_{\alpha\alpha}(\tau) = 1. \qquad \text{(at each } \tau=\tau_k) \tag{2.6}$$

There are a number of representations to express the term $Un_\uparrow n_\downarrow$, corresponding to various choices of the matrix $\hat{A}$. For example, $\hat{A}^{-1}$ is given by

$$\hat{A}^{-1}(\tau) = \begin{pmatrix} 0 & & \\ & 0 & \\ & & 1 \end{pmatrix}, \tag{2.7}$$

in the two-field method, while the representation by PRANGE AND KORENMAN [4] and HUBBARD [5] is given by

$$\hat{A}_i^{-1}(\tau) = \hat{e}_i(\theta,\phi)\hat{e}_i(\theta,\phi), \qquad \text{(in the sense of diadics)} \tag{2.8}$$

with $\hat{e}(\theta,\phi)$ being a unit vector.

We now rewrite (2.4) and (2.5) in a different form. First we introduce the Gaussian transformation:

$$\exp[-\frac{\beta}{4UN}\sum_i(\xi_i^{o2}+\vec{\xi}_i\hat{A}\vec{\xi}_i)] \tag{2.9}$$

$$= \int d\zeta^o d\vec{\zeta}\, \exp[-4U\pi^2\frac{\beta}{N}\sum_i(\zeta_i^{o2}+\vec{\zeta}_i\hat{A}^{-1}\vec{\zeta}_i) + 2\pi i\frac{\beta}{N}\sum_i(\zeta_i^o\xi_i^o+\vec{\zeta}_i\cdot\vec{\xi}_i)],$$

then we make use of the following changes of variables:

$$\zeta_i^o = -n_i/2\pi, \qquad \vec{\zeta}_i = i\vec{M}_i/2\pi,$$

$$i\xi_i^o = -2Un_i-\phi_i, \qquad \vec{\xi}_i = 2U\vec{M}_i\hat{A}_i^{-1}+\vec{h}_i. \tag{2.10}$$

In the new variables, the free energy ΔF is written as follows:

$$e^{-\beta\Delta F} \propto \int \delta\phi\delta n\delta\vec{h}\delta\vec{M}\, \exp[\sum_i U\int_o^\beta d\tau(n_i^2(\tau)-\vec{M}_i(\tau)\hat{A}_i^{-1}(\tau)\vec{M}_i(\tau))$$

$$+ \sum_i \int_0^\beta d\tau (\phi_i n_i - \vec{h}_i \cdot \vec{M}_i)] \langle T_\tau \exp[- \int_0^\beta d\tau \sum_i \{(2Un_i + \phi_i)\rho_i$$

$$- (2U\vec{M}_i \hat{A}_i^{-1} + \vec{h}_i) \cdot \vec{S}_i\}]\rangle . \tag{2.11}$$

The physical meanings of the new variables are apparent from the above expression. In particular, the variable $\vec{M}_i$ and n_i are the real spin and the charge densities (it should be noted that in the ξ-field method, the charge field becomes imaginary, while in (2.11) ϕ and $\vec{h}$ must be integrated along the imaginary axes).

The merit of expressing the functional in terms of variables $n_i(\tau)$ and $\vec{M}_i(\tau)$ may be ascribed to our hope that the low-lying excitation energies of the system are bodily well described in the form of the functional in terms of $n_i(\tau)$ and $\vec{M}_i(\tau)$. Actually, when there is a finite spin density amplitude, we expect the existence of the low-excitation modes described by the transverse motion of $\vec{M}_i(\tau)$. Therefore it is quite reasonable to split $\vec{M}(\tau)$ into two parts as follows:

$$\vec{M}_i(\tau) = \vec{M}_i^0(\tau) + \delta\vec{M}_i(\tau),$$
$$n_i(\tau) = n_i^0(\tau) + \delta n_i(\tau), \tag{2.12}$$

where $\vec{M}_i^0$ is assumed to represent the adiabatic slow transverse motion of the spin density and $\delta\vec{M}_i$ is defined as the deviation from $\vec{M}^0$. Correspondingly, the variables ϕ and $\vec{h}$ are also split into two parts and we employ the saddle point approximation for ϕ^0 and h^0. (It should be noted that we do not make saddle point approximation with respect fo $\vec{M}_i(\tau)$.) To begin with, if we neglect all the deviations δM, etc., the conditions to determine ϕ^0 and $\vec{h}^0$ in terms of n_i and $\vec{M}_i$ are given by

$$\vec{M}_i = \mathrm{Tr}\, \frac{\vec{\sigma}}{2} G_{ii} [\delta\mu, \vec{\Delta}],$$
$$n_i = \mathrm{Tr}\, \frac{1}{2} G_{ii} [\delta\mu, \vec{\Delta}], \tag{2.13}$$

with

$$\vec{\Delta}_i = U\vec{M}_i \hat{A}_i^{-1} + \frac{1}{2} \vec{h}_i . \tag{2.14}$$

The functional in terms of n_i and $\vec{M}_i$ is obtained, once the functional form of $\vec{h}_i$ and ϕ_i is obtained by the above relations through

$$\frac{\delta F}{\delta \vec{M}} = \vec{h}[\vec{M}, n]. \tag{2.15}$$

Because $\vec{\Delta}_i$ in (2.14) depends on the choice of the matrix $\hat{A}_i(\tau)$, $\vec{h}$ also depends on $\hat{A}_i(\tau)$. We here impose the condition that the difference in direction between $\vec{M}_i$ and $\vec{\Delta}_i$ becomes as small as possible. From this condition, it follows that the matrix $\hat{A}^{-1}$ must be chosen as follows:

$$A_i^{-1}(\tau) = \frac{1}{M_i^2(\tau)} \vec{M}_i(\tau)\vec{M}_i(\tau). \quad \text{(in the sense of diadics)} \tag{2.16}$$

Although (2.16) has the same form as (2.8), PRANGE and KORENMAN [4] and HUBBARD [5] considered the arbitrariness of choosing the direction of $\hat{e}(\theta,\phi)$ as the extra degrees of freedom, and roughly supposed that these extra angle average should be taken to retain the rotational symmetry. We show that such a overcounting of extra degrees of freedom is not necessary. The rotational symmetry is preserved automatically by taking into account the dynamical effect.

From the above construction, the transverse part of $\vec{h}$, $\vec{h}_t$, is now regarded as very small quantity, and therefore (2.13) can be expanded as follows

$$\vec{M}_i(\tau) - \mathrm{Tr}\, \frac{\vec{\sigma}}{2} G[\delta\mu, \vec{\Delta}_i] \cong \frac{\partial}{\partial \vec{h}} [\mathrm{Tr}\, \frac{\vec{\sigma}}{2} G] \cdot \vec{h}_t. \tag{2.17}$$

Thas is, the problem to obtain the functional is finally reduced to evaluate the difference of the left-hand side of (2.17) under the space-time-dependent exchange field $\vec{\Delta}_i(\tau)$. In the static approximation, (2.17) reproduces the same result as that obtained by CAPELLMANN and KORENMAN et al. if we make use of the gradient expansion method for the spatial coordinates (The systematic method to evaluate (2.17) is discussed in the next section). In our formalism, the time dependence of $\vec{M}_i(\tau)$ is also taken into account on the same footing as the spatial variables.

The important point to include the time-dependent effects is that the phenomenological classical equation of motion is derived in the saddle point approximation. Here it should be noted that the complete neglect of the time dependence is equivalent to the neglect of the commutation relations among quantum operators in (2.3). We know that the equations of motion of the classical mechanics are the direct consequences of the commutation relations from the classical to quantum correspondence. Therefore, the major effect to take into account the equation of motion for the spin densities is to introduce the effects of commutation relations or the effects of the quantization of the phase space. The most important effect to replace S^2 in the Curie constant with $S(S+1)$ comes from this quantization process. The formal procedure to perform the functional integral related to this part, is to use Feynman's path integral theory [6]. Besides the classical transverse modes, we have to consider the effects of longitudinal fluctuations of $\delta\vec{M}$, δn, etc. (The transverse part of $\delta\vec{M}$ does not contribute fo ΔF because $\delta\vec{M}$ is perpendicular to $\vec{A}^{-1}$). The functional is expanded in terms of their amplitudes. And also there are mode-mode coupling terms between the transverse and the longitudinal modes. These problems are very difficult many-body problems.

The static and the uniform components of the longitudinal modes must be determined by thermodynamic relations. Let us write the free energy ΔF as

$$\Delta F = \Delta F_{HF}(n,M) + \Delta F_1(n,M),$$

where ΔF_1 represents the deviation from the Hartree-Fock free energy. Then n and M must be determined by the saddle point of $\Delta F(n,M)$ as follows:

$$\delta\mu + \frac{1}{2}\frac{1}{N_o}\frac{\partial \Delta F_1}{\partial n} = Un,$$

$$\Delta - \frac{1}{2}\frac{1}{N_o}\frac{\partial \Delta F_1}{\partial M} = UM.$$

The Hartree-Fock theory corresponds to the complete neglect of ΔF_1. But in general, we expect the considerable contributions from ΔF_1. In fact, in the weakly ferro- and antiferromagnetic limit, this correction have played the significant role on the temperature dependence of various physical properties. How to estimate ΔF_1 is therefore an important problem and also one of the crucial points in the theory of itinerant electron systems.

3. Functional in the Long-Wavelength Limit

We present here the systematic method to evaluate the functional in the gradient expansion in the space-time coordinates. According to the preceding section, the problem is to obtain Green's function $G[\vec{\Delta}]$ under the exchange field $\vec{\Delta}(r,t)$. We employ the method of distribution function as given by KADANOFF and BAYM [7], which is particularly suited to obtain G in the space-time gradient expansions. In this formalism the equations of motion of Green's function $G(r_1t_1,r_1't_1')$ is written in terms of the center of mass coordinates (R,T) and relative coordinates (r,t):

$$R = (r_1 + r_1')/2, \qquad r = r_1 - r_1',$$

$$T = (t_1 + t_1')/2, \qquad t = t_1 - t_1'.$$

Equation of motion of Green's function in the lowest order is given by

$$L_oG_\rho + \vec{\Lambda}_1\cdot\vec{G} = 0, \tag{3.1}$$

$$L_o\vec{G} + \vec{\Lambda}_1G_\rho + \vec{\Lambda}_o\times\vec{G} = 0,$$

with

$$L_o = \frac{\partial}{\partial T} + v_k\cdot\frac{\partial}{\partial R}, \qquad (v_k \equiv \frac{\partial\varepsilon_k}{\partial k})$$

$$\vec{\Lambda}_o = 2\vec{\Delta}(R,T), \tag{3.2}$$

$$\vec{\Lambda}_1 = \frac{\partial\vec{\Delta}}{\partial R}\cdot\frac{\partial}{\partial k} - \frac{\partial\vec{\Delta}}{\partial T}\cdot\frac{\partial}{\partial\omega},$$

where G_ρ and $\vec{G}$ defined by

$$G = G_\rho(Rt,k\omega) + \vec{G}(RT,k\omega)\cdot\vec{\sigma}$$

are the Fourier transforms of G(RT,rt). We further rewrite $\vec{G}$ as follows

$$\vec{G} = \vec{m}(R,T)G_\ell + \vec{m}\times\vec{G}_t, \qquad (\vec{m} = \vec{\Delta}(R,T)/|\Delta|)$$

where G_ℓ and $\vec{G}_t$ represent the longitudinal and transverse components, respectively. In terms of G_ρ, G_ℓ and $\vec{G}_t$, (3.1) is written in the form

$$L_o G_\rho + (\vec{m}\cdot\vec{\Lambda}_1)G_\ell - \vec{m}\cdot(\vec{\Lambda}_1\times\vec{G}_t) = 0, \tag{3.3a}$$

$$L_o G_\ell + (\vec{m}\cdot\vec{\Lambda}_1)G_\rho + \vec{m}\cdot\{(L_o\vec{m})\times\vec{G}_t\} = 0, \tag{3.3b}$$

$$2\vec{\Delta}m\times\vec{G}_t - (\vec{m}\times L_o\vec{m})G_\ell - \vec{m}\times\vec{\Lambda}_1 G_\rho + L_o\vec{G}_t = 0. \tag{3.3c}$$

The zeroth order solutions of G_ρ and G_ℓ are obtained by solving (3.3a) and (3.3b) without the last terms, and are given by

$$G_\rho^o = (g_\uparrow + g_\downarrow)/2,$$

$$G_\ell^o = (g_\uparrow - g_\downarrow)/2, \tag{3.4}$$

with

$$g_\sigma(\omega) = [\omega - \varepsilon_k + \mu + \sigma\Delta(R,T)]^{-1}.$$

The amplitude M is determined by

$$M = \frac{1}{2}\int_{-\infty}^{\infty} d\omega\rho(\omega)[f_\uparrow(\omega) - f_\downarrow(\omega)], \tag{3.5}$$

where $\rho(\omega)$ and $f_\sigma(\omega)$ are the density of states and the Fermi distribution function, respectively. The first- and the second-order solution $G_t^{(1)}$ and $G_t^{(2)}$ are determined by

$$\begin{aligned} &2\Delta\vec{m}\times G_t^{(1)} - (\vec{m}\times L_o\vec{m})G_\ell^o - \vec{m}\times\vec{\Lambda}_1 G_\rho^o = 0, \\ &2\Delta\vec{m}\times G_t^{(2)} + L_o\vec{G}_t^{(1)} = 0. \end{aligned} \tag{3.6}$$

Eq.(3.6) can be easily solved to give

$$\vec{G}_t^{(1)} = \frac{1}{4\Delta}\,[(g_\uparrow - g_\downarrow)L_o\vec{m} + \vec{\Lambda}_1(g_\uparrow + g_\downarrow)], \tag{3.7}$$

$$\vec{G}_t^{(2)} = \frac{1}{4\Delta}\,\{\frac{1}{2}\frac{\partial}{\partial\varepsilon_k}\,(g_\uparrow + g_\downarrow) + \frac{1}{2\Delta}\,(g_\uparrow - g_\downarrow)\}\vec{m}\times(v_k v_k\,\frac{\partial^2\vec{m}}{\partial R\partial R}).$$

After carrying the summation with respect to k and ω in

$$\vec{M} - \frac{1}{2}\,\mathrm{Tr}\,\vec{\sigma}G(\vec{\Delta}) = \sum_{k,\omega} (\vec{G}_t^{(1)} + \vec{G}_t^{(2)} + \cdots)$$

we finally obtain the results

$$\delta\vec{h} = -\,\vec{m}\times\frac{\partial\vec{m}}{\partial T} + \vec{m}\times\vec{m}\times D\frac{\partial^2\vec{m}}{\partial R\partial R}\,, \tag{3.8}$$

with

$$D = \frac{1}{2M}\sum_k\{\frac{1}{2}\,\frac{\partial^2\varepsilon_k}{\partial k\partial k}\,(f_\uparrow - f_\downarrow) - \frac{v_k v_k}{2\Delta}\,(f_\uparrow - f_\downarrow)\}.$$

The spatial part of the free energy functional has the same form as that given by CAPELLMANN and KORENMAN et al. We also evaluate the effect of the time dependence in the lowest order based on the same theoretical frame as the spatial coordinates.

From (3.8), we see that the saddle point of the functional ($\delta\vec{h}=0$) gives rise to the classical equation of motion:

$$\frac{\partial m}{\partial T} = \vec{m}\times D\,\frac{\partial^2 m}{\partial R\partial R}\,. \tag{3.9}$$

This part of the functional may be regarded as the Lagrangian density in the field theory.

4. Effects of the Longitudinal Fluctuations

The functional Ψ_ℓ related to the longitudinal modes can be written down formally as follows:

$$\Psi_\ell = -\sum_{i,\sigma}(\frac{\sigma}{2}\,\delta n_{i\sigma}\delta n_{i-\sigma} + \delta\phi_{i\sigma}\delta n_{i\sigma}) - \mathrm{Tr}\,\log[1 - (\sum_\sigma \delta v_\sigma P_\sigma)\tilde{G}], \tag{4.1}$$

with

$$P_\sigma = \frac{1}{2}\,(1 + \sigma\hat{\sigma}_z) \quad \text{and} \quad \delta v_{i\sigma} = \delta\phi_{i\sigma} + U\delta n_{i-\sigma}\,.$$

In (4.1) Green's function $\tilde{G}[\vec{M}]$ can be expanded in terms of the gradient of the classical variable $\vec{M}(r,t)$. The zeroth-order terms Ψ_ℓ^0 represents the purely longitudinal contributions. The higher order terms with respect to the gradients represents the mode-mode coupling between the transverse and the longitudinal modes.

We here give the result of the lowest order mode coupling term.

$$\Psi' = -\frac{1}{16}\sum_{\substack{p,q,\\ \sigma,\sigma'}}\int dxdy\; w(q)w(p+q)\{(\vec{m}(x)\times\frac{\partial\vec{m}}{\partial x})\cdot(\vec{m}(y)\times\frac{\partial\vec{m}}{\partial y})$$

$$- \sigma\sigma' \frac{\partial\vec{m}}{\partial x}\cdot\frac{\partial\vec{m}}{\partial y}\} e^{-ip(x-y)} \delta v_\sigma(x)\delta v_{\sigma'}(y)$$

where

$$v_k = \frac{\partial\varepsilon_k}{\partial k} \quad \text{and} \quad w(k) = \frac{1}{2\Delta}(g_\uparrow - g_\downarrow) + \frac{1}{2}\frac{\partial}{\partial\varepsilon_k}(g_\uparrow + g_\downarrow).$$

5. Thermodynamics and Some Discussions

After obtaining the functional, we must perform the functional integration to discuss the thermodynamic properties. For the transverse modes, we must deal with the classical equation of motion, and the free energy contribution is given by the sum of all the classical paths. This is essentially equivalent to the Heisenberg model. For the longitudinal modes, we must treat the non-linear fluctuations. Moreover, the effects of the mode couplings between these modes have to be taken into account.

In the self-consistent renormalization theory, the functional Ψ_ℓ^o is approximated by the quadratic form given by

$$\bar{\Psi}_\ell^o = - \sum_{i,\sigma} (\frac{U}{2}\delta n_{i\sigma}\delta n_{i-\sigma} + i\delta\phi_{i\sigma}\delta n_{i\sigma})$$

$$- \frac{1}{2} \sum_{\sigma,i,j} \chi_{ij}^\sigma (i\delta\phi_{i\sigma} + U\delta n_{i-\sigma})(i\delta\phi_{j\sigma} + U\delta n_{j-\sigma}), \tag{5.1}$$

with χ_{ij} given by

$$\frac{1}{2}\chi_{ij}^{-1} = \frac{1}{2}\chi_{ij}^{o-1} + \frac{\partial}{\partial\chi_{ij}} \langle\delta\Psi_\ell[\delta n,\delta\phi]\rangle_c, \tag{5.2}$$

where χ_{ij}^o represents the contribution in the random phase approximation, $\delta\Psi_\ell$ is the non-linear part of Ψ_ℓ^o, and $\langle\cdots\rangle_c$ represents the cumulant average. The second term in (5.2) represents the renormalization effect from the non-linearity. In the weakly ferromagnetic limit, this second term has played the central role on the temperature dependence of the susceptibility. Therefore, at least in the case of the small amplitude spin density, the effects of the longitudinal modes should be very important.

On the other hand, the simplest way to deal with the transverse modes is to linearize (3.9) as follows:

$$\frac{\partial m_q^\pm}{\partial t} = \pm Dm_o^z q^2 m_q^\pm, \tag{5.3}$$

where $m_q^\pm = m_q^x \pm i m_q^y$. Then, according to the Feynman path integral theory [6], the correlation function between m_q^+ and m_q^- is evaluated as follows:

$$\langle m_q^- m_q^+\rangle = \frac{2m_o^z}{M} [\exp(\beta Dm_o^z q^2) - 1]^{-1}.$$

In this way, the boson charactor of the spin waves or the effect of the quantization is automatically taken into account. The temperature dependence of the magnetization is evaluated if we use the relation:

$$<S_z> = \frac{1}{2} <\{S_+, S_-\}_+> - <S^- S^+>,$$

and the correct $T^{3/2}$ dependence of the magnetization is recovered at low temperature, which can not be obtained by the saddle point approximation of the previous functional integral theories. In conclusion, we stress the significant role played by the equations of motion. Quantum mechanical features, i.e., quantization of the phase space due to commutation relations etc., are taken into account by the dynamical effects, as we have shown above.

Because of the presence of the mode couplings between the transverse and the longitudinal modes, we must generally deal with the complicated coupled equations of motion for both of these modes. The more elaborate treatment will be the subject of our future investigations.

Acknowledgement

The author would like to express his sincere thanks to Professor T. Moriya for reading the manuscript and for many valuable discussions.

References

1. For a review of the recent development, see T. Moriya, J. Magn. Magn. Mat. 14, 1 (1979).
2. H. Capellmann, J. Phys. F4, 1466 (1974).
3. V. Korenman, J. L. Murray and R. E. Prange, Phys. Rev. B16, 4032 (1977).
4. R. E. Prange and V. Korenman, Phys. Rev. B19, 4691 (1979).
5. J. Hubbard, Phys. Rev. B19, 2626 (1979).
6. R. P. Feynman and A. R. Hibbs, *Quantum Mechanics and Path Integrals* (McGraw-Hill, New York, 1965); J. Schwinger, *Quantum Kinematics and Dynamics* (W. A. Benjamin, Inc., New York, 1970).
7. L. P. Kadanoff and G. Baym, *Quantum Statistical Mechanics* (W. A. Benjamin, Inc., Menlo Park, California, 1962).

Part II

d Metals and Compounds – Ground State Properties

The Screening of Excess *d* Holes by *s* and *p* Electrons in Nickel

Junjiro Kanamori

Department of Physics, Faculty of Science, Osaka University, Toyonaka
Osaka 560, Japan

Abstract

The screening of d holes by electrons in s and p states in nickel is discussed by adapting our previously published theory of sp valence impurities where the impurity charge is screened apparently mainly by d electrons at surrounding nickel atoms. The smallness of the phase shifts of s and p states at the Fermi energy which results from the presence of an antiresonance dip near the top of the d band leads us to the conclusion that a d hole in nickel is accompanied by an electron deficiency of about 0.4 in total at surrounding nickel atoms. Implications of this fact, in particular the inter-atomic interaction between holes and a narrowing of the d band are discussed.

I. Introduction

The purpose of this paper is to examine the applicability of the Hubbard model to transition metals, in particular to nickel by focusing attention on the screening of excess d holes by electrons in nearly free states, or in other words those in the s and p symmetry atomic states. We shall conclude that a difference in the screening mechanism may be a possible origin of characteristics of nickel found recently by photoemission experiments.

The Hubbard model assumes that d electrons interact with each other only when they happen to be at the same atom. This neglect of the inter-atomic interaction between two d electrons or holes in the model has been justified usually by invoking s and p electrons for the screening of an excess or a deficiency of d electrons within the same atom [1]. We shall show, however, that the socalled s-d mixing affects the s and p states in such a way that a disturbance in the charge distribution extends inevitably to near neighboring atoms in nickel. In the case of iron, on the other hand, d holes in other orbitals than those occupied by holes can screen the excess charge by going out from the atom. As will be discussed in more details below, the fact that the screening cannot be completed within the same atom distinguishes nickel from other metals.

Before going into the discussion of the screening mechanism, we review briefly relevant problems. Recently the satellite at 6 eV below the Fermi level in the photoemission spectra of nickel [2-6] has attracted much attention. Though the origin of the satellite seems to be still controversial, a majority of authors suggest that it is somehow associated with the two-hole bound state at the same atom [5-10]. When analyzed in terms of the Hubbard model without taking into account the screening due to s electrons explicitly, it yields $U \leq 2$ eV according to PENN [7] and TREGLIA, DUCASTELLE and SPANJAARD [8]. On the other hand if it is localized with an extra s electron screening it, as was suggested by IWAN, HIMPSEL and EASTMAN [6], we may conclude $U \leq 5$ eV. In the former analysis the bound state is an itinerant one in which the band energy as well as U tends to increase its energy. In the latter estimate we assume that U of the Hubbard model is the interaction between two holes screened fully by s (and p) electrons and the band energy is irrelevant.

IGARASHI [11] shows at this symposium that a value of around 5 eV or 1.3 times the width of the d band is required for U in the single-band Hubbard model to keep the whole majority spin states below the Fermi energy, i.e. to obtain the strong ferromagnetism in nickel. In order to explain the persistence of the strong ferromagnetism in Ni-Fe alloys, we need an even larger value of U [11, 12]. Since IGARASHI's many-body theory should be extended to include the Hund rule coupling, i.e. the exchange interaction between two holes at different orbitals of the same atom, we cannot draw any definite conclusion for the value of U at present. LIEBSCH [9] has shown within the t matrix approximation, however, that if one insists on the itinerant bound state picture without taking into account the screening by s electrons explicitly, it is difficult to reconcile the satellite position with the observed exchange splitting [13] even when the Hund rule coupling is taken into account; the observed value of the exchange splitting leads us to a deeper position of the satellite. We conclude that we need to consider the screening by s and p electrons for the two-hole bound state on one hand and to look for a possible contribution to the exchange splitting made by the inter-atomic interaction on the other hand.

Apart from the Hubbard model problem, we may say that if the inter-atomic shielding is important in nickel, a doubt can be cast on the validity of the local density functional approach in the band calculation. HIMPSEL, KNAPP and EASTMAN [13] have shown that the experimental band width of a d hole in nickel is narrower by 30 percent than the calculated value. A narrowing of the d-band width is expected from the hole-hole correlation effect as has been discussed by several authors [7-9,11]. We do not know, however, whether the bare band width, i.e. the one in the absence of the hole-hole correlation in nickel is correctly given by the local density functional approach or not. As will be discussed below, a d hole in nickel is accompanied by an electron deficiency 0.4 in total at surrounding atoms. The inter-atomic screening mechanism is quite possible to result in a narrower d band than that calculated by the local density functional approach which may be justified if the screening is always intra-atomic.

In the next section (section II) we discuss the screening of an impurity charge in nickel. Section III deals with the discussion of the inter-atomic interaction between two d hole in nickel. In the final section (section IV) we discuss further implications of the screening by s and p electrons.

II. The shielding of an impurity charge in nickel

In the following discussions the s or p symmetry of a wave function is defined with respect to the nuclear position of a given impurity. An experimental evidence for the fact that the shielding of an impurity charge can not be achieved within the states of s and p symmetries in nickel is given by the data on the decrease of the saturation magnetization due to those impurities whose valence states are of s and p symmetries. We list in Table 1 the data measured by CRANGLE and MARTIN [14], which show an approximate relation $d\mu/dc = -Z\mu_B$ per impurity atom for a given impurity of valence Z. TERAKURA and KANAMORI [15] gave a theory for the mechanism underlying the relation between $d\mu/dc$ and Z and verified it by carrying out a detailed calculation with a muffin-tin potential model for the case of Al in nickel. We summarize the conclusions of our discussions below.

Table I Decrease of magnetization of Ni caused by an impurity of non-transition element [14]

Solute element	Valency	$-d\mu/dc$ (μ_B)
Cu	1	1.14
Zn	2	2.11
Al	3	2.80
Si	4	3.77
Ge	4	3.70
Sn	4	4.22
Sb	5	5.31

First of all we can show that s and p valence impurities remain practically nonmagnetic when embedded in ferromagnetic nickel (and also in other ferromagnetic metals); this fact was verified by our calculation and also by neutron diffraction experiment [16]. Since a nonmagnetic impurity atom replaces a nickel atom having the magnetic moment approximately equal to 0.6 μ_B, we can expect a contribution of -0.6 μ_B to $d\mu/dc$ from the impurity site. The above-mentioned experimental data indicates that the rest of $d\mu/dc$, $-(Z - 0.6)\mu_B$ reflects a decrease of magnetic moments of surrounding nickel atoms. This was confirmed by neutron diffraction [16]. On the other hand, the impurity atoms remain electrically almost neutral, or in other words they keep their s and p valence electrons even in nickel. This was concluded by our calculation of the local electron density and supported by the agreement between the calculation and a variety of hyperfine interaction data [17]. Since this means that the impurity charge may look as screened within its atomic cell, the disturbance in the magnetic moments of surrounding nickel atoms demands an explanation.

The answer to the above-mentioned question is given by the fact that the phase shifts of the s and p symmetry states at the Fermi energy, or in other words the change of the number of occupied states due to the impurity are very small in nickel. The underlying mechanism is the same as the Fano effect producing the antiresonance in optical absorption [18]. Figure 1 reproduces our old calculation of the phase shifts in the case of Al in nickel. We show there the phase shifts ΔN_s and ΔN_p as function of energy which show characteristic dips near the Fermi energy. We show also the self-energy Σ_ℓ (ℓ=s or p) which an electron in a given s or p valence orbital of the impurity atom acquires through the mixing with the host metal states; the phase shifts can be shown to be approximately proportional to $-\mathrm{Im}\Sigma_\ell$.

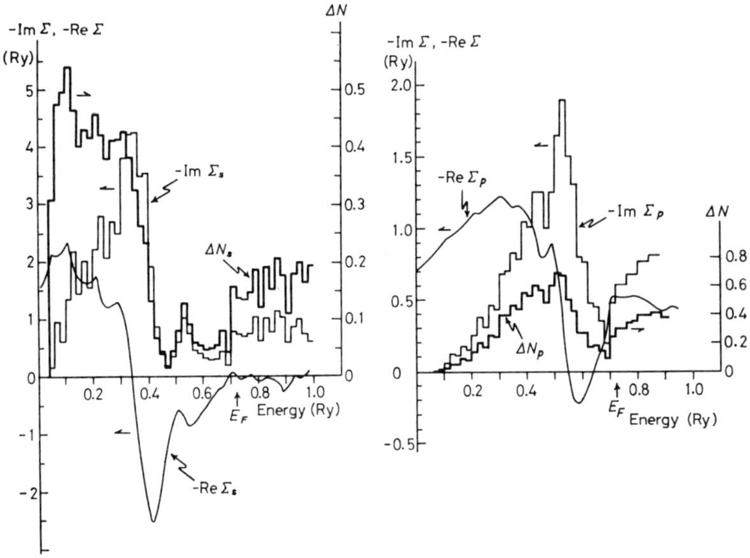

Fig. 1 The phase shift ΔN_ℓ and the self-energy Σ_ℓ (ℓ=s or p) of the majority spin states in the case of a substitutional Al atom in nickel [15].

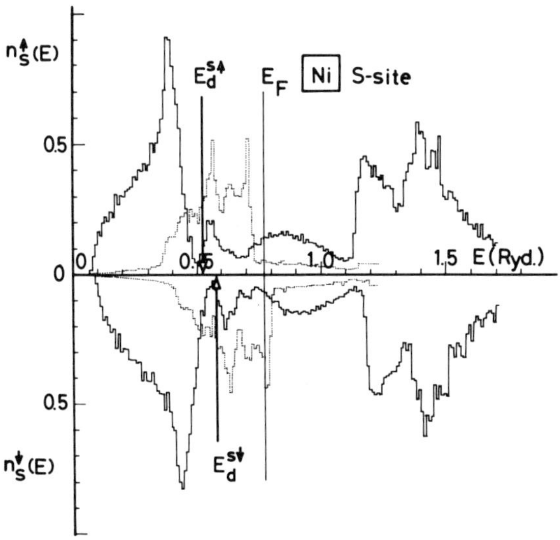

Fig. 2 The local density of states within a muffin-tin sphere for the s symmetry in nickel; a profile of the d band and the approximate position of the antiresonance are shown also [17].

The self-energy consists of the term arising from direct mixings with nearly free states and the term from indirect mixings with the d-band states via nearly free states; we can expect the Fano antiresonance from this structure of the self-energy irrespective of the impurity potential. We can show also by a general argument that the dip position should be near the Fermi energy in nickel. We refer for further details of the discussions to [15, 17, 19].

Since the phase shift at the Fermi energy corresponds to the spatial integration of an excess or deficiency of electrons, the local neutrality of the impurity which corresponds to an excess Z - 0.6 compared with a nickel atom must be compensated by a deficiency of the same amount outside the impurity to result in the small phase shifts. The deficiency can be ascribed to a decrease of the amplitudes of host metal states in the bonding states between the impurity valence orbitals and host metal states, since the dip separates the bonding states from the antibonding states [20]. In the impurity problem the deficiency will be shielded by the d holes of minority spin. The observed $d\mu/dc$ is thus explained, since the number of d holes of minority spin decreases by (Z - 0.6) per impurity.

Turning to pure nickel, we can expect that if a nickel atom has a d hole, it produces an electron deficiency (1 - 0.6)=0.4 at near-neighboring nickel atoms. We discuss this problem in the next section.

III. The effective interaction between two d holes in nickel

Figure 2 shows the energy dependence of the local density of states of s symmetry in a nickel atom. When a d hole is present at a given atom, the local density in the energy region below the d band increases due to the attractive potential. Since the screening is done by electrons in the states lying rather well below the Fermi energy, we may neglect the band energy of d holes when we discuss the interaction between two holes in the states lying near the Fermi energy. In other words we treat d holes as if they are static impurities.

The importance of the screening due to the s electrons in estimating the intra-atomic coulomb integral U between two d holes was emphasized by HERRING [21]. He has shown that if the energies of the states of the configurations $4s^0\ 3d^{10}$, $4s^1\ 3d^9$ and $4s^2\ 3d^8$ of a free nickel atom are compared, U is estimated to be around 2 eV, while U $\sim$ 8 eV is obtained by the comparison of $3d^{10}$, $3d^9$ and $3d^8$ neglecting the s electrons. As we have seen in the preceding section, however, the screening due to the s and p electrons in nickel metal is quite different from that in the case of a free atom. Though we intend to calculate U in near future by carrying out an electronic structure calculation of an impurity nickel atom having two d holes embedded in otherwise pure nickel atom, we do not know at present any reliable estimate of U. In the following discussion we treat U as a phenomenological parameter.

We turn now to the inter-atomic interaction between two d holes. As was mentioned in the preceding section, a d hole will produce an electron deficiency of 0.4 at near-neighboring nickel atoms. The deficiency is expected to occur mostly in the nearest neighbors, though so far only a somewhat indirect verification has been made by the electronic structure calculation. We may suppose that a d hole is accompanied by 0.4/z holes at each nearest-neighboring atom with z $\sim$ 12. Suppose another d hole comes to one of the nearest-neighboring atom. Since the electron deficiency is produced in the s and p symmetry states lying rather low in energy, the states of the second d hole near the Fermi energy can be assumed to be independent of these s and p states participating in the shielding of the first hole. Then the second hole will be subject to an intra-atomic interaction $(0.4/z)U$ at the nearest-neighboring atom, U denoting the intra-atomic coulomb integral. Taking into account the fact that the first hole also interacts with the electron deficiency produced by the second hole, we obtain the magnitude of the resulting inter-atomic interaction as $U_1 = (0.8/z)U$.

If we assume the three dε band model proposed by GAUTIER [22] (see also [1]), we can write down the inter-atomic interaction as

$$H_{int} = (U_1/2N)\sum_{m,n}\sum_{\rho}\sum_{k,k',q}\sum_{\sigma,\sigma'} \exp(i\vec{q}\cdot\vec{\rho}) \times a^*_{mk+q}\, a^*_{nk'-q\sigma'}\, a_{nk'\sigma'}\, a_{mk\sigma} \qquad (1)$$

where m denotes one of the dε bands, σ the spin direction, ρ one of the vectors connecting to the nearest neighbors, k,k' and q are wave vectors.

IV. Implications of the extended screening

Adopting the three-dε-band model, we can see that the intra-band term m=n in Eq.(1) can contribute to the exchange splitting, since the exchange hole extends to near neighbors in the case of low density. In the Hartree-Fock approximation we obtain a rough estimate of the contribution as 0.27 U times the number of holes in a given dε band for low density, where we take into account the two-dimensional character of the model [22]. Thus the inter-atomic interaction may enhance the intra-atomic coulomb interaction by a factor ∿1.3 in the Hartree-Fock approximation. In order to judge its importance, we need a many-body theoretical treatment such as Igarashi's [11]. In general, however, we doubt that a low-density approximation such as the t-matrix one is applicable to nickel when the interaction range is extended. Anyhow, one cannot neglect the inter-atomic interaction easily in such a delicate problem as the stability of the strong ferromagnetism in nickel.

As was mentioned before, an excess of d holes in the case of iron can be easily screened by d holes in other orbitals of the same atom which will keep the electrical neutrality by going out of it. In other words the phase shift in the d states which cannot contribute to the local screening in nickel is important in iron. It is known in the calculation for transition metal alloys that the charge disturbance outside the atom is very small when an impurity charge is screened by d electrons [23]. Thus the inter-atomic interaction will be negligible in iron. The value of U for the two-hole state will be also rather small because of the intra-atomic screening by d electrons. The case of cobalt is similar to the case of iron as far as the inter-atomic interaction is concerned. The U value will be intermediate between the cases of nickel and iron, since the two-d hole cannot be screened by other d holes.

In connection with these differences among ferromagnetic transition metals produced by the screening due to s and p electrons we discuss the possibility that the band calculation based on the local density functional approach may not be quantitatively correct in estimating the band energy of a d hole in nickel. The screening by s and p electrons becomes inter-atomic in nickel because these states are strongly mixed with the d states of surrounding nickel atoms. We need a non-local exchange-correlation potential to take into account the situation in the density functional. We can suppose by analogy with the polaron case that the effective mass of a d hole will become heavier than that obtained by the band calculation based on the local density functional approach. Then the bare band width itself in the Hubbard model might be narrower. In fact, if U is as small as 2 eV, the narrowing in the t-matrix approximation [11] seems to be not enough to explain the observed width in nickel [13]. Finally we mention the fact that the s and p phase shifts at the Fermi energy in copper are free from the Fano effect. Thus the screening of the impurity charge produced by a d hole may be achieved more normally than in nickel. This is consistent with the fact that the band width is not narrowed in copper [6]. We have no intention to deny the possibility of the interpretation of the band narrowing as the manifestation of the many-body effect. However, the screening effect can be partially responsible for the narrowing.

Acknowledgements

The author thanks Dr. Igarashi for valuable discussions and Mrs. A. Egusa for preparing the manuscript. The present work is financially supported partially by a Grant-in-Aid for Scientific Research D and by a Grant-in-Aid for Cooperative Research A from Ministry of Education, Science and Culture.

References

1. J. Kanamori: Prog. Theor. Phys. 30, 275 (1963)
2. S. Hüfner and G. K. Wertheim: Phys. Letter 51A, 299 (1975)
3. P. C. Kemeny and N. J. Shevchik: Solid State Commun. 17, 255 (1975)
4. R. J. Smith, J. Anderson, J. Hermanson and G. J. Lapeyre: Solid State Commun. 21, 459 (1977)
5. C. Guillot, Y. Ballu, J. Paigné, J. Lecante, K. P. Jain, P. Thiry, R. Pinchaux, Y. Pétroff and L. M. Falicov: Phys. Rev. Letters 39, 1632 (1977)
6. M. Iwan, F. J. Himpsel and D. E. Eastman: Phys. Rev. Letters 43, 1829 (1979)
7. D. Penn: Phys. Rev. Letters 42, 921 (1979)
8. G. Treglia, F. Ducastelle and D. Spanjaard: J. Physique 41, 281 (1980)
9. A. Liebsch: Phys. Rev. Letters 43, 1431 (1979)
10. L. A. Feldkamp and L. C. Davis: Phys. Rev. Letters 43, 151 (1979)
11. J. Igarashi: in this book
12. J. Kanamori and T. Jo: in this book
13. F. J. Himpsel, J. A. Knapp and D. E. Eastman: Phys. Rev. B 19,2919 (1979)
14. J. Crangle and M. J. C. Martin: Phil. Mag. 4,1006 (1955)
15. K. Terakura and J. Kanamori: Prog. Theor. Phys. 46, 1007 (1971)
16. J. B. Comly, T. M. Holden and G. G. Low: J. Phys. C 1, 458 (1968)
17. H. Yoshida (formerly Katayama), K. Terakura and J. Kanamori: Jour. Phys. Soc. Japan 49, 972 (1980) and papers referred therein
18. U. Fano: Phys. Rev. 124, 1866 (1961)
19. J. Kanamori, H. Yoshida and K. Terakura: Proceedings of the Fifth Intl. Conf. on Hyperfine Interactions (North-Holland, Amsterdam 1981)
20. C. Herring, Magnetism IV, ed. by G. T. Rado and H. Suhl (Academic Press, New York 1966)
21. F. Gautier: J. Phys. Chem. Solids 24, 387 (1963)
22. J. Kanamori: J. Applied Phys. 36, 929 (1965)

The Ground State of fcc Nickel-Base Ferromagnetic Disordered Alloys

Junjiro Kanamori and Takeo Jo

Department of Physics, Faculty of Science, Osaka University, Toyonaka, Osaka 560, Japan

Abstract

Possible causes of the instability of the strong ferromagnetism in fcc nickel-base disordered alloys are discussed. In the case of Ni-Mn alloys it is caused by the appearance of Mn atoms with magnetic moments antiparallel to the bulk magnetization. Although the same situation is expected in the case of Ni-Fe alloys in the Hartree-Fock approximation, the many-body effect arising from the electron-magnon scattering is more likely to be responsible for the instability of the strong ferromagnetic state in this alloy system. The reasoning leading to this conclusion is discussed in detail. The spin fluctuation in the ground state in Ni-Fe alloys after the onset of the instability is expected then to become large.

I. Introduction

It is generally agreed that the socalled strong ferromagnetism, i.e. the state where the whole majority spin states of the d band are occupied is realized in nickel. The strong ferromagnetism seems to persist to a few percent of Mn in Ni-Mn alloys, to about 50 percent in Ni-Fe alloys and in the whole concentration range in Ni-Co alloys. The weak ferromagnetic state in which some of the majority spin states are unoccupied follows the strong ferromagnetic one in both Ni-Mn and Ni-Fe alloys, although the boundary concentration between the two states is not well defined. The purpose of this paper is to discuss possible causes of this change of the ferromagnetic state in fcc Ni-base disordered alloy systems.

The stability of the strong ferromagnetism in fcc Ni-base alloys was discussed first by SHIMIZU and HIROOKA [1] by use of the rigid band Stoner model. They concluded a first-order-like transition to the nonmagnetic state in Ni-Fe alloys. A more realistic treatment of the alloying effect on the electronic structure was given by HASEGAWA and KANAMORI [2]. They combined the Hartree-Fock approximation for the intra-atomic interaction between electrons of the Hubbard model with the coherent potential approximation (CPA) for the alloying effect. This Hartree-Fock CPA treatment yields the result that some of the majority spin states are pushed up across the Fermi level at a critical concentration by the alloying effect on the Hartree-Fock energy level of Fe or Mn. The critical concentration can be adjusted by a suitable choice of parameters to reproduce approximately the observed behavior of the saturation magnetization which starts to deviate from the Slater-Pauling curve at the onset of the weak ferromagnetic state. Later JO and MIWA [3] pointed out the possibility of finding two stable Hartree-Fock solutions corresponding to the state of a given solute atom with a parallel magnetic moment and that with an antiparallel one to the bulk magnetization. They showed that Mn impurities in Ni-Co alloys were in either of these two states, depending upon the local environment. JO [4] extended the theory to concentrated Ni-base alloys within the single-site CPA by extending Hasegawa and Kanamori's treatment to include the possibility of the antiparallel moment. He applied this theory successfully to Ni-Mn alloys, explaining a variety of experimental data [5]. The existence of Mn atoms with antiparallel moments in concentrated alloys has been confirmed by NMR experiment [6].

Although JO [4] discussed also the Ni-Fe system to predict the appearance of Fe atoms with antiparallel moments in the weak ferromagnetic state, the relation between Hasegawa and Kanamori's picture of the onset of the weak ferromagnetism and the corresponding one in Jo's theory has not been very clear. The fact that no experimental evidence for the appearance of Fe atoms with antiparallel moments has been found in the weak ferromagnetic phase between 50 percent and 65 percent of Fe (the invar concentration) [7] has complicated the situation furthermore. This paper will give a discussion on what we can expect from the Hartree-Fock approximation. We then point out that the many-body effect due to the electron magnon scattering first discussed by HERTZ and EDWARDS [8] and more quantitatively by IGARASHI [9] and EDWARDS [10] at this symposium is probably the more effective cause of the instability of the strong ferromagnetism in Ni-Fe alloys. From this point of view we conclude that the ferromagnetic state in alloys of Fe concentration higher than 50 percent is characterized by a large amplitude of the spin fluctuation even in the ground state.

In the next section (section II) we summarize the Hartree-Fock approximation to the ground state of Ni-base alloys. In section III we discuss the inadequacy of the Hartree-Fock approximation for Ni alloys and the possibility of the ground state with a large spin fluctuation.

II. The instability of the strong ferromagnetic state in the Hartree-Fock approximation

We assume the alloy Hamiltonian defined by

$$H= \sum_{i,\sigma} \varepsilon_i a^*_{i\sigma} a_{i\sigma} + \sum_i U_i a^*_{i\sigma} a_{i\sigma} a^*_{i-\sigma} a_{i-\sigma} + \sum_{i,j,\sigma} t_{ij} a^*_{i\sigma} a_{j\sigma} \quad , \tag{1}$$

where ε_i is the random variable representing the atomic energy level, being equal to ε_{Ni} or ε_{Fe} (ε_{Mn}) according to the species of atom occupying the i-th lattice site, U_i the intra-atomic coulomb integral being equal to either U_{Ni} or U_{Fe}(U_{Mn}), σ denotes the spin state, and the transfer integral t_{ij} is assumed to be independent of the atomic species. In the Hartree-Fock approximation we define the effective atomic energy $\varepsilon_{i\sigma}$ by

$$\varepsilon_{i\sigma} = \varepsilon_i + U_i \langle n_{i-\sigma} \rangle \quad , \tag{2}$$

where $\langle n_{i-\sigma} \rangle$ is the self-consistently determined average value of $n_{i-\sigma} = a^*_{i-\sigma} a_{i-\sigma}$.

We adopt the single-site CPA in which each atom is treated as an impurity embedded in the effective medium described by the coherent potential [2]. The CPA equation determining the coherent potential and the Hartree-Fock equation are simultaneously solved. Hereafter we discuss Ni-Fe alloys. We show in Fig.1 Hasegawa and Kanamori's calculation of the density of states

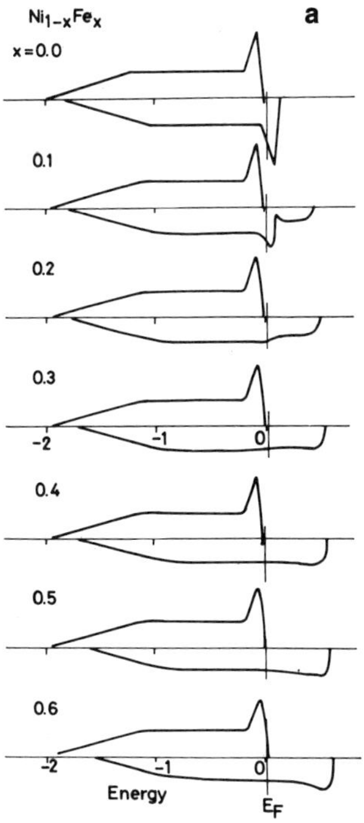

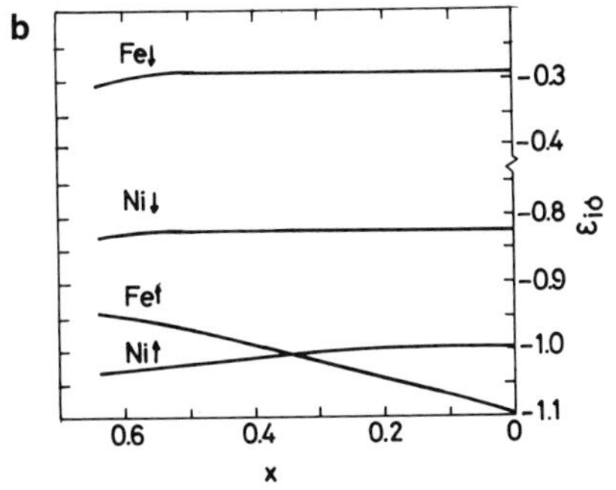

Fig. 1a The density of states of the d band in Ni-Fe fcc alloys for each spin direction [2].

Fig. 1b The concentration dependence of the Hartree-Fock atomic energy levels $\varepsilon_{i\sigma}$'s [2].

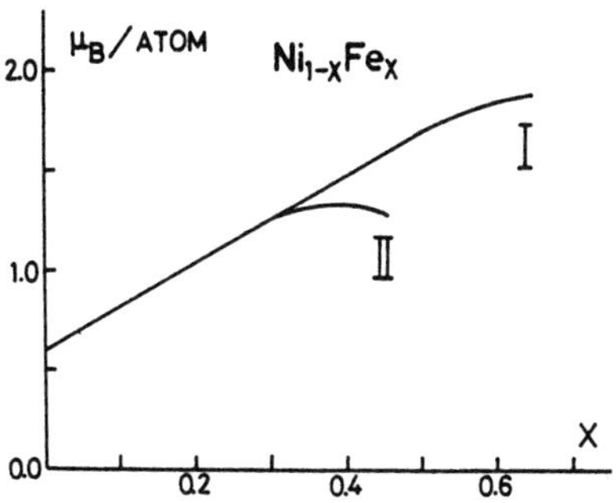

Fig. 2 The calculated saturation magnetization. I represents Hasegawa and Kanamori's calculation with U_{Ni} = 0.7W and U_{Fe} = 0.625W and II is the calculation by JO [4,5] with U = 0.7W for both Ni and Fe. If the same U_{Fe} as in I is used, the saturation magnetization in II deviates from the linear increase at an even smaller x.

and $\varepsilon_{i\sigma}$ of Ni-Fe alloys which does not take into account the possibility of the Hartree-Fock solution with an antiparallel moment for Fe (or Mn) atom. In the calculation the relative value of ε_{Ni} to ε_{Fe} is determined by the condition that the Fermi energies of pure nickel and pure iron in the paramagnetic state should coincide with each other. The U values are chosen such that the strong ferromagnetic state becomes unstable near 50 percent. We refer to [2] for more details. The mechanism giving rise to the instability can be summarized as follows. As an impurity in nickel, a Fe atom loses its minority spin electrons to surrounding Ni atoms to some extent because of the potential difference. It recovers the electrons with increasing Fe concentration, because it will have some Fe atoms on its near-neighboring sites.

Thus $\varepsilon_{Fe\uparrow}$ (↑ denotes the majority spin) increases with the increase of $<n_{Fe\downarrow}>$ to push up some majority spin states across the Fermi energy at a critical concentration.

The above-mentioned calculation is essentially to solve the Hartree-Fock equation for an impurity atom embedded in a metal described by the coherent potential. In the cases of Fe and Mn in Ni-base alloys we find two stable solutions, while we find only one solution for Ni atoms [3,4]. As was mentioned before, the two solutions for Fe (and also for Mn) atom correspond to the states with a parallel moment and an antiparallel one. As a first step to take into account this fact, we adopt the pseudo-ternary alloy approach in which a given Ni-Fe alloy is regarded as an alloy of three atomic species, i.e. Ni and FeI with a parallel moment and FeII with an antiparallel one [4,5]. The concentrations of FeI and FeII are determined by the condition of minimum energy which turns out to be the equality of the energies of FeI and FeII ; note that the energies are functions of the concentrations [4,5]. Figure 2 compares the pseudo-ternary alloy approximation with Hasegawa and Kanamori's treatment in the calculation of the saturation magnetization, where the same values are assumed for the parameters in the two calculations. We can see that the strong ferromagnetic state becomes unstable by the appearance of FeII at a Fe concentration smaller than the onset of the weak ferromagnetic state in Hasegawa and Kanamori's calculation. Since this situation does not depend on the choice of parameters, we conclude that within the Hartree-Fock approximation combined with the single site CPA the weak ferromagnetic state anticipated by Hasegawa and Kanamori will not be realized.

The energies of FeI and FeII in a given alloy will depend on the atomic composition of near-neighboring atoms, or in other words on the local environment [3]. Thus the appearance of FeII atoms with antiparallel magnetic moments will be more gradual than what the pseudo-ternary alloy theory predicts. We discuss this point in the next section.

III. Possibility of the ground state with a large spin fluctuation in Ni alloys

Though the onset of the weak ferromagnetism in Ni-Fe alloys predicted in [2] has been contradicted by JO [4], it is apparently more consistent with the experimental data than the state with FeII with an antiparallel moment. For example, it can explain a steep increase of the electronic specific-heat coefficient with Fe concentration in the range more than 50 percent, while the state with FeII fails, since FeII atoms produce a split band above the Fermi level in the majority spin states without increasing the density of states at the Fermi level appreciably. Furthermore, as was mentioned before, no experimental evidence has been so far found for the existence of FeII [7] which is predicted to have a magnetic moment as large as FeI near the critical concentration [4].

In our opinion the many-body theory presented by IGARASHI [9] is capable of solving the above-mentioned dilemma, although there remain many points to be elucidated in future. He shows that the electron(hole)-magnon scattering pushes up the majority spin states to result in a smaller exchange splitting between the two spin directions than in the Hartree-Fock approximation. Also he develops a theory which unifies the t matrix theory of the hole-hole correlation with the theory of the electron-magnon scattering. He applies the theory to Ni-base alloys, assuming the rigid band approximation. He concludes that the electron-magnon scattering contribution to the self-energy of a majority spin hole is responsible for the instability of the strong ferromagnetism in the alloy system.

We can adapt Igarashi's theory to the CPA at least qualitatively. As can be seen from Fig.1, the majority spin states are only slightly affected by the alloying effect, while the density of states for minority spin is flattened out by the effect. We can modify Hasegawa and Kanamori's treatment by including the self-energy correction due to the electron-magnon scattering which will push up the majority spin states to some extent. Thus we can conclude that the top of the majority spin states reaches the Fermi level at a smaller Fe concentration than in the original Hartree-Fock approximation, though there will be some correction to the Hasegawa and Kanamori mechanism arising from the concentration dependence of the self-energy correction. The problem is now whether this modified mechanism gives rise to the instability of the strong ferromagnetism at a smaller Fe concentration than the appearance of Fe atoms with antiparallel moments or not.

The formation energy of a Fe atom with an antiparallel moment is also affected by the electron-magnon scattering. This problem is left to future investigation. We may say, however, that the electron-magnon scattering energy will not affect particularly favorably the appearance of FeII. Thus we can suppose that the self-energy correction reverses the situation in the Hartree-Fock approximation in Ni-Fe alloys, favoring the onset of the weak ferromagnetism of the type described in Fig.1. We comment also that we need a larger U than that assumed in the calculation of Figs.1 and 2 (0.7 times the band width) when we include the electron-magnon scattering correction. In fact Igarashi suggests U= W with W representing the band width [9]. With a larger U it is quite possible that the formation of a Fe atom with an antiparallel moment is suppressed in the concentration range 50 to 65 percent of Fe even when we take into account the local environment effect, since the nature (either ferromagnetic or antiferromagnetic) of the interaction between two Fe atoms depends quite delicately on the situation [11]. We need a further investigation on this point, too.

Turning to Ni-Mn alloys where the appearance of MnII has been established experimentally, we point out the importance of the local environment effect and the electron-magnon scattering correction for a more detailed understanding than the pseudo-ternary approach; we emphasize, however, the fact that the latter approach is quite useful for the understanding of overall situations. The pseudo-ternary alloy approach tends to conclude a smaller U value phenomenologically, since it averages the local environment effect. In the case of Ni-Mn alloys it is suggested [12] that a given Mn atom may have an antiparallel moment if it has more than two Mn atoms on its nearest-neighbor shell. Such a cluster will appear with an appreciable probability at an average concentration smaller than the one corresponding to the local concentration of the cluster. The pseudo-ternary alloy approach always assumes the environment of the average concentration in the discussion of the appearance of an antiparallel Mn. Thus by use of a smaller U value, it adjusts the calculated critical concentration to the observed concentration range where the deviation of the saturation magnetization from the linear increase starts. Like the case of Ni-Fe alloys, therefore, we conclude that a larger U value than that assumed in our previous calculation [4,5] can be used if the local environment effect is taken into account. We note also that even in the case of a single Mn atom embedded in nickel the electron magnon self-energy correction might produce some majority spin states above the Fermi level; this would explain the small magnetic moment of Mn in dilute concentration i.e. around 3.0 μ_B. In the Hartree-Fock approximation it was difficult to explain this, because again the appearance of an antiparallel moment should occur prior to the onset of the weak ferromagnetic situation suggested above. Finally we mention the fact that in the phase with MnII or in the weak ferromagnetic state the effective U value tends to decrease with increasing number of holes in the majority spin states due to the electron-magnon scattering, as is shown by Igarashi. This will explain the early disappearance of the ferromagnetic phase in Ni-Mn alloys which presents a difficulty in the pseudo-ternary approach with a larger U value.

The competition between the weak ferromagnetic state and the state with FeII or MnII with antiparallel moment is an interesting theoretical problem left to future. The former state is characterized by a large spin fluctuation, i.e. by the presence of virtual magnons (including the Stoner excitations). In order to understand the state, an extension of the CPA from the single site to many sites may be useful. In the many-site CPA combined with the Hartree-Fock approximation we search for solutions of various configurations of atomic moments. In fact a preliminary calculation of a two-atom cluster shows that the atoms may have canted magnetic moments with respect to each other and also to the bulk magnetization. The state may correspond to the ground state with a spin fluctuation which is frozen in the adiabatic treatment.

Acknowledgements

The authors would like to thank Dr. J. Igarashi for helpful discussions and Mrs. A. Egusa for preparing the manuscript. This research is supported partially by Grant-in-Aid for Cooperative Research A and a Grant-in-Aid for Scientific Research D given by Ministry of Education, Science and Culture.

References

1. M. Shimizu and S. Hirooka: Phys. Letters 27A, 53 (1968)
2. H. Hasegawa and J. Kanamori: J. Phys. Soc. Japan 31, 382 (1971); 33, 1599 (1972)
3. T. Jo and H. Miwa: J. Phys. Soc. Japan 40, 706 (1976)
4. T. Jo: J. Phys. Soc. Japan 40, 715 (1976)
5. T. Jo: J. Phys. Soc. Japan 48, 1482 (1980)
6. Y. Kitaoka and K. Asayama: J. Phys. Soc. Japan 40, 1521 (1976)
7. B. Window: J. Phys. F4, 329 (1974)
8. J. A. Hertz and D. M. Edwards: J. Phys. F 3, 2174 (1973); D. M. Edwards and J. A. Hertz: J. Phys. F 3, 2191 (1973)
9. J. Igarashi: in this book
10. D. M. Edwards: in this book
11. Y. Teraoka and J. Kanamori: Physica 91B, 199 (1977)
12. J. W. Cable and H. R. Child: Phys. Rev. B10, 4607 (1974)

Many-Body Effect on the Ferromagnetism in Transition Metals

Jun'ichi Igarashi

Department of Physics, Faculty of Science, Osaka University, Toyonaka, Osaka 560, Japan

Abstract

A many-body theory, which takes into account simultaneously the electron-magnon interaction and the electron-electron scattering in the d band of transition metals is developed. On the basis of it, the exchange splitting and the band width of nickel are calculated in a fair agreement with the angle-resolved photoemission experiment, by using the single-orbital Hubbard model. It is concluded by numerical calculation that the U value (intra-atomic coulomb interaction) should be greater than 5-6 eV in order to maintain the strong ferromagnetism of nickel. Furthermore it is found that the strong ferromagnetism of nickel becomes unstable in decreasing the electron number. Being based on this result, a new mechanism is proposed to explain the deviation of the magnetization from the Slater-Pauling curve in Ni-Fe alloys with iron concentration higher than 50 percent.

I. Introduction

The many-body effect in transition metals has been recently recognized to be important in understanding photoemission experiments [1-6]. Especially the recent angle-resolved photoemission spectra (ARPS) experiment of nickel [1] shows disagreements with the band calculation based on the local spin density formalism (LSDF) [7-8]: the experimental exchange splitting is only half the value of the calculated one, and the occupied d-band width determined experimentally is about 25 percent smaller than the calculated one. It should be noted here that in the case of iron the experiment [2] shows a fair agreement with the band calculation.

These facts suggest to us that in nickel the inter-atomic d-electron correlation, which is believed to be important, is inadequately taken into account in LSDF, since the correlation energy is replaced by that of the homogeneous electron gas of the corresponding local density. On the other hand, in Fe, the intra-atomic screening by d electrons, which is believed to be sufficiently included in LSDF, is so effective in reducing the effective interaction between d electrons, which makes consequently the inter-atomic correlation small.

One purpose of the present paper is to give the many-body theory of nickel based on the Hubbard model and to explain thereby the ARPS, the special attention being paid to the stability of the strong ferromagnetism of nickel. The degenerate d bands in nickel are treated as three independent bands, as was discussed by KANAMORI [9].

There exist two important mechanisms to be considered in studying many body effects [10]. One is the electron-electron scattering, which is treated by means of the multiple scattering (t-matrix) theory [9], and the other is the electron-magnon interaction [10-13]. For the electron-magnon interaction, HERTZ and EDWARDS [12] proposed an approximate vertex function, which satisfied the Ward-Takahashi Identity (WTI) with an ad hoc modification, in order to guarantee the rotational invariance in spin space. They pointed out that the exchange splitting became small by this interaction, although the estimation was qualitative. Later MATSUMOTO, UMEZAWA, SEKI and TACHIKI [13] studied the same problem by means of the one loop approximation, although the practical calculation was confined to the parabolic band model with a contact interaction. Their formulation has various merits as is compared with that in [12]. For example, it naturally satisfies the WTI, owing to being based on an invariant expansion. As we intend to carry out a more quantitative calculation than that in [12], we adopt MATSUMOTO et. al.'s formulation [13] in treating the electron-magnon interaction.

Recently, LIEBSCH [14] calculated the self-energy correction to the one-electron energy based on the t-matrix theory, in which the five degeneracy of d bands in nickel and the exchange interaction as well as the coulomb one are taken into account, and obtained the result that the exchange splitting had a reasonable value and the band width was reduced considerably. His theory, however, ignores completely the electron-magnon interaction, which has a considerable effect on the exchange splitting. So it is quite necessary to treat the electron-magnon interaction and the electron-electron scattering simultaneously. As is well known, however, there exists a redundacy between the self-energy terms coming from the two mechanisms mentioned above. We propose an approximate method of removing the redundancy within the framework of one loop approximation by extending the theory of MATSUMOTO et. al. [13].

The important consequence of our calculation is that the U value (intra-atomic coulomb interaction) should be larger than 5-6 eV in order to maintain the strong ferromagnetism of nickel. This conclusion is in conflict with $U \leq 2$ eV concluded by PENN [15] and TREGLIA, DUCASTELLE and SPANJAARD [16], who try to interpret as arising from the two-hole bound state the satellite of x ray (XPS) [3,4] and ultraviolet (UPS) photoemission spectra [5,6] at 6 eV below the Fermi energy, within the framework of the Hubbard model. From the discovery of resonating shake-up 3d-4s satellites in copper, IWAN, HIMPSEL and EASTMAN [17] discussed recently that the two 3d holes are localized with an extra 4s electron screening it. With this picture, the U value of nickel is estimated to be $\sim$4 eV, which is consistent also with Auger experiments [18]. The full and penetrating discussion about the role of s and p electrons and about the validity of the Hubbard model is given by KANAMORI [19].

Another purpose of the present paper is related to Ni-Fe alloy. The coherent potential approximation (CPA), which is combined with the Hartee-Fock (HF) approximation for the intra-atomic interaction between electrons of the Hubbard model, was first applied to various transition metal alloys by HASEGAWA and KANAMORI [20]. For Ni-Fe alloy, they pointed out that the top of the majority spin states of iron was pushed upward across the Fermi level as increasing the iron concentration by the alloying effect. Subsequently JO and MIWA [21] and JO [22] extended Hasegawa and Kanamori's treatmnet by taking into account the possibility of the atom with the reverse moment to the bulk magnetization. JO [22] pointed out that for Ni-Fe alloy the Fe atom with the moment antiparallel to the bulk magnetization appeared as increasing the iron concentration. Recently KANAMORI and JO [23] concluded that the Fe atom with the moment antiparallel to the bulk magnetization always appeared before the top of the majority spin states crossed the Fermi level, as increasing the iron concentration, within the HF-CPA theory. Contrary to the theoretical prediction, the Fe atom with the reverse moment has not been observed. (The Ni-Mn alloy system is not the case, where Mn atom with the reverse moment is observed [24].) In this paper we remove the inconsistency between the HF-CPA calculation and experiment by taking into account the many-body effect. (See also [23].)

In section II we discuss the basic formalism of the present calculation. In section III the calculated results are discussed. The final section is devoted to the supplementary discussion.

II. Formulation

The Hamiltonian employed by us is given by

$$H = \sum_{k\sigma} \varepsilon(k)\, a_{k\sigma}^{+} a_{k\sigma} + U\sum_{i} a_{i\uparrow}^{+} a_{i\uparrow} a_{i\downarrow}^{+} a_{i\downarrow}, \tag{1}$$

where k refers to the Bloch states, $a_{i\sigma}$ is the annihilation operator of an electron with spin $\sigma(=\uparrow$ or $\downarrow)$ of the atomic orbital at the i-th lattice site, and U is coulomb interaction at the same site.

In this section, we present the many-body theory, which combines the t-matrix theory with that of the electron-magnon interaction, being based on the loop expansion method [13]. The detailed discussion of it is omitted here and will be published elsewhere [25]. The same notations as in [13] are used in this paper. In Fig.1, we show diagramatically the equations for the propagators and vertex functions, where the solid line represents the electron propagator, the wavy one the magnon propagator, the dotted one the coulomb interaction and the black circle the vertex part of the electron-magnon interaction. Figure 1(a) represents the diagrams for the down spin electron propagator, where $\Lambda^{0}(p)$ is defined as $\{p_0 - \varepsilon(p) + \mu^2\}$ with μ^2 the chemical potential. (The diagrams for the up spin electron propagator are omitted.) The last diagram of Fig.1(a) is decomposed into three types of diagrams, which are shown in Fig.1(b). The first term represents the process of the electron-electron scattering, the second one that of the electron-magnon interaction and the third one the correction one to the double counting of the second-order diagram. (The origin of the correction term will be discussed later in this section.) Figure 1(c) expresses the vertex part for the electron-magnon interaction, where the third term also represents the correction one to the double counting. Figure 1(d) expresses the equation for the magnon propagator.

Within the one-loop diagrams [13], the vertex parts of the second and third terms of Fig.1(b) can be replaced by $\Gamma^{(3)}_{-}(000)$ $(\equiv -g)$, since the momentum dependence of it is regarded to be a contribution from a higher order loop. From Fig.1(c), we obtain the following equations for the vertex function, by applying the WTI to the first term and by using the relation that $\Gamma^{(3)}_{-}(p+q;\, p,q) = \mp M\Gamma^{(4)}_{+}(p+q,0;p,q)$ with q_0 the bound state energy of an electron-hole pair (spin-wave), and by neglecting the momentum dependence of the vertex part in the second and third term [13]:

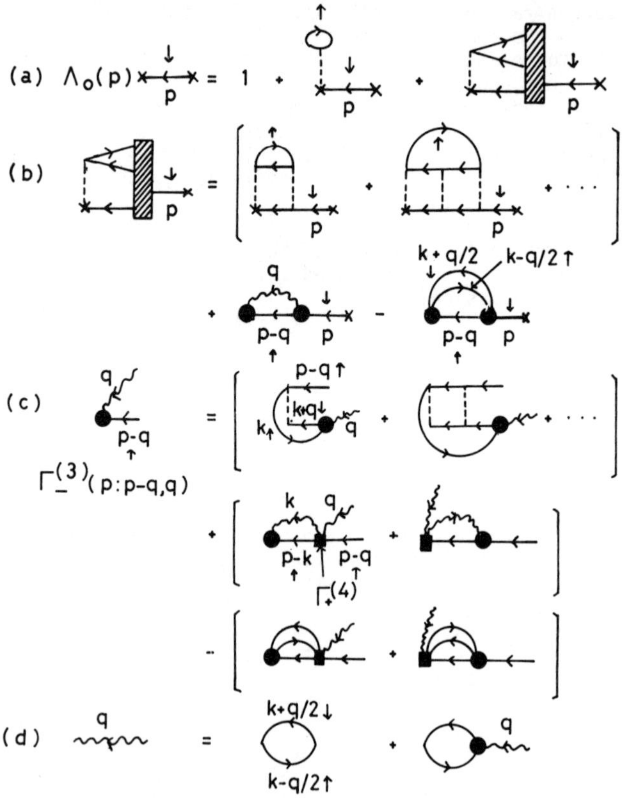

Fig. 1 Diagramatic equations for the propagators and vertex functions. The solid line represents the electron propagator, the wavy one the magnon propagator and the broken one the coulomb interaction.

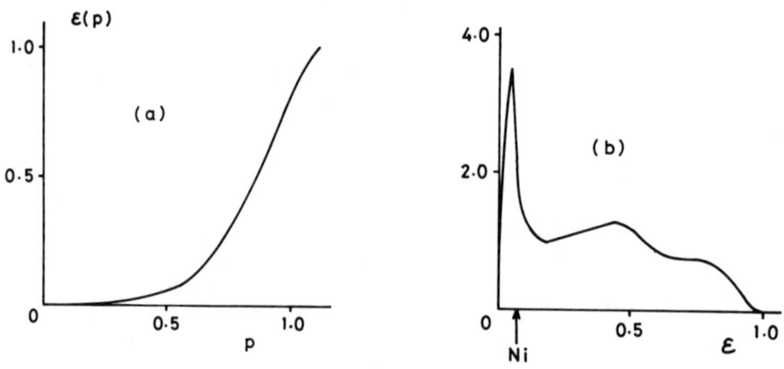

Fig. 2 (a) The assumed energy vs. momentum relation. The momentum p is scaled by the $2\pi/a$ with a a lattice constant. (b) The assumed density of states. The arrow indicates the Fermi level for pure nickel.

$$g = \overline{U} + g^2 M^{-1}(i/2\pi)N^{-1} \int d^4q\Delta_B(q)\ [S_+(-q)-S_-(q)]$$

$$- g^2 M^{-1}(i/2\pi)N^{-1} \int d^4qD(q)\ [S_+(-q)-S_-(q)]. \qquad (2)$$

$$\overline{U} = M^{-1}(-i/2\pi)N^{-1} \int_+ d^4k\ U\{1 - U\ G_2(k)\}^{-1}\{S_+(k)-S_-(k)\}. \qquad (3)$$

$$D(q) = (-i/2\pi)N^{-1} \int_+ d^4k\ S_-(k+q/2)\ S_+(k-q/2). \qquad (4)$$

$$G_2(q) = (i/2\pi)N^{-1} \int d^4k\ S_-(q/2+k)\ S_+(q/2-k). \qquad (5)$$

Here $S_\pm(p)$ denotes the up and down spin electron propagators and $\Delta_B(q)$ the magnon propagator. M represents the magnetization, which is defined by

$$M = (-i/2\pi)N^{-1} \int_+ d^4k\ [S_+(k) - S_-(k)]. \qquad (6)$$

The symbol + in Eqs.(3)(4) and (6) means that the integration path for k_o is taken along the half circle in the upper plane for complex k_o. From Eq.(2), it is found that the t-matrix term (the first term of Fig.1(b)) with p=o is expressed by g and the contribution from the one-loop diagram. Thus we can rewrite the electron propagators in terms of the renormalized quantity as

$$S_\pm(p) = p_0 - \{\varepsilon(p) - \tilde{\mu}^2 \mp gM/2 + (-i/2\pi)N^{-1}\int_+ d^4k[u(p+k)-u(k)]S_\mp(k)$$

$$+ g^2(i/2\pi)N^{-1}\int d^4q\Delta_B(q)[S_\mp(p\pm q) - S_\mp(\pm q)]$$

$$- g^2(i/2\pi)N^{-1}\int d^4qD(q)[S_\mp(p\pm q) - S_\mp(\pm q)]\}, \qquad (7)$$

where

$$u(p) = U[1 - UG_2(p)]^{-1}\ G_2(p)U\ , \qquad (8)$$

and $\tilde{\mu}^2$ denotes the renormalized chemical potential. From Fig.1(d), we obtain the magnon propagator in the lowest order approximation, as

$$\Delta_B(q) = D(q)/(1 + gD(q)), \qquad (9)$$

by replacing the vertex part by -g. It is the difference between the above equations and those in [13] that only the former contains the t-matrix terms (the first one in Eq.(2) and the one containing u(p) in Eq.(7)) and the correction ones to the double counting (the third one in Eq.(2) and the last one in Eq.(7)).

These equations are solved by the following procedure. First g is assumed to be a certain value. $\overline{U}$, D(q), $G_2(q)$ and $\Delta_B(q)$ are evaluated by use of Eqs.(3),(4),(5) and (9), with the assumption that the electron propagators $S_\pm(p)$ are given by the lowest order approximation, i.e., $\{p_0 - \varepsilon(p) + \tilde{\mu}^2 \pm gM/2\}^{-1}$. The use of Eq.(9) for the magnon propagator is appropriate in this case. g is recalculated by use of $\overline{U}$, D(q) and $\Delta_B(q)$ in the right-hand side of Eq.(2). We iterate the above procedure until Eq.(2) is satisfied. Then $S_\pm(p)$ are evaluated to the next order by use of Eqs.(7) and (8). Thus we can obtain the correct result for $S_+(p)$ within the limit of the one-loop diagram, although the momentum dependence of the t-matrix term in Eq.(7) may contain a higher order loop. On the basis of the loop expansion method, we may suppose that the t-matrix term is expanded by the renormalized coupling constant g. So the double counting term is considered to be the second-order diagram with the renormalized coupling constant in the lowest order approximation. This is why we adopt the diagram shown in Fig.1(b) as the correction term to the double counting. This iterative calculation is expected to be fairly correct, since the electron propagator in the lowest order approximation has been expressed by the renormalized value. This is especially so when an electron lies near the Fermi level, since the self-energy correction is zero at p=o.

We point out here that our formulas tend to the exact atomic limit. The proof is given in the following, where we assume the strong ferromagnetic state in order to simplify the discussion. When $U\to\infty$, g remains a finite value (Eq.(2)) and u(k) tends to -U with k_0 a finite value (Eq.(8)). Thus we obtain from Eqs.(7) and (8),

$$S_-(p) \sim \{p_0-UM/[1-U(1-M)/p_0]\}^{-1}=(1-M)/p_0+M/(p_0-U) \tag{10}$$

with $p_0 \sim U$.

III. Results of the calculation

In the following calculation, we adopt the hole picture and assume a spherical symmetric function for $\varepsilon(p)$, as is shown in Fig.2(a), where the energy and the momentum are scaled by the band width and $2\pi/a$ with $\underline{a}$ a lattice constant, respectively. The band width is assumed to be 4.5 eV. $\varepsilon(\vec{p})$ is determined to be consistent with the density of states, which simulates that of the t_g band of nickel, as is shown in Fig.2(b). In pure nickel, each of three t_g bands is assumed to have 0.18 electrons per atom. We follow the procedure given in section II. The integrations over the momentum space were performed in the first Brillouin zone of fcc lattice, where the Umklapp process is taken into account. In the calculation of $G_2(\vec{q})$ and D(q), we approximate them by the spherical average value over the $\vec{q}$ direction.

Let us first discuss the results of the electron-magnon interaction. We can obtain the required formulas, identical with those in [13], by replacing the first term in Eq.(2) by U and by neglecting the terms containing u(p) and D(q) in Eqs.(2) and (7). In Fig.3, g is shown by the solid line as a function of the hole number, where U is assumed to be the band width. The broken line stands for the quantity defined by the Fermi energy divided by the hole number, which is the smallest value of g to maintain the strong ferromagnetism. It is observed from the crossing point of the solid line over the broken one that the both spin states become partially filled by holes, i.e., the strong ferromagnetism becomes unstable, at the hole number 0.32 [26]. It is also found that the renormalized coupling constant g tends to decrease fairly rapidly with increasing the hole number after the onset of the <u>weak</u> ferromagnetic state. The exchange splitting between the p=0 holes is nearly equal to gM. The large value of it obtained by the Hartree-Fock approximation is reduced by the electron-magnon interaction, and the bottom of the down spin hole band approaches quite near the Fermi level. (In nickel it is reduced from 0.81 eV to 0.40 eV when U is assumed to be the band width.) The magnon propagator is also renormalized. In the case of nickel, the stiffness constant of the spin wave dispersion is reduced

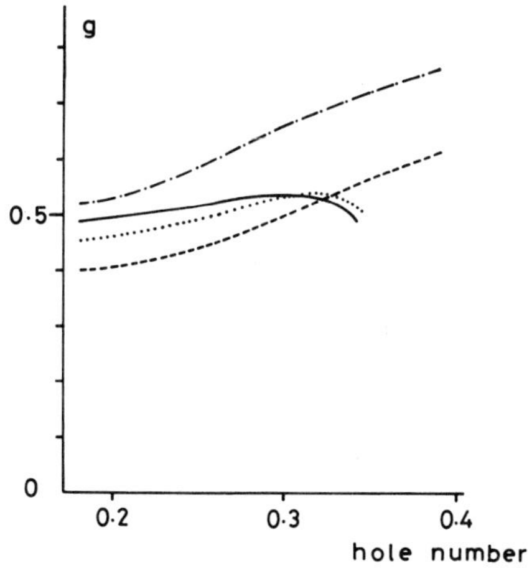

Fig. 3 The renormalized coupling constant g as a function of the hole number. The solid line represents the g value evaluated by taking into account only the electron-magnon interaction. (U is equal to the band width.) The broken dotted line represents the $\bar{U}$ (Eq.(3)) evaluated by the t-matrix theory. The dotted line represents the g value evaluated by the theory proposed in this paper. (U is equal to twice the band width in both cases.) The broken line represents the minimum value of g required to maintain the strong ferromagnetism. The unit of energy is the band width.

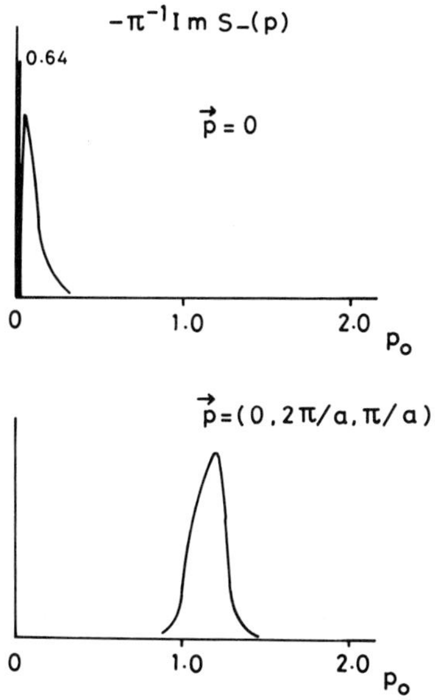

Fig. 4 The spectra for the down spin hole in Ni evaluated by taking into account only the electron-magnon interaction. The origin of the energy corresponds to the Fermi level. The spectra weight in the δ-function is indicated. P_0 is in the unit of the band width.

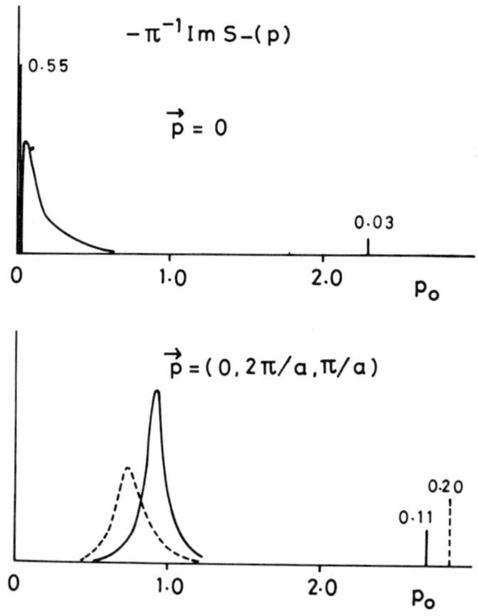

Fig. 5 The spectra for the down spin hole in Ni evaluated by taking into account the electron-electron scattering and the electron-magnon interaction simultaneously (solid line). The broken line represents the results obtained by taking into account further the electron-electron interaction in the different orbitals. (See text.)

from 420 meV A^2 to 270 meV A^2. (Here the lattice constant $\underline{a}$ is assumed to be 3.46 A.) Its value is fairly small in comparison with experiment (∿500 meV A^2) [27]. This disagreement, however, is not important by the following reason: (1) we can obtain a more reasonable value of it by changing $\varepsilon(k)$ slightly, because it sensitively depends on the shape of $\varepsilon(k)$. (2) the spin-wave branch has not any considerable effect on the estimate of g [28]. In Fig.4, the spectrum for the down spin hole of nickel with a definite momentum is shown, where U is also assumed to be the band width. It is noted that the band width becomes broad by the electron-magnon interaction.

Next let us discuss the calculated results of the theory proposed in the preceding section. In the following calculation, U is assumed to be twice the band width. In Fig.3, the calculated value of g is shown by the dotted line. From the figure, it is found that the strong ferromagnetism becomes unstable at the hole number 0.32 [26], and the exchange splitting is reduced from the large Hartree-Fock value. In the same figure, $\overline{U}$, which is defined by Eq.(3), is shown by the broken-dotted line, when the electron-magnon interaction is neglected (t-matrix theory). We calculated $\overline{U}$ self-consistently by assuming the electron propagators to be the lowest order approximation with the exchange splitting $\overline{U}M$ and by using Eq.(3). From the figure, it is found that the t-matrix theory does not cause the instability of the strong ferromagnetism.

In Fig.5 the down spin hole spectrum of nickel with a definite momentum is shown, where the same U value (twice the band width) is assumed. The band width reduces little due to the cancellation effect of the electron-magnon term and the t-matrix one.

The further reduction of the band width is expected, if we take into account the coulomb interaction between electrons in different orbitals. (We assume three orbitals.) By use of the t-matrix theory, we obtain the self-energy correction to the up and down spin hole propagators as

$$\Delta\Sigma_{\pm}(p) = (-i/2\pi)N^{-1}\cdot 2\int_{+} d^4k\ U/[1-UG_2^{+\pm}(p+k)]S_{+}(k) \tag{11}$$

with

$$G_2^{+\pm}(p) = (i/2\pi)N^{-1}\int d^4k\ S_{+}(p/2-k)\ S_{\pm}(p/2+k) \tag{12}$$

in the strong ferromagnetic state. The factor 2 in Eq.(11) means the number of the other orbitals. Note that there appears the self-energy correction to the up spin hole propagator as well as the down spin one. In order to estimate the band width, we calculate $\Delta\Sigma_{\pm}(p)-\Delta\Sigma_{\pm}(0)$ by use of $S_{\pm}(p)$ previously calculated. In this paper we do not discuss the contribution to the exchange splitting, since a self-consistent calculation of $\Delta\Sigma_{\pm}(p)$ will be required to that purpose. The spectrum for the down spin hole thus obtained is also shown in Fig.5, by the broken line, in a fair agreement with the experiment.

As is shown in the same figure, we have obtained the result that the two-hole bound states are located about 2.5 (∿11eV) above the Fermi level. If U is assumed to be smaller than the one used in our calculation, the satellite position will become closer to the Fermi level. However it has been found from the numerical calculation that U can not be smaller than 1.3 (5∿6 eV) in order to maintain the strong ferromagnetism of nickel, within the single-orbital Hubbard model. The two-hole bound state in our calculation is the itinerant one. As is suggested by IWAN et. al.[7], the 6 eV satellite may correspond to the two-hole bound state localized with an extra 4s electron screening it. Thus we may not have to make the two-hole bound state in our calculation correspond to the 6 eV satellite.

IV. Supplementary discussion

In this paper we presented the many-body theory, which took into account simultaneously the electron-electron scattering and the electron-magnon interaction. The well-known redundancy between the self-energy parts coming from the two mechanisms mentioned above was removed within the limit of the one-loop approximation. By use of this theory, the self-energy correction to the one-electron propagator was evaluated. As for the exchange splitting, both of the two mechanisms made it much smaller than the Hartree-Fock value. They had a tendency to destroy the strong ferromagnetism. So the U value had to be greater than 5-6 eV in order to be consistent with the strong ferromagnetism of nickel. With increasing the hole number from the nickel case, the role of the electron-magnon interaction for the reduction of the exchange splitting became more and more important. (See Fig.4.) We expect that the inclusion of the electron-magnon interaction improves to some extent the low density approximation in the t-matrix theory. As for the band narrowing phenomena, it was found that the two mechanisms worked in the opposite direction: the electron-electron scattering made the band width narrower and the electron-magnon interaction broader. Within the single-band model simulating nickel, the cancellation is fairly complete. So we had to take into account the coulomb interaction between d electrons in different orbitals in order to get the reasonable reduction of the band width. In order to make a definite comparison with the experiment [1], however, we have to know the bare hand width in the Hubbard model or to what extent the band calculation contains the many-body effect mentioned above. These points are not fully understood. Nevertheless we think that the many-body effect given in this paper is the main origin of the band narrowing.

What value of U is permitted in the Hubbard model so as to be consistent with the ferromagnetism ? Within the single-band model, a fairly large value is required. If we depart from it and take into account the Hund rule coupling between d electrons in the different orbitals, U may be permitted to be small, since the Hund rule coupling assists the ferromagnetism. Moreover, it is suggested by KANAMORI [19] that the inter-atomic interaction is important in nickel. This mechanism may assist ferromagnetism. Our theory can be extended to include these effects. However it is rather complicated, since the magnon propagator contains then optical modes as well as the spin-wave mode. The numerical estimate of these effects remains as a future problem.

It was also found that the strong ferromagnetism became unstable due to the electron-magnon interaction as increasing the hole number. This comes from the fact that the effect of the electron-magnon interaction on the renormalized coupling constant g is nearly independent of the hole number in the strong ferromagnetic state and thereby the top of the majority spin electron state approaches to the Fermi level. On the other hand, the t-matrix theory did not cause the instability of the strong ferromagnetism, since the effective U value calculated by this theory increased with the hole number increasing. After the onset of the weak ferromagnetism, the g value decreased fast with the hole number increasing. This means that the contribution of the electron-magnon interaction increases rapidly. It should be noted here that the state obtained in this paper is quite different from that given by the Hartree-Fock approximation. On the basis of the rigid band approximation, we can make the many-body state like this correspond to that in Ni-Fe alloys.

In order to discuss the alloying effect, however, we have to combine our theory with the CPA theory. This is a difficult problem left in future. We now merely comment that in the HF-CPA theory, the many-body effect is roughly taken into account by replacing the large bare coulomb interaction by a small adjustable parameter, which is assumed not to depend on the hole number. From our point of view, this assumption is not valid, especially in the concentration range where the saturation magnetization deviates from the Slater-Pauling curve in Ni-Fe alloy. So the HF-CPA theory is considered not to be reliable especially in this concentration range.

Acknowledgements

The author would like to express his sincere thanks to Professor J. Kanamori for suggesting him the present work and for valuable discussions. Thanks are also due to Mrs. A. Egusa for preparing the manuscript.

References

1. F. J. Himpsel, J. A. Knapp and D. E. Eastman: Phys. Rev. B 19, 2919 (1979).
2. D. E. Eastman, F. J. Himpsel and J. A. Knapp: Phys. Rev. Lett. 44, 95 (1980).
3. S. Hüfner and G. K. Wertheim: Phys. Lett. 51A, 299 (1975).
4. P. C. Kemeny and N. J. Shevchik: Solid State Commun. 17, 255 (1975).
5. R. J. Smith, J. Anderson, J. Hermanson and G. J. Lapeyre: Solid State Commun. 21, 459 (1977).
6. C. Guillot, Y. Ballu, J. Paigné, J. Lecante, K. P. Jain, P. Thiry, R. Pinchaux, Y. Pétroff and L. M. Falicov: Phys. Rev. Lett. 39, 1632 (1977).
7. V. L. Moruzzi, J. F. Janak and A. R. Williams: Calculated Electronic Properties of Metals (Pergamon, New York, 1978)
8. C. S. Wang and J. Callaway: Phys. Rev. B15, 298 (1977).
9. J. Kanamori: Prog. Theor. Phys. 30, 275 (1963).
10. P. W. Anderson: Phil. Mag. 24, 203 (1971).
11. U. Brandt: Z. Phys. 244, 217 (1971).
12. J. A. Hertz and D. M. Edwards: J. Phys. F3, 2174 (1973); D. M. Edwards and J. A. Hertz: J. Phys. F3, 2191 (1973).
13. H. Matsumoto, H. Umezawa, S. Seki and M. Tachiki: Phys. Rev. B 17, 2276 (1978).
14. A. Liebsch: Phys. Rev. Lett. 43, 1431 (1979).
15. D. Penn: Phys. Rev. Lett. 42, 921 (1979).
16. G. Treglia, F. Ducastelle and D. Spanjaard: J. Physique 41, 281 (1980).
17. M. Iwan, F. J. Himpsel and D. E. Eastman: Phys. Rev. Lett. 43, 1829 (1979).
18. E. Antonides, E. C. Janse and G. A. Sawatzky: Phys. Rev. B 15, 1669 (1977).
19. J. Kanamori: in this book.
20. H. Hasegawa and J. Kanamori: J. Phys. Soc. Japan 31, 382 (1971); 33, 1599 (1972).
21. T. Jo and H. Miwa: J. Phys. Soc. Japan 40, 706 (1976).
22. T. Jo: J. Phys. Soc. Japan 40, 715 (1976).

23. J. Kanamori and T. Jo: in this book.
24. Y. Kitaoka and K. Asayama: J. Phys. Soc. Japan 40, 1521 (1976).
25. J. Igarashi: in preparation.
26. 0.32 is nearly equal to the number of the 3d holes included in the one orbital per atom at the iron concentration 50 percent, if the five d bands are equally occupied.
27. H. A. Mook, J. W. Lynn and R. M. Nicklow: Phys. Rev. Lett. 30, 556 (1973).
28. This is easily seen from Eq.(2), where the momentum contributing to the integral in the second term has to be greater than the Fermi momentum. Except for the case when the Fermi momentum is fairly small, the Stoner excitation has a more important role than the bound state of an electron-hole pair.

Electronic Band Structure and Helical Spin Density Wave of MnSi

Osamu Nakanishi, Akira Yanase*, and Mitsuo Kataoka

The Research Institute for Iron, Steel and Other Metals
Tohoku University, Sendai, Japan

* Department of Physics, Faculty of Science, Tohoku University, Sendai, Japan

1. Introduction

The intermetallic compounds with the B20 type of structure have been received continual attention and studied in detail in the past decades [1-3]. This is because their electrical and magnetic properties show a variety, depending on the constituent elements. Among them, MnSi has been studied most extensively and revealed to be a helical magnet whose Néel temperature T_N is 29 K [4]. According to the neutron diffraction experiments performed by ISHIKAWA et al., this helical spin density wave (HSDW) has the peculiar properties as follows [4,5]: (i) The period is very long (180 Å). (ii) The equilibrium direction of the wave vector $\vec{Q}$ is <111>. When an external magnetic field is applied, the direction of $\vec{Q}$ follows that of the magnetic field. (iii) The period does not depend on the direction of $\vec{Q}$, but depends weakly on temperature. (iv) The HSDW is changed into an induced ferromagnetic state (IFMS) by a weak external magnetic field (6.2 KOe). This means that the energy difference between the HSDW and the ferromagnetic state (FMS) is much small. The IFMS is supposed to be a typical example of itinerant weak ferromagnetism on the basis of the following experimental facts: (i) The induced ferromagnetic moment 0.4 μ_B/Mn is much less than 1.4 μ_B/Mn obtained by using the Curie-Weiss constant in the paramagnetic phase [2,6]. (ii) The magnetization curve does not saturate even at 150 KOe [7]. (iii) The spin-lattice relaxation time T_1 of ^{55}Mn in MnSi is independent of temperature at temperatures much higher than T_N[8,9].

MORIYA and MAKOSHI applied the self-consistent renormalization theory for itinerant weak ferromagnetism to MnSi in order to clarify the properties of the HSDW [10-12]. In their theory, the origin of the HSDW was attributed to a maximum of the susceptibility $\chi_0(\vec{Q})$ of the non-interacting electron system at a wave vector $\vec{Q}$ corresponding to the period of the HSDW. As will be shown in the next section, the calculated band structure [13] shows that there exists no nesting between the electronic states near the Fermi surface by the wave vector $\vec{Q}$ corresponding to the period of the HSDW. Then we consider another mechanism for the origin of the long-period spin structure.

We note that, since MnSi has the crystal structure of low symmetry (Fig.1), the antisymmetric exchange interaction or the Dzyaloshinsky-Moriya (DM) interaction [14,15] can be important to spin structures. This interaction causes a parallel alignment of neighboring spins to deviate toward a perpendicular configuration. This implies that a HSDW is a spin structure favorable to the DM interaction in a certain class of crystal structure [16-20]. In §3, it will be shown that the observed peculiar properties of the HSDW in MnSi can be explained by assuming that the DM interaction causes the long-period spin structure [20].

2. Self-Consistent APW Calculation for the Electronic Band Structure

Figures 1 and 2 show the crystal structure of MnSi and its Brillouin zone, respectively. The space group is $P2_13(T^4)$. A unit cell contains four molecules. We have made the self-consistent APW calculation for the electronic band structure of this compound [13]. The $X\alpha$ parameter α is taken to be 2/3 and both muffin-tin radii of Mn and Si are determined to be 1.12 Å. The sum over the angular momentum, ℓ, in the APW functions is truncated after $\ell_{max}=6$. The magnitudes of the wave vectors are smaller than $3.6(2\pi/a)$, where a (= 4.558 Å) is the lattice constant. The number of augmented plane waves is about 200. The starting charge density in the crystal is a superposition of the charge densities for the neutral atoms $Mn(3d)^5(4s)^2$ and $Si(3s)^2(3p)^2$. The four symmetry points, Γ, R, X and M, are chosen as the sampling points, at which the energies and the wave functions are calculated. The crystal charge density is obtained by a summation over occupied states on the sampling points. A sum of 80% of the input charge density and 20% of the output charge density is used to construct a new potential for the next step of iteration. After 15 iterations, the calculated energies become stable within 0.002 Ryd. Using the obtained self-consistent potential, we calculate the energy bands along the five symmetry lines, T, Λ, Δ, Z and Z'. In order to obtain the band structure at general points in the Brillouin zone, we fit the LCAO bands to the calculated ones whose energies lie between 0 and 1.2 Ryd.

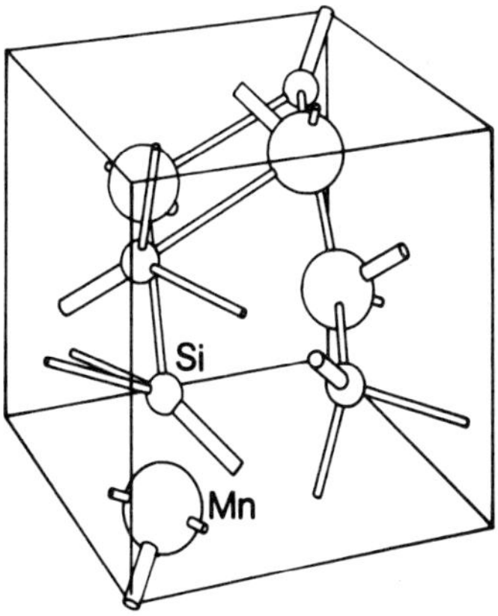

Fig.1 Crystal structure of MnSi. There are four Mn ions and four Si ions in a unit cell. Large and small circles show Mn and Si, respectively. The positions of Mn and Si ions in a unit cell are given by (u,u,u), (1/2+u,1/2-u,-u), (-u,1/2+u,1/2-u) and (1/2-u,-u,1/2+u) where u_{Mn} and u_{Si} are 0.138 and 0.845, respectively.

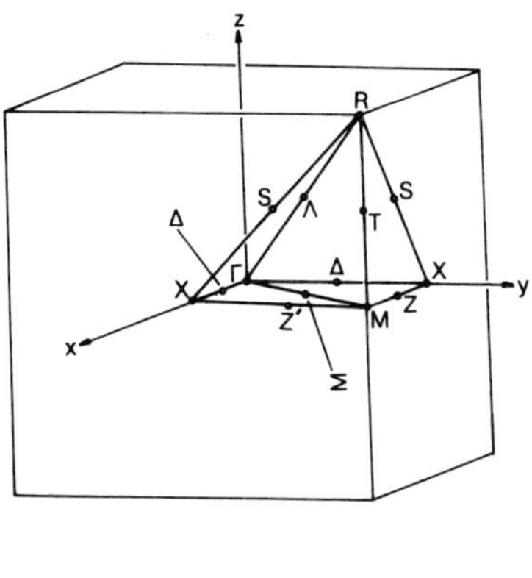

Fig.2 Brillouin zone of MnSi. The solid and open circles show the symmetry points and the symmetry lines, respectively. Since the isogonal point group is T, the representation domain is 1/12 of the Brillouin zone. The inclusion of time reversal symmetry causes degeneracies between $\vec{k}$ and $-\vec{k}$. Therefore, it is sufficient to calculate the energy levels in 1/24 of the Brillouin zone. Any symmetry operation does not transform the Z line into the Z' line.

Figures 3 and 4 show the calculated energy bands and the density of states, respectively. The dashed line in Fig.3 shows the Fermi level E_f (=0.93 Ryd.). The band states with energies near E_f consist mainly of the 3d orbitals of Mn. The following two properties of this band structure should be noted. (i) The band structure has a narrow gap at about 0.97 Ryd., below which 48 electrons/cell can be piled up. If we adopt the rigid band approximation, the Fermi level of FeSi should lie in the energy gap. This can explain the

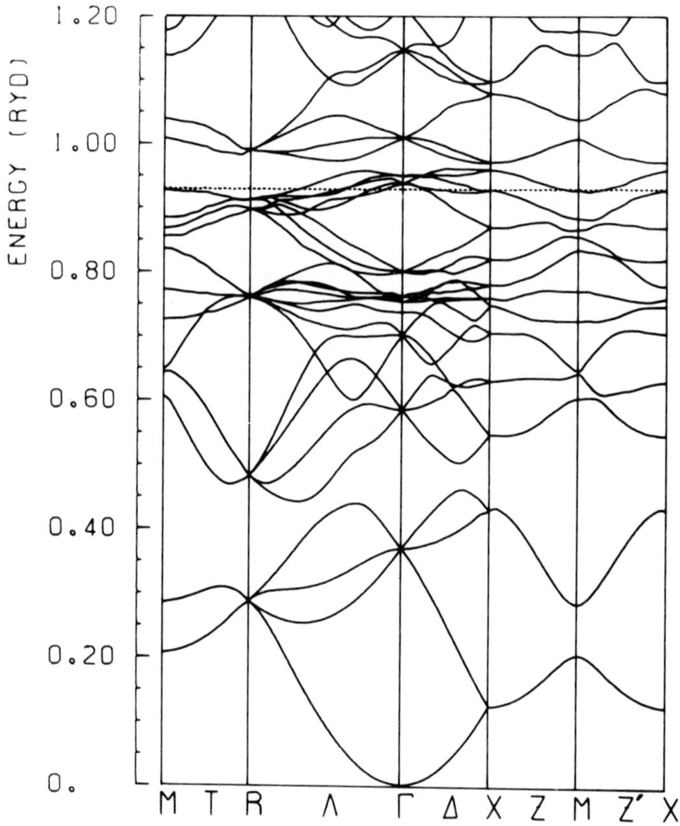

Fig.3 Energy bands of MnSi. The dashed line shows the Fermi level E_f (= 0.93 Ryd.). It is noted that there exists a narrow gap at about 0.97 Ryd. The bands along the Λ and Δ lines are not degenerate, while those along the T , Z and Z' lines are doubly degenerate.

Fig.5 Perspectives of the five Fermi surfaces of MnSi.

(a) Electron sheet of the 24-th band. The center of figure is the R point. The corners are the Γ points.
(b) Electron sheet of the 23-rd band. The center is the R point.
(c) Hole sheet of the 22-nd band. The center is the Γ point. The corners are the R points.
(d) Hole sheet of the 21-st band. The center is the Γ point.
(e) Hole sheet of the 20-th band. The center is the Γ point.

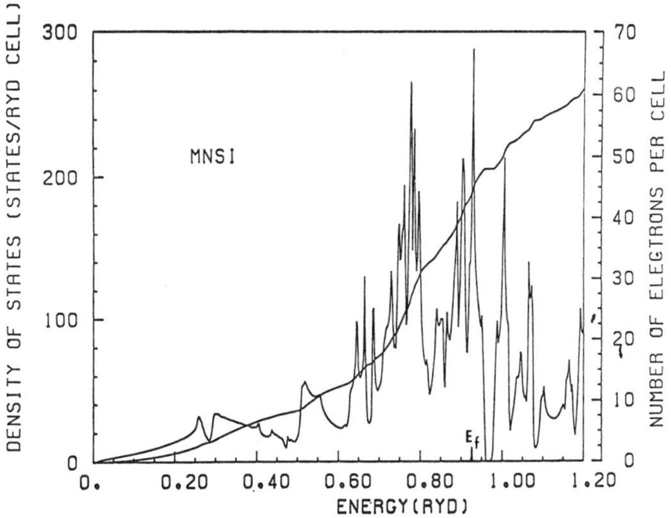

Fig.4 Density of states of MnSi. The number of electrons/cell is also shown. These curves are obtained by using the LCAO bands fitted to the APW ones calculated along the high symmetry lines. E_f shows the Fermi level.

Fig.5.
Caption see opposite page

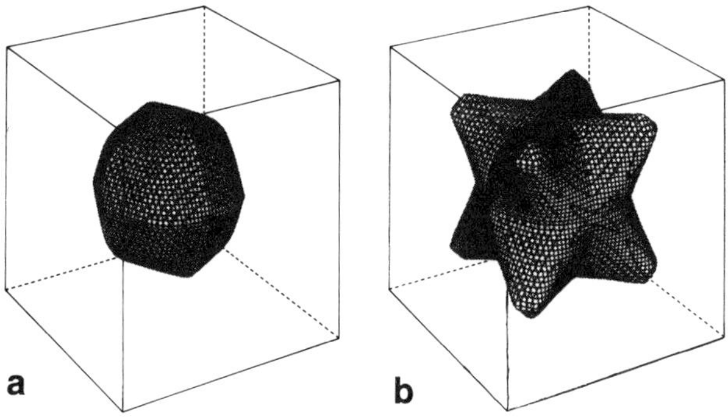

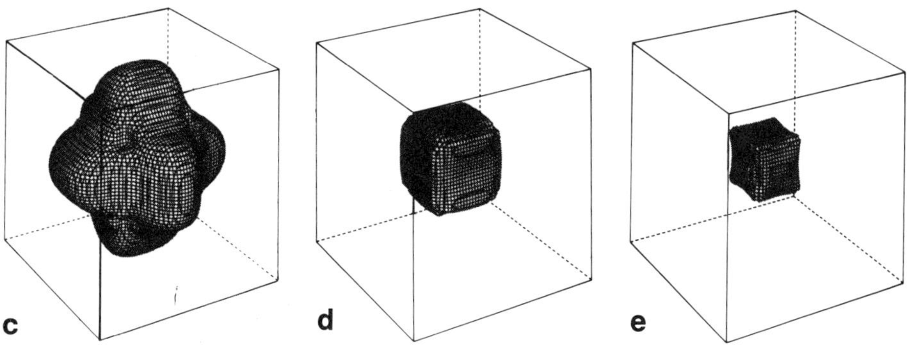

observed fact that FeSi is a narrow gap semiconductor [3,21]. (ii) When we shift the up- and down-spin bands so that MnSi has the induced ferromagnetic moment 0.4 μ_B/Mn, both the up- and down-spin bands have large densities of states at the Fermi level. This agrees with the experimental fact that the IFMS in MnSi is not saturated [7].

To visualize the shapes for the Fermi surfaces, we try to draw their perspectives by the aid of a computer. Since the shapes for the Fermi surfaces are very sensitive to the energy dispersions, the global fitting of the LCAO bands is inadequate to drawing of the shapes for the Fermi surfaces. Therefore we fit again the LCAO bands only to the 8 bands below the gap at 0.97 Ryd. to reproduce the energy dispersions near the Fermi level more precisely. The energy differences between the original and fitted bands are within 0.01 Ryd. This precise fitting gives 148 states/Ryd. cell for the density of states at E_f. Since the Fermi level crosses the five bands as shown in Fig. 3, MnSi has the five Fermi surfaces, which are shown in Fig.5(a) to 5(e). We note that there exists no nesting between electronic states by a small $\vec{Q}$ corresponding to the period of the HSDW in MnSi.

3. Origin of the Helical Spin Density Wave

We know that the antisymmetric exchange interaction between a pair of magnetic ions with no inversion symmetry can affect an antiparallel alignment of localized spins to be canted [14,15]. In the following, we see how a ferromagnetic spin structure stabilized by the usual symmetric exchange interaction is modified by the antisymmetric one [20].

It is assumed that the free energy of the spin system, F, is expressed by using the spin density $\vec{S}(\vec{r})$ as

$$F = \sum_n \sum_{i_1 i_2 \cdots i_{2n}} \int \tilde{g}_{i_1 i_2 \cdots i_{2n}} (\vec{r}_1, \vec{r}_2, \cdots, \vec{r}_{2n}) \times S_{i_1}(\vec{r}_1) S_{i_2}(\vec{r}_2) \cdots S_{i_{2n}}(\vec{r}_{2n}) d\vec{r}_1 d\vec{r}_2 \cdots d\vec{r}_{2n} , \tag{1}$$

where i's specify the x, y and z components of $\vec{S}(\vec{r})$. The neutron diffraction experiments on MnSi do not show any multiple spin structure of the sublattices consisting of four Mn ions in a unit cell [22], so that the spin structure may be well defined in terms of a single $\vec{Q}$. Therefore the spatial variation of the spin density $\vec{S}(\vec{r})$ is expressed by

$$\vec{S}(\vec{r}) = \vec{S}_1 \cos(\vec{Q}\cdot\vec{r}) + \vec{S}_2 \sin(\vec{Q}\cdot\vec{r}) = [\vec{S}^+ \exp(-i\vec{Q}\cdot\vec{r}) + \vec{S}^- \exp(i\vec{Q}\cdot\vec{r})]/2, \tag{2}$$

$$\vec{S}^{\pm} = \vec{S}_1 \pm i\vec{S}_2, \tag{3}$$

where $\vec{S}_1$ and $\vec{S}_2$ are arbitrary spin vectors. Substituting (2) into (1) and expanding F in powers of $\vec{Q}$ whose magnitude is assumed to be small, we obtain

$$F = \sum_n \sum_{\sigma_1+\sigma_2+\cdots+\sigma_{2n}=0} \sum_{i_1 i_2 \cdots i_{2n}} [g(i_1,i_2,\cdots,i_{2n}) \prod_{\nu=1}^{2n} S_{i_\nu}^{\sigma_\nu}$$

$$+ \sum_m g^{\sigma_1\sigma_2\cdots\sigma_{2n}}(i_1,i_2,\cdots,i_{2n};\ j_1,j_2,\cdots,j_m) \prod_{\nu=1}^{2n} S_{i_\nu}^{\sigma_\nu} \prod_{\mu=1}^{m} Q_{j_\mu}], \quad (4)$$

$$g(i_1,i_2,\cdots,i_{2n}) = \frac{1}{2^{2n}} \int \tilde{g}_{i_1 i_2 \cdots i_{2n}}(\vec{r}_1,\vec{r}_2,\cdots,\vec{r}_{2n}) d\vec{r}_1 d\vec{r}_2 \cdots d\vec{r}_{2n}, \quad (5)$$

$$g^{\sigma_1\sigma_2\cdots\sigma_{2n}}(i_1,i_2,\cdots,i_{2n};\ j_1,j_2,\cdots,j_m)$$

$$= \frac{(-i)^m}{2^{2n} m!} \int \prod_{\mu=1}^{m} (\sum_{\nu=1}^{2n} \sigma_\nu \vec{r}_\nu)_{j_\mu} \tilde{g}_{i_1 i_2 \cdots i_{2n}}(\vec{r}_1,\vec{r}_2,\cdots,\vec{r}_{2n}) d\vec{r}_1 d\vec{r}_2 \cdots d\vec{r}_{2n}, \quad (6)$$

where j's specify the x, y and z components of $\vec{Q}$; σ's are +1 or -1, and S^σ means S^+ or S^-. We have taken into account that $(\sigma_1+\sigma_2+\cdots+\sigma_{2n})\vec{Q}$ with a small n can not be equal to nonvanishing reciprocal lattice vectors because of the smallness of Q.

Eq.(4) shows that

$$g(i_1,i_2,\cdots,i_{2n}) \quad \text{and} \quad g^{\sigma_1\sigma_2\cdots\sigma_{2n}}(i_1,i_2,\cdots,i_{2n};\ j_1,j_2,\cdots,j_m)$$

are, respectively, transformed as

$$(r_{i_1} r_{i_2} \cdots r_{i_{2n}}) \quad \text{and} \quad (r_{i_1} r_{i_2} \cdots r_{i_{2n}} r_{j_1} r_{j_2} \cdots r_{j_m})$$

by the symmetry operations constructing the isogonal point group of $P2_13(T^4)$, T. The number of the independent g's or $g^{\sigma_1\sigma_2\cdots\sigma_{2n}}$'s is equal to that of the bases of the identity representations. We derive, as an example, the term quadratic in $\vec{S}$ and linear in $\vec{Q}$. The symmetry of T requires that

$$g^{+-}(x,y;z) = g^{+-}(y,z;x) = g^{+-}(z,x;y),$$
$$g^{+-}(y,x;z) = g^{+-}(z,y;x) = g^{+-}(x,z;y), \quad (7)$$

and the other g^{+-}'s are all zero. In addition to these relations, (6) gives

$$g^{+-}(i_1,i_2;j_1) = -\ g^{-+}(i_1,i_2;j_1) = -\ g^{+-}(i_2,i_1;j_1). \quad (8)$$

Substituting (7) and (8) into (4), we obtain the antisymmetric form $D[\vec{S}_1\times\vec{S}_2]\cdot\vec{Q}$, where

$$D = -4ig^{+-}(x,y;z)$$

$$= -\int (z_1 - z_2)\tilde{g}_{xy}(\vec{r}_1,\vec{r}_2)d\vec{r}_1 d\vec{r}_2. \tag{9}$$

For a discussion of the stability of the spin state, we should take the terms in (4) at least up to the fourth and the second orders of S's and Q's, respectively. The similar procedures to the above lead us to the total free energy as follows.

$$\begin{aligned} F = & \{\tfrac{1}{2}[\chi(0)^{-1} + cQ^2]\langle S^2\rangle_p + \tfrac{1}{4}\gamma\langle S^4\rangle_p\} + D[\vec{S}_1\times\vec{S}_2]\cdot\vec{Q} \\ & + \tfrac{1}{2}\{A_1[\langle 3S_z^{\,2} - S^2\rangle_p(3Q_z^{\,2} - Q^2) + 3\langle S_x^{\,2} - S_y^{\,2}\rangle_p(Q_x^{\,2} - Q_y^{\,2})] \\ & \quad + A_2[\langle 3S_z^{\,2} - S^2\rangle_p(Q_x^{\,2} - Q_y^{\,2}) - \langle S_x^{\,2} - S_y^{\,2}\rangle_p(3Q_z^{\,2} - Q^2)] \\ & \quad + 4A_3[\langle S_xS_y\rangle_pQ_xQ_y + \langle S_yS_z\rangle_pQ_yQ_z + \langle S_zS_x\rangle_pQ_zQ_x]\} \\ & - \tfrac{1}{2}K_1[\langle S_x^{\,4}\rangle_p + \langle S_y^{\,4}\rangle_p + \langle S_z^{\,4}\rangle_p], \end{aligned} \tag{10}$$

where the coefficients are to be expressed by linear combinations of g's and

$$\begin{aligned} \langle S^2\rangle_p &= \tfrac{1}{2}(S_1^{\,2} + S_2^{\,2}), \\ \langle S_iS_{i'}\rangle_p &= \tfrac{1}{2}(S_{1i}S_{1i'} + S_{2i}S_{2i'}), \\ \langle S^4\rangle_p &= \tfrac{3}{8}(S_1^{\,2} + S_2^{\,2})^2 - \tfrac{1}{2}|\vec{S}_1\times\vec{S}_2|^2, \\ \langle S_i^{\,4}\rangle_p &= \tfrac{3}{8}(S_{1i}^{\,2} + S_{2i}^{\,2})^2. \end{aligned} \tag{11}$$

It is noted that $\langle\cdots\rangle_p$ in (10) and (11) corresponds to an average over a period, i.e.,

$$\langle f\rangle_p = \frac{Q}{2\pi}\int_0^{2\pi/Q} f(\vec{r})dr_{/\!/}\,, \tag{12}$$

where $r_{/\!/}$ is the parallel component of $\vec{r}$ to $\vec{Q}$. An arbitrary translation of the origin of $\vec{r}$ in (2) does not change F given by (10). This means that (10) does not contain the effects of the crystal surface. The first term on the right-hand side of (10) is the symmetric exchange energy. If both c and γ are positive, this term stabilizes a FMS. The third and fourth terms give the anisotropy energies of spin and $\vec{Q}$. We can expect that the first term is of the order of k_BT_N, while the other terms are at most of the order of magnetic anisotropy energies.

The second term can become lowest when $\vec{S}_1$, $\vec{S}_2$ and $\vec{Q}$ are perpendicular to each other. As seen from (2), this configuration corresponds to a HSDW whose helical plane is perpendicular to the direction of $\vec{Q}$. The clockwise or counterclockwise rotation of spins is energetically favorable for a negative or positive D, respectively. The distortion from the perpendicular configuration is small even if the direction of $\vec{Q}$ deviates from <111> or <001>. Substituting $|\vec{S}_1|=|\vec{S}_2|=S$ and $\vec{S}_1 \perp \vec{S}_2$ into (10), we obtain the free energy of the HSDW, F_{HSDW}, as

$$F_{HSDW} = \frac{1}{2}[\chi(0)^{-1} - 2|D|Q + (c + A_1 - A_3)Q^2]S^2 + \frac{1}{4}(\gamma - \frac{3}{4}K_1)S^4$$
$$+ \frac{1}{2}[(A_3 - 3A_1)S^2Q^2 - \frac{3}{8}K_1S^4](\beta_1{}^4 + \beta_2{}^4 + \beta_3{}^4), \quad (13)$$

where β's are the direction cosines of $\vec{Q}$. By minimizing F_{HSDW} with respect to S, (13) is written as

$$F_{HSDW} = -\frac{1}{4\gamma\chi(0)^2}\{[1 + \left(\frac{3K_1}{4\gamma}\right) - 4|D|\chi(0)Q + 2(c + A_1 - A_3)\chi(0)Q^2]$$
$$+ [\left(\frac{3K_1}{4\gamma}\right) + 2(A_3 - 3A_1)\chi(0)Q^2](\beta_1{}^4 + \beta_2{}^4 + \beta_3{}^4)\}. \quad (14)$$

The free energy of a FMS, F_{FMS}, is easily obtained by replacing $\langle S^2\rangle_p$, $\langle S^4\rangle_p$ and $\langle S_i{}^4\rangle_p$ in (10) by S^2, S^4 and $S_i{}^4$, respectively. On the other hand, the free energy of a sinusoidal spin density wave (SSDW), F_{SSDW}, is obtained by assuming $\vec{S}_1 /\!/ \vec{S}_2$. Figure 6 shows the free energies thus obtained. When Q approaches to zero, F_{HSDW} is higher than F_{FMS} by a part of the magnetic anisotropy energy. This is because the directions of spins in a HSDW rotate sacrificing the anisotropy energy. If a lowering in energy due to the antisymmetric exchange energy overcomes the loss of the magnetic anisotropy energy, the HSDW becomes more stable than the FMS. Eq.(14) shows that F_{HSDW} has its minimum at $\vec{Q}=\vec{Q}_0$ where Q_0 $(=|\vec{Q}_0|)$ is given by

$$Q_0 = \frac{|D|}{c}[1 + \left(\frac{A_3 - A_1}{c}\right) - \left(\frac{A_3 - 3A_1}{c}\right)(\beta_1{}^4 + \beta_2{}^4 + \beta_3{}^4)]. \quad (15)$$

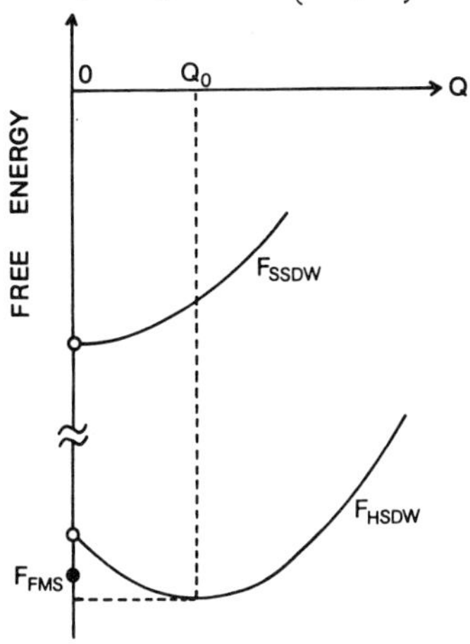

Fig.6 Q dependence of the free energy of the HSDW, F_{HSDW}, given by (14). The solid circle represents the free energy of the FMS, F_{FMS}. The free energy of the SSDW, F_{SSDW}, is also shown. The open circles mean that both F_{HSDW} and F_{SSDW} have no meaning at Q=0. F_{HSDW} has its minimum at Q_0 $(\simeq|D|/c)$. Only when the magnetic anisotropy energy is sufficiently small, the HSDW can become most stable.

Since $|D|/c$ and $|A_3 - 3A_1|/c$ are much small compared with unity, $\vec{Q}_0$ has a very small magnitude hardly dependent on its direction. Furthermore, Q_0 is independent of S and therefore only weakly dependent on temperature. From (14) and (15), the anisotropy energy of the direction of $\vec{Q}$ is obtained as

$$-\frac{1}{4\gamma\chi(0)}\left[\frac{3K_1}{4\gamma\chi(0)} + 2\left(\frac{D}{c}\right)^2(A_3 - 3A_1)\right](\beta_1{}^4 + \beta_2{}^4 + \beta_3{}^4). \tag{16}$$

An external magnetic field induces a magnetization by distorting the HSDW to a conical spin density wave (CSDW). If the anisotropy energy (16) is smaller than the difference between F_{FMS} and the minimum F_{HSDW} and is exceeded by the Zeeman energy, the direction of $\vec{Q}$ follows that of a magnetic field. The CSDW crushes to the IFMS with increasing the magnetic field further. The critical field is of the order of D^2/c as seen from (10).

The characteristic properties of the HSDW obtained above agree with those observed in MnSi [4]. We conclude that the antisymmetric exchange interaction can cause the HSDW, of which a typical example is observed in MnSi.

4. Discussion

Once the HSDW due to the antisymmetric exchange interaction is found to be realized in real materials, it is desirable to investigate in more detail the various properties of the HSDW.

On the basis of the Landau theory, we can discuss the thermodynamic properties of the HSDW at a temperature T near T_C which is the Curie temperature of the FMS in the absence of $D[\vec{S}_1 \times \vec{S}_2]\cdot\vec{Q}$. We assume that $\chi(0)^{-1} = c_0(T/T_C - 1)$ where c_0 is a positive constant and that the other coefficients in (10) are independent of T. The Néel temperature of the HSDW, T_N, is given by

$$T_N = T_C\left[1 + \left(\frac{D^2}{c_0 c}\right)\right], \tag{17}$$

where small terms have been neglected. It is noted that T_N is very little higher than T_C. If an external magnetic field H in the direction of $\vec{Q}$ is increased, the HSDW is distorted to a CSDW and finally changed into an IFMS. The transition between the CSDW and the IFMS is of the second order. By extending the calculation in §3 to the case of a spin system under external magnetic fields, the magnetization S_0 is shown to be given by

$$c_0(T_N/T_C - 1)S_0 = H \tag{18}$$

in the CSDW and by

$$c_0(T/T_C - 1)S_0 + \gamma S_0{}^3 = H \tag{19}$$

in the IFMS. This magnetization process is the same as that expected in a HSDW caused by a maximum $\chi_0(\vec{Q})$ [10]. The magnetization in the CSDW phase increases steeply with a magnetic field, since $(T_N/T_C - 1)$ is much small as seen from (17). These results agree well with the observation in MnSi [5]. On the other hand, the phase boundary between the CSDW and the IFMS phases in the T-H plane is expressed by

$$H = c_0\sqrt{c_0/\gamma}\,(T_N/T_C - 1)\sqrt{(T_N - T)/T_C}\ . \qquad (20)$$

The observed phase diagram of MnSi [5,23] shows that the critical field increases more abruptly than $\sqrt{(T_N - T)/T_C}$ in (20) with lowering temperature from T_N. More detailed investigations on these properties will be reported elsewhere.

We are interested also in the microscopic origin of $D[\vec{S}_1\times\vec{S}_2]\cdot\vec{Q}$. As in the case of a localized spin system [15], the spin-orbit coupling plays an important role in the antisymmetric exchange interaction in an itinerant electron system. However, the existence of the Fermi surfaces leads us to a different situation from that in a localized electron system. Spin degeneracies of energy bands are split by the spin-orbit coupling in the crystal with no inversion symmetry. When the split bands are occupied by electrons up to the Fermi level, the energy of the electron system involves the antisymmetric exchange energy. The calculation of D on the basis of the electronic band structure of MnSi will be reported elsewhere.

Acknowledgements

The authors would like to express their sincere thanks to Professor A. Hasegawa, Professor Y. Ishikawa and Professor T. Moriya for valuable discussions.

1. D. Shinoda and S. Asanabe, J. Phys. Soc. Jpn. 21, 555 (1966).
2. J. H. Wernick, G. K. Wertheim and R. C. Sherwood, Mat. Res. Bull. 7, 1431 (1972).
3. F. A. Sidorenko, E. A. Dmitriev and P. V. Gel'd, Izvestiya Vysshikh Uchebnykh Zavedenii, Fizika 8, 15 (1972).
4. Y. Ishikawa, K. Tajima, D. Bloch and M. Roth, Solid State Commun. 19, 525 (1976).
5. Y. Ishikawa, T. Komatsubara and D. Bloch, Physica 86-88B, 401 (1977).
6. H. J. Williams, J. H. Wernick, R. C. Sherwood and G. K. Wertheim, J. Appl. Phys. 37, 1256 (1966).
7. D. Bloch, J. Voiron, V. Jaccarino and J. H. Wernick, Phys. Lett. 51A, 259 (1975).
8. H. Yasuoka, V. Jaccarino, R. C. Sherwood and J. H. Wernick, J. Phys. Soc. Jpn. 44, 842 (1978).
9. T. Moriya and K. Ueda, Solid State Commun. 15, 169 (1974).
10. T. Moriya, Solid State Commun. 20, 291 (1976).
11. K. Makoshi and T. Moriya, J. Phys. Soc. Jpn. 44, 80 (1978).
12. K. Makoshi, J. Phys. Soc. Jpn. 46, 1767 (1979).
13. O. Nakanishi, A. Yanase and A. Hasegawa, J. Magn. Magn. Mat. 15-18, 879 (1980).
14. I. Dzyaloshinsky, J. Phys. Chem. Solids 4, 241 (1958).
15. T. Moriya, Phys. Rev. 120, 91 (1960).
16. I. Dzyaloshinsky, Sov. Phys. JETP 19, 960 (1964).
17. R. L. Melcher, Phys. Rev. Lett. 30, 125 (1973).
18. L. L. Liu, Phys. Rev. Lett. 31, 459 (1973).
19. K. Adachi, N. Achiwa and M. Mekata, J. Phys. Soc. Jpn. 49, 545 (1980).
20. O. Nakanishi, A. Yanase, A. Hasegawa and M. Kataoka, Solid State Commun. 35, 995 (1980).
21. Y. Takahashi and T. Moriya, J. Phys. Soc. Jpn. 46, 1451 (1979).
22. Y. Ishikawa (private communication).
23. S. Kusaka, K. Yamamoto, T. Komatsubara and Y. Ishikawa, Solid State Commun. 20, 925 (1976).

Part III

Some Theoretical Aspects of Narrow-Band Problems

Stoner and Kondo Spin Glasses

J.A. Hertz

NORDITA, Blegdamsvej 17, DK-2100 Copenhagen Ø, Denmark

Abstract

A mean-field description is given for spin glass systems without ideal classical local moments. The starting point is a moderately dilute set of Anderson model impurities, each with a finite Kondo or spin fluctuation temperature, coupled by indirect exchange. The results are consistent with a picture in which the Kondo and spin glass interactions compete, leading to the existence of a finite concentration threshold for $T = 0$ spin glass behavior.

1. Introduction and Model

The usual theoretical models for spin glasses [1] assume the existence of well-defined local moments and allow for spin glass freezing by virtue of the random exchange interactions between moments. However, just as in the case of itinerant ferromagnetism, there are interesting questions that arise in experimentally realizable systems where one has to understand the relation between the generation of the moments and their ordering. In the highly localized limit ($T_K \to 0$ for the single impurity) the two problems are decoupled, while in the opposite or pure itinerant limit the moments are only generated collectively as the order parameter of the system. As in the ferromagnetic case, moreover, most experimentally accessible systems lie awkwardly in an intermediate region between these simple limits.

In a previous paper [2] I formulated a theory that might be applicable to the more itinerant alloys, such as RhCo. It was essentially the adaptation to spin glasses of the Moriya-Kawabata theory for itinerant ferromagnetism [3], formulated for a finite concentration of Wolff model [4] impurities. Here I give a different formulation, in terms of a set of Anderson resonant level impurities [5]. It is thus a picture of transition metal impurities in a noble metal host, and more appropriate to, say, AuCo than RhCo. The formalism can, in principle, describe the whole range of coupling strengths from an uncorrelated Friedel resonance to the Kondo limit and of concentrations from an isolated impurity to 100%. In this latter regime, we would have the picture of transition metal magnetism given by Anderson and McMillan [6]. However, we shall concentrate here on fairly dilute situations, where intersite interactions are easily handled in perturbation theory, and on the case of strong local spin fluctuations, where the Kondo aspects of the problem can be characterized fairly simply.

The model may be written

$$H = \sum_{k\sigma} \epsilon_{k\sigma} c^+_{k\sigma} c_{k\sigma} + \sum_{i\in S,\sigma} \Big[\epsilon_d d^+_{i\sigma} d_{i\sigma} + \tfrac{1}{2}U\, d^+_{i\sigma} d_{i\sigma} d^+_{i,-\sigma} d_{i,-\sigma} + \sum_k \Big(V_{dk}\, e^{i\vec{k}\cdot\vec{r}_i}\, d^+_{i\sigma} c_{k\sigma} + \text{h.c.} \Big) \Big] \quad (1)$$

where S is a set of impurity sites $\vec{r}_i$. We treat only the symmetric case $\epsilon_d = -U/2$, although strongly asymmetric cases might be of interest in situations with mixed-valent impurities. We try to solve it in the functional integral formulation of Stratonovich and Hubbard [7]. The partition function is then written

$$Z = Z_0 \int \delta\vec{\xi}\, \delta\eta\, \exp\Big\{ -\frac{1}{2U} \sum_{i\in S} \int_0^\beta d\tau \Big[\vec{\xi}^{\,2}_i(\tau) + \eta^2_i(\tau) \Big] \Big\} Z[\vec{\xi},\eta] \quad (2)$$

where $Z(\vec{\xi},\eta)$ is the partition function for the noninteracting system in the presence of magnetic and scalar fields $\vec{\xi}_i(\tau)$ and $i\eta_i(\tau)$:

$$Z[\vec{\xi},\eta] = \Big\langle T \exp\Big\{ -\tfrac{1}{2} \sum_{i\in S} \int_0^\beta d\tau [\vec{\xi}_i(\tau)\cdot d^+_i(\tau)\vec{\sigma} d_i(\tau) + i\eta_i(\tau) d^+_i(\tau) d_i(\tau)] \Big\} \Big\rangle . \quad (3)$$

In the succeeding sections we try to reduce (2) to a tractable form by making a set of physically sensible approximations and then try to study the spin glass transition, first in the simplest mean-field theory and then in a better approximation.

2. Toward a Simple Model

Because the system is fairly dilute, we begin by ignoring interactions between impurities, and reviewing the single-impurity problem [8,9]. If we are near the local-moment or Kondo limit, this is a highly non-linear problem and it is essential to preserve this nonlinearity in any sensible approximate treatment. Actually, this problem has never been solved satisfactorily in this approach in a rotationally invariant formalism; the only successful account of the Kondo limit from this viewpoint is that of Hamann [9], who broke rotational invariance, using the so-called two-field functional integral, in which $\vec{\xi}$ in (2) and (3) is replaced by a scalar coupled only to $d^+\sigma_z d$. Since we would like to maintain a rotationally invariant description in handling the spin glass freezing later on, we will have to make a patch-up job of this: We will have to take results from the single-impurity problem as done by Hamann and artificially put them into the formalism with the vector $\vec{\xi}$. Fortunately, the best way we know of to treat the spin glass freezing just happens to allow a clean separation between the single-impurity and interaction effects, so this artificial step can be carried out without any ambiguities.

Following Hamann, then, we begin by separating the exponent in the functional integral into two pieces, one local in time (the adiabatic piece or effective potential) and the other containing all terms in $\ln Z(\vec{\xi},\eta)$ which involve time derivatives or nonlocality in time. The first of these is easy to calculate, since it just requires evaluating (2) for constant $\vec{\xi}$ and η . Making the saddle-point approximation for the scalar potential η , we find an effective potential

$$V(\vec{\xi}) = \frac{1}{2U}\xi^2 - \int_0^{\xi} d\xi' \int_{\mu(\xi')-\xi'/2}^{\mu(\xi')+\xi'/2} d\epsilon \rho_0(\epsilon) \tag{4}$$

where $\mu(\xi)$ is the chemical potential, adjusted with magnetic field to preserve charge neutrality, and ρ_0 is the Anderson resonant level density of states $\pi^{-1}\Delta/[(\epsilon-\epsilon_d-Un/2)^2+\Delta^2]$. In the symmetric case $\epsilon_d = -U/2$, $n = 1/2$, charge and spin degrees of freedom decouple completely, and we may set $\mu = 0$, independent of ξ . At finite temperature, ρ_0 must be convoluted with the derivative of the Fermi function to give the correct static effective potential [10], but we deal here with temperatures so low compared to the level width as to be able to ignore this [2,8,9]. Then the integral in (4) may be evaluated exactly, giving

$$V(\vec{\xi}) = \frac{1}{2U}\xi^2 - \frac{2\Delta}{\pi}\left\{\frac{\xi}{2\Delta}\tan^{-1}\left(\frac{\xi}{2\Delta}\right) - \frac{1}{2}\ln\left[1+\left(\frac{\xi}{2\Delta}\right)^2\right]\right\} \tag{5}$$

V has the familiar two-well structure for $U \gg \Delta$. As discussed at length by Hamann, in order to get the Kondo limit right, it is also necessary to retain nonlinearity to all orders in ξ in the remaining, nonadiabatic, part of the exponent of (2), although only the asymptotic long-time limit of the response to $\xi(\tau)$ is necessary. Fortunately, however, we will never need to use these complicated terms explicitly here. We will only need to know the susceptibility $\chi(T)$ of the Kondo problem, which we can take from other calculations. In fact, in the end we will not even need to know the adiabatic effective potential (4) or (5) explicitly, although for the simple mean-field discussion we give first it is essential.

We turn now to the inter-impurity interactions, which are responsible for the spin glass behavior. We could, in principle, make a cluster expansion of $\ln Z(\vec{\xi},\eta)$ to describe these. However, in the dilute region with which we are concerned here their effects are very simple: There is an effective exchange between the impurity and conduction band electrons calculable in perturbation theory [11], and thus a mechanism for indirect exchange between impurities, of magnitude

$$J_{ij} = J_{SW}^2 \chi^0(\vec{r}_i - \vec{r}_j) \tag{6}$$

where J_{SW} is the Schrieffer-Wolff exchange $8|V_{dk}|^2/U$. Rather than obtain this fact by cluster expansion of the full partition function, which would be slightly tedious, we will just put it in by hand. That is, we add terms $J_{ij}\ \vec{S}_i \cdot \vec{S}_j$ to the original Hamiltonian explicitly. Then, writing Z in terms of the functional integral and extracting the adiabatic piece gives an effective potential just like (4), generalized to a finite range exchange:

$$V[\vec{\xi}] = \tfrac{1}{2} \sum_{i,j\in S} \vec{\xi}_i \cdot J_{ij}^{-1} \vec{\xi}_j - \sum_{i\in S} \frac{2\Delta}{\pi} \left\{ \frac{\xi_i}{2\Delta} \tan^{-1}\left(\frac{\xi_i}{2\Delta}\right) - \tfrac{1}{2}\ell\left[1 + \left(\xi_i/2\Delta\right)^2\right] \right\} \tag{7}$$

where the matrix J_{ij} is defined by (6) for $i \neq j$ and $J_{ii} = U$. Then, to complete the specification of the simplified model, we just add the nonadiabatic terms in the form given by Hamann, and taken as site diagonal:

$$Z = \int \delta\vec{\xi}\ \exp\left\{-\int_0^\beta d\tau \left[V[\vec{\xi}(\tau)] + \sum_{i\in S} \int_0^\beta d\tau'\, T(\xi_i(\tau), \xi_i(\tau'), \dot{\xi}_i(\tau'))\right]\right\}. \tag{8}$$

As mentioned above, the exact form of T is unimportant to us here. All that really matters about it is that it allows "hops" in (imaginary) time between the minima of $V(\xi)$ which quench the local moment of the impurity, provided that the time interval is long enough for a few hops to take place during it. Therefore we identify the mean hopping rate with T_K[9].

The inter-impurity interactions are thus treated as weak in the sense that $\vec{\xi}_i \cdot \vec{\xi}_j$ terms in V occur only quadratically. This is because the big nonlinearity in the problem is a local one; the intersite terms are a weak perturbation, at least in the limit $U >> \Delta$, where the Schrieffer-Wolff exchange is small. The other very simple and convenient property of the present approximate partition function, namely, the fact that the local spin fluctuation dynamics are just those of an isolated impurity, is a consequence of our looking at dilute alloys. At any finite concentration, interimpurity interactions will actually modify the local density of states which goes into calculating the effective potential (4) or (5), and so on. This situation requires an effective medium approach such as has been used for pure Hubbard model ferromagnets by Hubbard [12] and Hasegawa [13], and we will not attempt that here. Thus our results will be sensible only to lowest order in the concentration.

It is now a fairly straightforward matter to set up the formal apparatus of perturbation theory and study the spin glass transition. It is similar in many respects to the standard theories of Edwards-Anderson classical Heisenberg-Ising spin glass because the effective Lagrangians are very similar. The classical models would have a different nonlinear term (log cosh in the Ising case) and no T term (it is a quantum effect).

3. A Mean-Field Theory

The simplest thing we can do formally to generate a mean-field theory is to look for static stationary points of the effective Lagrangian (the exponent in (8)), i.e. solutions of $\partial V/\partial\vec{\xi} = 0$. We get

$$\vec{\xi}_i = \sum_j J_{ij} \, (\vec{\xi}_j/\xi_j) \, \pi^{-1} \tan^{-1}(\xi_j/2\Delta) \tag{9}$$

which could have been obtained quite directly without any functional integrals: It is just that statement that the molecular field on site i comes from the magnetizations on other sites j (π^{-1} tan ($\xi/2\Delta$) is just the magnetization of an Anderson resonant level in a field ξ). Thus the $i \neq j$ terms in (9) correspond exactly to the Weiss mean-field equations for a classical spin system (there the $\pi^{-1}\tan^{-1}$ would be a tanh) and the $i=j$ term is just Anderson's equation describing local moment formation in the impurity problem. (Equivalently, (9) are just the counterparts of the Stoner mean-field equations of ordinary band magnetism.) In order to look for a spin glass phase, we try to calculate the Edwards-Anderson order parameter

$$q \equiv \overline{m_i^2} \equiv \overline{[\pi^{-1} \tan^{-1} (\xi_i/2\Delta)]^2} \,. \tag{10}$$

We average over the configurations of impurities in the approximation used by Edwards and Anderson [1] and by Sherrington and Kirkpatrick [14]. It can be described simply as follows: Suppose that there are enough impurities interacting with a given one that, by virtue of a central limit theorem argument, the total molecular field (9) has a Gaussian distribution. Then

$$q = \int dh \, [2\pi \, \overline{\xi^2}]^{-1/2} \exp(-h^2/2\overline{\xi^2}) \; [\pi^{-1} \tan^{-1}(h/2\Delta)]^2 \tag{11}$$

and we determine the variance $\overline{\xi^2}$ from (9)

$$\begin{aligned} \overline{\xi^2} &= \sum_{jk} \overline{J_{ij} J_{ik} m_j m_k} \\ &\approx \sum_j \overline{J^2_{ij}} \, q = \Big[U^2 + c J^2_{SW} \sum_{j \neq i} \Big(\chi^0(\vec{r}_i - \vec{r}_j) \Big)^2 \Big] q \end{aligned} \tag{12}$$

where c is the impurity concentration.

The threshold condition for a nontrivial solution of (11) and (12) is

$$U^2 + c \, J^2_{SW} \sum_{j \neq i} (\chi^0(\vec{r}_i - \vec{r}_j))^2 > (2\pi\Delta)^2 \tag{13}$$

so in this approximation the indirect exchange can lead to a spin glass even when the Anderson condition for local moments $U > 2\pi\Delta$ is not satisfied. This sort of situation is what might be called a Stoner glass, as described for a slightly different sort of model in ref. [2]. To obtain a transition to the normal state at some finite temperature, it is necessary to take into account the smearing of the density of states in (4) by the convolution with the Fermi function, as mentioned above. Then the $\tan^{-1}$ in equation (11) for q will be similarly smeared out, leading to the transition at some sufficiently high T_g. This is the usual Stoner mechanism for breakdown of magnetic order. This approximation may be described as SK [14] theory for the spin glass combined with a Stoner description of the magnetization. As such, it has the vices of both these kinds of theories. In particular, the spin glass phase has unphysical divergent fluctuations, in the sense that the quantity

$$\chi_{EA} = \sum_j \overline{\chi_{ij}^2} \tag{14}$$

which diverges as the spin glass transition temperature T_g is approached from above, remains divergent in the spin glass phase [15,16]. This is in contrast with what happens to the ordinary susceptibility $\chi = \partial m/\partial h$ in a ferromagnet. There (in an Ising model) χ blows up at the Curie temperature but is rendered finite below T_c because the presence of the finite magnetization reduces the fluctuations of which χ is a measure. In a Heisenberg model, χ^{-1} goes to zero and stays zero below T_c, just on the edge of instability, because uniform rotations cost no energy for a rotationally invariant system. In the SK spin glass theory, by contrast, although χ_{EA} is affected by the presence of a nonzero q below T_g, the corrections are not enough to restore stability - χ_{EA}^{-1} remains negative when calculated in that theory.

The other obvious defect of this approximation is the fact that it does not contain the proper physics of local spin fluctuations for $U >> 2\pi\Delta$: It is a completely static theory and cannot describe the Kondo quenching of local moments. Since it neglects these quantum fluctuation effects, it obviously cannot treat the competition between them and spin glass freezing. Thus it predicts, spuriously, a spin glass phase whenever the Anderson criterion for local moments is satisfied, no matter how weak the intersite interactions are.

The alert reader will have also noticed that this theory has a third difficulty. Its mean-field instability against Anderson local moment formation is not even given correctly, the threshold is off by a factor 2. This is one of the well-known problems of the rotationally invariant functional integral representation [17]. No satisfactory remedy is known, within the functional integral method.

4. Improved Mean Field Theories

One way to improve on this theory and include the suppression of local moments was given in ref. [2]. That paper used a Wolff model description of the impurities rather than an Anderson one, but the general structure of the theory carries over directly to the present model. Such an approximation might be sensible for the intermediate region $U = O(\pi\Delta)$, where nonlinear effects in the local spin fluctuations are important but one is still far from the Kondo limit. Then one truncates the expansion of $V(\xi)$ at fourth order and approximates the T term in (8) by its quadratic part. Written in terms of Matsubara Fourier components this has a simple form [8,9]

$$T \approx \sum_{\omega, i\in S} \frac{|\omega|}{4\pi\Delta^2} \left|\vec{\xi}_{i\omega}\right|^2 . \tag{15}$$

One then has a quantum field theory with free propagators $\langle\xi^\alpha_{i\omega}\ \xi^\alpha_{i,-\omega}\rangle \equiv G_{ij}(\omega)$ which satisfy

$$G^{-1}_{ij}(\omega) = J^{-1}_{ij} - (2\pi\Delta)^{-1}\ (1 - |\omega|/\Delta)\,\delta_{ij} \tag{16}$$

and a ξ^4 interaction of strength of order Δ^{-3}. Treating these interactions in one-loop approximation then gives a qualitatively correct account of both the suppression of Anderson local moment formation and of Moriya-Kawabata fluctuation effects on the spin glass transition. However, this description is not appropriate when one approaches the Kondo limit $U \gg \Delta$. Furthermore, in its current form, it avoids the Almeida-Thouless disease only at the cost of restricting itself to small (strictly speaking, infinitesimally small) values of the EA order parameter. While there is no fully understood cure for Almeida-Thouless disease in any model (except in the limit where the number of spin components goes to infinity), there is a theory, due to Parisi [18], which appears to be quite successful in several respects, although we do not yet understand how the cure works. In the remainder of this section, we shall apply this theory to the model functional (8).

First, we summarize the features of the Parisi theory which are essential to the present calculation. Suppose we try to calculate the "order parameter susceptibility" χ_{EA} (14), first of all above T_g. In mean-field theory, we obtain χ_{ij} from

$$m_i = \chi^0_{ii}\ (h_i + \sum_j J_{ij}\ m_j) \tag{17}$$

(which is just the Curie-Weiss equation when J_{ij} is translationally invariant), so that

$$\chi_{ij} = \partial m_i/\partial h_j = [\chi^{0^{-1}} - J]^{-1}_{ij}$$

$$= \chi^0_{ii}\delta_{ij} + \chi^0_{ii} J_{ij} \chi^0_{jj} + \sum_k \chi^0_{ii} J_{ik} \chi^0_{kk} J_{kj} \chi^0_{jj} + \cdots \quad (18)$$

This series is represented graphically in fig. 1a. In this discussion, we can imagine that all <u>local</u> interactions from the anharmonic parts of V and T in (8), as well as the contributions from the site-diagonal exchange J_{ii} have been taken account of in advance in calculating χ^0_{ii}. That is, χ^0_{ii} is just the exact local susceptibility for the one-impurity problem. While we do not know this quantity exactly, there are very accurate calculations of it available [19], so we shall take it as known. Then the J_{ij}'s in (17) and (18) are only intersite. To calculate χ_{EA}, we must average the square of (18) over impurity configurations, and at this level of approximation that is accomplished by the set of "ladder" graphs of fig. 1b [16]. They form a geometric series, so

$$\chi_{EA} = \frac{(\chi^0_{ii})^2}{1 - (\chi^0_{ii})^2 \sum_{j \neq i} \overline{J^2_{ij}}} . \quad (19)$$

The divergence of this quantity signals the spin glass transition. It is evident how we have suppressed the previous spurious instability (13), since all intrasite interactions have been absorbed into the presumed correct calculation of χ^0_{ii}.

Now for the spin glass phase. In SK theory, we calculate χ_{ij} in the same way (18), but now χ^0_{ii} is the local susceptibility in the presence of the local frozen magnetization. Thus, for example, for the Ising model,

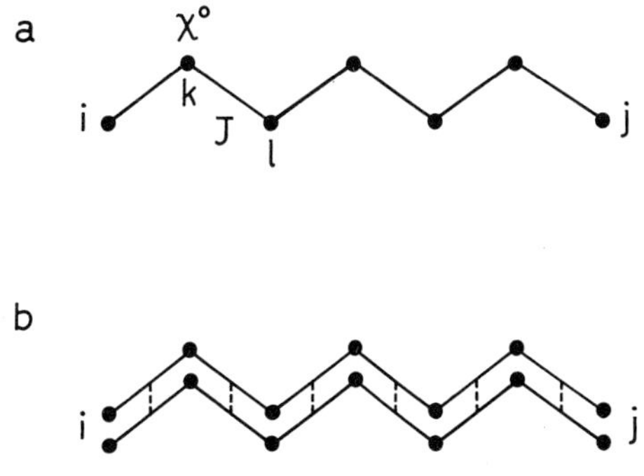

Fig. 1(a) Mean-field series for χ_{ij}. A dot stands for the local susceptibility and a line from site k to site ℓ for $J_{k\ell}$.
(b) Mean-field series for the EA order parameter susceptibility χ_{EA}. A dotted line connecting two $J_{k\ell}$ bonds stands for $\overline{J^2_{k\ell}}$.

$$\chi_{ii}^{0} = \frac{\langle (S_i - \langle S_i \rangle)^2 \rangle}{T} = \frac{1-\overline{\tanh^2 \beta \xi_i}}{T} \tag{20}$$

and at each site in the ladder series for χ_{EA} , instead of just two factors of $\chi_{ii}^{0} = 1/T$ we have the average of the square of (20) over the distribution of internal fields ξ_i . But this leads to a divergent series which, if formally summed by analytic continuation, gives a negative number [16] --- a ridiculous result, since χ_{EA} is a positive definite quantity by definition. This is the Almeida-Thouless disease. Parisi's cure for it (although formulated in a very different way) can be obtained, in our simple way of looking at it, by replacing the average of $(\chi_{ii}^{0})^2$ by the square of the average χ_{ii}^{0} , and determining the latter quantity by the condition that the series (19) for χ_{EA} just diverge, as it does at T_g . Thus, everywhere in the spin glass phase,

$$\chi = \chi_{ii} = \left(\sum_{j\neq i} \overline{J_{ij}^2}\right)^{-1/2} = J_{SW}^{-1}[c \sum_{j\neq i} (\chi^0(\vec{r}_i - \vec{r}_j))^2]^{-1/2} \; . \tag{21}$$

This χ is manifestly temperature independent. This way of fixing χ and χ_{EA} is obviously ad hoc; it is rather a mystery how it falls out of the Parisi formulation. In an n-component spin model, it actually occurs naturally in a straightforward calculation in the limit $n \to \infty$ [20]. We have not discussed the temperature dependence of the Edwards-Anderson order parameter q , since that is an ambiguous quantity in the Parisi theory. However, it is not an observable quantity in general, so we content ourselves with knowing χ .

It is thus very simple to calculate χ as a function of c and T . It is just the lesser of two quantities: the local χ of the normal state and the spin glass χ given by (21) Thus, at a small impurity concentration, χ rises as we lower T , first according to a Curie law for $T \gg T_K$, and then saturating at T_K^{-1} as $T \to 0$ (we take this as the definition of T_K for the purposes of this discussion), just as in the single-impurity problem. But if χ should get as large as (21), further temperature dependence suddenly stops as it finds itself in the frozen spin glass state. This implies a finite threshold concentration for a spin glass state to exist at $T = 0$, since if χ does not ever reach the value given by (21), no transition will occur. The threshold is thus given by equating (21) to T_K^{-1} . This can be written

$$c_{crit} = (T_K/T_g^0)^2 \tag{22}$$

where T_g^0 is just the T_g we would have if T_K were zero. The result (22) has a very simple interpretation: we simply compare the characteristic condensation energies of the Kondo system and the local-moment spin glass; the system assumes the low-temperature state with the greater binding. There is an analogy with the situation in the Kondo lattice [21], but the systems under discussion here obviously allow a greater varia-

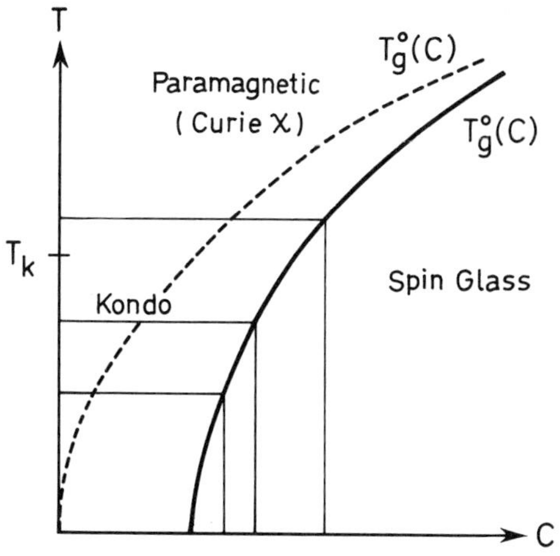

Fig.2 Phase diagram of the Kondo glass. Thick line: spin glass transition T_g. Dotted line: spin glass transition temperature T_g for $T_k = 0$. Thin lines: lines of constant χ per impurity.

tion in the characteristic ratio T_g^0/T_K than does the usual Kondo lattice, since the concentration can be varied. The phase diagram is indicated in fig. 2. We note also that in the spin glass region, the susceptibility per impurity is proportional to $c^{-\frac{1}{2}}$; thus the measured χ is expected to be proportional to c for $c \leq c_{crit}$ and to $c^{\frac{1}{2}}$ for $c \geq c_{crit}$.

We close this section with a few final comments: (1) Notice how the Parisi trick neatly separated the spin glass interactions from the Kondo effect in the local χ . This feature, together with the hypothesis of small concentration, saved us from having to deal with the two kinds of physics simultaneously. At larger c , of course, the local electronic properties, and thus the local χ , are affected by inter-impurity couplings. Hence our analysis can be correct only to leading order in c , as mentioned earlier. (2) It is well to add a warning that this is only a mean-field theory of the spin glass state. It is therefore quite unlikely to survive the large fluctuation effects expected for spin glasses on account of their order parameter fluctuations [22]. Nevertheless, as always, it may well be qualitatively correct for appropriate time and length scales. (3) The validity of this theory is not restricted to the Kondo limit $U \gg \Delta$. If we have good calculations of χ_{ii}^0 and the Schrieffer-Wolff effective exchange for intermediate coupling, the whole procedure can be carried through just as well there. (We still require low concentration, however.)

5. Alternative Approach to the Kondo Glass

It may be useful to think about the $U \gg \Delta$ limit, which may be called the Kondo glass, in a different way, which complements the one given above. The description is in terms of the interplay of the probability distribution of local molecular fields with the Kondo quenching of local moments which generate these fields, calculated using the Kondo Hamiltonian.

We start out by ignoring the Kondo effect. In the SK mean field theory, then, the local molecular field distribution is Gaussian, with a width

$$q\Delta = cq\, J_{SW}^2 \sum_{j\neq i} (\chi^0(\vec{r}_i-\vec{r}_j))^2 \equiv q\, T_g^{0\,2} \,. \tag{23}$$

In a better theory [23] it is probably not Gaussian, but (23) still serves as an estimate of the width. But we know that the Kondo effect on a given impurity depends on the local magnetic field; a field suppresses the Kondo divergences encountered in perturbation theory. So if all the spins lie in big molecular fields, $((q\Delta)^{\frac{1}{2}} \gg T_K)$ the Kondo effect will be completely suppressed by the presence of the spin glass state. However, to the extent that there are spins in very low fields $(\ll T_K)$, the Kondo effect can quench their moments. These spins should then not be counted as giving rise to the spin glass molecular fields; that is, the effective c in (22) should be reduced to an effective concentration c' given roughly by

$$c' = c\left[1 - \int_{-T_K}^{T_K} P(h)\, dh\right] . \tag{24}$$

This is an idealization, since spins in fields just a bit less than T_K are not quenched completely, and even those that are have some susceptibility. Similarly, even those in fields greater than T_K suffer some Kondo suppression. Thus $P(h)$ in (24) should really be multiplied by a smooth function of h that has a fairly strong but continuous rise toward saturation for h near T_K, instead of the step function implied by our approximation. However, if $P(h)$ is not too rapidly varying near $h = T_K$, the step function approximation will not be bad.

We now make this argument self-consistent by requiring that the new (lower) T_g calculated by putting the corrected effective concentration into (23) give the width of the distribution in (24). We then obtain (for the Gaussian field distribution), at $T = 0$,

$$(T_g/T_g^0)^2 \doteq 2\int_{T_K}^{\infty} \frac{dh}{\sqrt{2\pi T_g{}^2}}\, e^{-h^2/2T_g{}^2} = \operatorname{erfc}\,(T_K/T_g\sqrt{2}) \,. \tag{25}$$

This can be solved graphically, and one finds, in qualitative agreement with the arguments of the preceding section, that there is a finite value of $c \propto (T_K/T_g^0)^2$ necessary for a spin glass state to exist at $T = 0$. Since this argument is somewhat different from those given earlier, in that it looks <u>explicitly</u> at the effects of the local molecular fields on the dynamics of the local moments rather than simply determining the local χ indirectly by requiring that the series for χ_{EA} be at the threshold for divergence, it is reassuring that the results agree qualitatively.

References:

1. S. F. Edwards and P. W. Anderson, J. Phys. F 5, 965 (1975)
2. J. A. Hertz, Phys. Rev. B 19, 4796 (1979), see also J. Mathon, J. Phys. F 8, 1783 (1978)
3. T. Moriya and A. Kawabata, J. Phys. Soc. Japan 34, 693 (1973), 35, 669 (1973); T. Ramakrishnan, Phys. Rev. B 10, 4014 (1974); G. Gumbs and A. Griffin, Phys. Rev. B 13, 5054 (1976); see also ref. 10 and K. Murata and S. Doniach, Phys. Rev. Lett. 29, 285 (1972)
4. P. A. Wolff, Phys. Rev. 124, 1030 (1961)
5. P. W. Anderson, Phys. Rev. 124, 41 (1961)
6. P. W. Anderson and W. L. McMillan, Varenna Lectures 1967, Ital. Phys. Soc., International School of Physics "Enrico Fermi", Course 37 (Academic Press, New York 1967) pp. 50-86
7. R. L. Stratonovich, Dokl. Akad. Nauk. SSSR 115, 1097 (1957) [Soviet Physics Doklady 2, 416 (1957)]; J. Hubbard, Phys. Ref. Letters 3, 77 (1959)]
8. S. Q. Wang, W. E. Evenson and J. R. Schrieffer, Phys. Rev. Letters 23, 92 (1969)
9. D. R. Hamann, Phys. Rev. B 2, 1373 (1970)
10. J. A. Hertz and M. A. Klenin, Phys. Rev. B 10, 1084 (1974)
11. J. R. Schrieffer and P. A. Wolff, Phys. Rev. 149, 491 (1966)
12. J. Hubbard, Phys. Rev. B 19, 2626, 20, 4584 (1979)
13. H. Hasegawa, J. Phys. Soc. Japan 46, 1504 (1979); 49, 178, 963 (1980); Solid State Commun. 31, 597 (1979)
14. D. Sherrington and S. Kirkpatrick, Phys. Rev. Lett. 36, 415 (1976); M. Feigelman, Soviet Phys. JETP, to appear
15. J. R. L. de Almeida and D. J. Thouless, J. Phys. A 11, 983 (1978)
    E. Pytte and J. Rudnick, Phys. Rev. B 19, 3603 (1979)
16. J. Hertz and A. Khurana, J. Phys. C 13, 2715 (1980)
    J. Rudnick, Phys. Rev. B 22, 3356 (1980)
17. See, e.g., R. E. Prange and V. Korenman, Phys. Rev. B 19, 4691, or J. Hubbard, ref. 12
18. G. Parisi, Phys. Lett. A 73, 203 (1979); Phys. Rev. Lett. 23, 1754 (1979); J. Phys. A 13, 1101, L 115, 1887 (1980); Phil. Mag. 41, 677 (1980); see also G. Parisi and G. Toulouse, J. Physique-Lettres 41, L 361 (1980)
19. H. R. Krishna-murthy, J. W. Wilkins and K. G. Wilson, Phys. Rev. Lett. 35, 1101 (1975); Phys. Rev. B 21, 1003, 1044 (1980)
20. J. M. Kosterlitz, D. J. Thouless and R. C. Jones, Phys. Rev. Lett. 36, 1217 (1976); also proved for the random-axis model by A. Khurana (unpublished)
21. S. Doniach, Physica B 91, 231 (1977), R. Jullien, J. Fields and S. Doniach, Phys. Rev. Lett. 38, 1500 (1978)
22. A. J. Bray and M. A. Moore, J. Phys. C 12, 79 (1979); J. A. Hertz, L. Fleishman and P. W. Anderson, Phys. Rev. Lett. 43, 942 (1979); R. Fisch and A. B. Harris, Phys. Rev. Lett. 38, 785 (1977); R. V. Ditzian and L. P. Kadanoff, Phys. Rev. B 19, 4631 (1979); P. Reed, M. A. Moore and A. J. Bray, J. Phys. C 11, L139 (1978)
23. D. J. Thouless, P. W. Anderson and R. G. Palmer, Phil. Mag. 35, 593 (1977)

Itinerant Magnetism Approach for Understanding One-Dimensional Antiferromagnets

F.D.M. Haldane

Institut Laue-Langevin, 156X, F-38042 Grenoble, France

Abstract

One-dimensional antiferromagnets, and quantum fluids in general, have ground-state properties resembling itinerant fermion systems. An approach to the Heisenberg antiferromagnet based on referring it back to an itinerant spin-1/2 fermion system is outlined. One advantage of this description is that full spin rotational invariance can be retained at all stages of a calculation.

1. Introduction

One-dimensional models occupy a special niche in the theory of highly correlated systems: on the one hand, essentially all the non-trivial models that can be exactly solved are one-dimensional, but on the other hand, the physics that such solutions reveal is of a special one-dimensional kind, without parallel in the higher-dimensional world. Indeed, although the original motivation for studying such models was as a "warming-up" exercise before tackling the real, three-dimensional many-body problem, the rich variety of specifically one-dimensional features that have emerged - such as "solitons" or mobile topological defects, for example - have stimulated a great deal of recent experimental interest in systems composed of loosely coupled molecular chains which to some extent, at least, may be described as "quasi-one-dimensional".

One dimension is special because fluctuation effects become very important. As is well known, in the absence of long-range forces, long-range order cannot develop at any finite temperature in one dimension. At zero temperature too, quantum zero-point fluctuations limit possible long-range order. A ground state with a discrete broken symmetry remains possible, as does one with a continuous, but conserved, order parameter, such as ferromagnetism. However, the standard broken continuous symmetry with non-conserved order parameter is not possible, because of zero-point fluctuations of the collective "Goldstone" mode with a linear dispersion at long wavelengths that always accompanies such a state. Typically, one-dimensional analogues of systems that show such ordering in higher dimensions do have a corresponding "Goldstone-mode"-like feature in their spectrum of elementary excitations, but have correlations that asymptotically decay with various, often non-universal, power laws. The canonical example is the 1-D harmonic lattice, usually used in the student texts to introduce the phonon concept. Long-wavelength phonons are of course just the collective Goldstone mode associated with the presence of broken translational symmetry, or long-range order of the lattice phase variable $\theta_i = 2\pi u_i/a$, where u_i is the displacement at site i, and a is the lattice spacing. In one dimension, the spectrum of course has the characteristic linear Goldstone form, but a simple calculation of the phase coherence along the chain shows that correlations fall off with inverse power law exponent $\eta = 2\pi\hbar/(mav_S)$, where m is the atomic mass, and v_S the phonon sound velocity. In this case, it is seen that the quantum fluctuations are suppressed if $h \to 0$, or if the atoms become infinitely heavy.

Analogous effects can be found in 1-D magnetic systems. In this talk, I will concentrate on the 1-D Heisenberg antiferromagnet, though similar behaviour will occur in both ferro- and antiferromagnetic easy-plane systems, and even easy-axis antiferromagnets in the "spin-flop" phase occuring in finite field. It turns out that by a very complicated method due to Bethe [1], the $S = \frac{1}{2}$ chain with nearest-neighbour exchange J can be exactly diagonalised. In zero field, its excitations resemble the familiar doubly periodic Néel spin-wave type: $\omega_k = \frac{1}{2}\pi J\sin|ka|$: linear long-wavelength behaviour reflecting broken spin-rotational symmetry, with a doubled periodicity reflecting the additional breaking of the discrete lattice translational symmetry to give a doubled magnetic unit cell. In fact, this is not at all the case, and it is known (and will be rederived here) that the Néel correlations fall off with inverse power law $\eta = 1$. Perhaps stranger still is the behaviour in finite magnetic field: the periodicity of the spectrum decreases from $\pi/a = \frac{1}{2}G$ to $\frac{1}{2}G(1-|\langle S^z\rangle|/S)$, quite unlike the higher dimensional case, which retains the Néel periodicity. This incommensurate periodicity in the spectrum is strongly reminiscent of the Kohn anomaly at $2k_F$ in a <u>fermion</u> system, which also varies linearly with changes in density.

The exact solutions [1] are not very helpful in providing physical insight into the physics of the problem. In fact only recently has a systematic method for calculating correlation-function exponents been developed [2], and despite the availability of exact wavefunctions, direct calculation of correlation functions has not yet been achieved. One way of investigating the $S = \frac{1}{2}$ chain is to use the existence of a canonical transformation that relates it to a <u>spinless fermion</u> system, in this way making the identification of its "Kohn anomaly" feature with a fermion $2k_F$ more direct. I will briefly describe the <u>Jordan-Wigner</u> [3] transformation in the next section; however there are several disadvantages of the corresponding formalism: it is special to $S = \frac{1}{2}$ systems, and it hides the spin-rotational symmetry in such a way that it becomes lost when approximations are made. This is unfortunate, because the Jordan-Wigner form is only a stage on the way to a more tractable description of the low-energy physics of the quantum fluctuations: the first real progress in this direction came when Luther and Peschel [4] developed a linearised low-energy form that could be handled with techniques developed in connection with the soluble <u>Luttinger</u> [5,6] 1-D fermion model. In the rest of this talk, I will develop an alternative way of bringing the Heisenberg chain into correspondance with a fermion model, and hence using the Luttinger model techniques: this will first relate it to a "precursor" itinerant spin-$\frac{1}{2}$ electron that develops a magnetic insulating state. This approach will allow full spin-rotation symmetry to be explicitly retained throughout the linearisation process, and furthermore has the advantage that the extension to $S > \frac{1}{2}$ magnetic systems will become obvious (if not quite as straightforward to handle).

2. Various Representations of the $S = \frac{1}{2}$ Heisenberg Antiferromagnet

The obvious representation of the model is in terms of $S = \frac{1}{2}$ spin operators:

$$H = J \sum_i \vec{S}_i \cdot \vec{S}_{i+1} \; ; \quad \vec{S} = \sum_i \vec{S}_i \; ; \quad S_i = \tfrac{1}{2} \, , \quad \vec{S}_{i+N_a} \equiv \vec{S}_i \, . \tag{1}$$

Not so obvious or well known is the representation of the momentum operator, but results of Baxter [7] allow it to be written as

$$P = G\,(2\pi i)^{-1} \ln\{\mathrm{Tr.} \prod_{i=1}^{N_a} [\tfrac{1}{2}\underline{\underline{I}} + \underline{\underline{\vec{\sigma}}} \cdot \vec{S}_i]\} \quad , \qquad (S_i = \tfrac{1}{2}) \; , \tag{2}$$

where G is the reciprocal lattice vector, and the trace is over the 2X2 Pauli and identity matrices in the product. The two global symmetry generators, $\vec{S}$ and P (modulo G) are thus explicitly given.

The second representation is the Jordan-Wigner form [3] in terms of spinless fermion operators $c_i^\dagger$:

$$S_i^z = (n_i - \tfrac{1}{2}) ; \quad S_i^+ = \prod_{j \neq i} [\mathrm{sgn}(j-i)]^{(n_j)} c_i^\dagger$$

$$H = \tfrac{1}{2} J [\sum_i c_i^\dagger c_{i+1} + \mathrm{h.c.} + 2(n_i - \tfrac{1}{2})(n_{i+1} - \tfrac{1}{2})] ; \quad c_{i+N_a} = -(-1)^{(S^z + \frac{1}{2}N_a)} c_i \; ;$$

$$P = \sum_k k c_k^\dagger c_k \; ; \quad c_k^\dagger = N_a^{-\frac{1}{2}} \sum_i e^{ikx_i} c_i^\dagger \, , \quad e^{ikN_a} = -(-1)^{2S^z} , \tag{3}$$

where $x_n = na$. In this formulation, the momentum operator is quite tractable, but the expressions for S^+ and S^- become quite complex due to the prefactors that change the fermion anticommutation of operators on different sites into the commuting algebra of spin operators. Following Luther and Peschel [4], it is quite easy to develop a linearised form of the Hamiltonian suitable for discussing the anisotropic system with a small degree of exchange J_z between axial spin components. The only somewhat non-standard (but very typical) aspect of the spectrum is the alternation of boundary conditions with total azimuthal spin S^z. This ensures that low-energy excitations between states in even and odd sectors of S^z involve multiples of $2k_F$, rather than odd multiples of k_F, as in a "normal" fermion system. The main problem in developing the linearisation is to avoid losing the spin-rotation symmetry which is a crucial aspect of the isotropic Heisenberg magnet itself; on the other hand, (3) is a perfectly suitable starting point for discussion of the easy-plane anisotropic systems.

Various other representations in terms of bosons are possible: there is the representation in terms of the hard-core Bose lattice gas due to Matsubara and Matsuda [8] and the arbitrary-spin Schwinger coupled-boson scheme, recently used [9] in connection with the ferromagnetic chain. However, the one I want to introduce here is based on the Hubbard model [10]. This uses spin-$\frac{1}{2}$ fermions with the usual periodic boundary conditions:

$$H = \text{Lim. } U \to \infty\ [H(U)]$$

$$H(U) = \tfrac{1}{2}(JU)^{\frac{1}{2}} \sum_{i\sigma} c^{\dagger}_{i\sigma} c_{i+1\sigma} + \text{h.c.} + U\sum_i (n_{i\uparrow} - \tfrac{1}{2})(n_{i\downarrow} - \tfrac{1}{2})$$

$$P = \tfrac{1}{2}G + \sum_{k\sigma} k c^{\dagger}_{k\sigma} c_{k\sigma}\ , \quad c^{\dagger}_{k\sigma}\ N_a^{-\frac{1}{2}} \sum_i e^{ikx_i} c^{\dagger}_{i\sigma}\ , \quad e^{ikL} = +1\ ,$$

$$\vec{S} = \tfrac{1}{2}\sum_{i\sigma\sigma'} \vec{\sigma}_{\sigma\sigma'} c^{\dagger}_{i\sigma} c_{i\sigma'}\ . \qquad (4)$$

It is well known that in the limit of large U, this Hamiltonian develops a gap of order U for excitations that have charge different from one electron per site, and that the remaining low-lying states become those of the $S = \frac{1}{2}$ antiferromagnet. This system is isotropic because, although the interaction term in (4) has been written in a way that is non-manifestly isotropic, the Pauli principle allows it to be rewritten in an isotropic way. An at first sight surprising detail is the necessity to adjust the definition of the momentum P by $\frac{1}{2}G$: this is the momentum of the state with a filled band of up-spin electrons and an empty band of down-spin electrons. In the magnetic picture, this must correspond to the zero-momentum ferromagnetic state; in the electronic picture, the occupied states have momenta that cancel in pairs, except for the zone center and zone edge, with a net momentum $\frac{1}{2}G$. This representation is valid for two-sublattice bandstructures, with hopping only between different sublattices, in arbitrary dimensions; it gives rise to a completely "unfrustrated" antiferromagnet, with no competing exchange interactions. Among its advantages are the "clean" representations of <u>both</u> the spin and momentum operators; its disadvantage lies in the introduction of the extra charge degrees of freedom that must be suppressed by the limit $U \to \infty$.

Like the Jordan-Wigner form (3), this is an <u>interacting</u> fermion model. In general, the transition to the magnetic insulating ground state only occurs above some critical coupling $U > U_c \sim J$; when this gap has opened, the character of the remaining low-lying states with energies much less than the gap will be that of the antiferromagnet. However, in <u>one dimension</u>, there is an important simplification: $U_c = 0$. This is known both from studies of this "Umklapp instability" [11] of the spin-$\frac{1}{2}$ 1-D Fermi gas, and from the exact solution (again using the method of Bethe [1]) of the 1-D Hubbard chain found by Lieb and Wu [12]. This means that the low-energy character of the 1-D antiferromagnet is already established in (4) for <u>arbitrarily weak U</u>, and a truly weakly interacting Fermi gas can be studied. In contrast, the Jordan-Wigner form (3) only generates the low-lying states of the antiferromagnet at an intermediate dimensionless coupling of order unity.

3. The Weak-U Limit and its Linearised Form: the Spin-$\frac{1}{2}$ Luttinger Model

In the weak-U limit of (4), the Bloch orbital basis becomes appropriate: it is convenient to make the transformation $c_{i\sigma} \to (-1)^i c_{i\sigma}$ to invert the electron dispersion; then

$$H = -(v_F/a)\sum_{k\sigma} \cos(ka) c^{\dagger}_{k\sigma} c_{k\sigma} + \tfrac{1}{2}U \sum_{kk'q\sigma\sigma'} c^{\dagger}_{k+q\sigma} c^{\dagger}_{k'-q\sigma'} c_{k'\sigma'} c_{k\sigma}\ . \qquad (5)$$

For weak U, only the structure near the Fermi level will be important; it is now convenient to replace the non-linear Hubbard chain dispersion $\varepsilon_k = (v_F/a)\cos(ka)$ by the linearised Luttinger model form [5,6] $\varepsilon_{kp} = v_F pk$, $p = \pm 1$. The spectrum of the Luttinger model consists of two kinds of fermion: electrons, here with a Fermi sea filled up to a Fermi vector $k_F = \frac{1}{4}G$ corresponding to a half-filled band, plus a filled Dirac sea of negative energy states, or positrons. These extra, unphysical particles are required for "bookkeeping" purposes in developing the formalism: states in which they are excited have energies of at least $v_F k_F$ and do not significantly affect the low-energy states in the weak-coupling limit. Instead of the electron + positron basis, it is more convenient to work with two alternative fermion fields, $\psi^{\dagger}_{p\sigma}(x)$, $p = \pm 1$, one associated with each linear branch of the spectrum. The Luttinger model kinetic energy term is then

$$H^{o} = v_F \sum_{pk\sigma} (pk-k_F)(n_{kp\sigma} - \theta(k_F-pk)) \quad . \tag{6}$$

The advantage of working with the Luttinger model is that the alternative "boson representation" [13-16] of fermion states can then be used. This uses spin-labelled bosons $b_{q\sigma}$ ($q=2\pi n/L$, $n \neq 0$), plus integer charge operators $N_{p\sigma}$ and their conjugate phase operators $\bar{\theta}_{p\sigma}$, which represent $q = 0$ modes. In this basis, the kinetic energy and momentum become

$$H^{o} = v_F[\sum_{q\sigma} |q| b^{\dagger}_{q\sigma} b_{q\sigma} + (2\pi/L) \sum_{p\sigma} (N_{p\sigma} - \tfrac{1}{4}N_o)^2]$$

$$P = \tfrac{1}{2}G + \sum_{p\sigma} (\tfrac{1}{2}G + 2\pi N_{p\sigma}/L) p N_{p\sigma} + \sum_{q\sigma} q b^{\dagger}_{q\sigma} b_{q\sigma} \tag{7}$$

where N_o is the ground-state charge. The extra factor $\frac{1}{2}G\, \Sigma_{p\sigma} N_{p\sigma}$ results from the inversion of the Hubbard band structure. The fermion fields have the representation

$$\psi^{\dagger}_{p\sigma}(x) = L^{-\frac{1}{2}} A_{p\sigma} e^{i\phi^{\dagger}_{p\sigma}(x)} e^{i\bar{\theta}_{p\sigma}} e^{i\phi_{p\sigma}(x)} \quad ,$$

$$\phi_{p\sigma}(x) = 2\pi p (x/L) N_p + \lim_{\varepsilon\to 0} i \sum_{q\neq 0} \theta(pq) \left|\frac{2\pi}{Lq}\right|^{\frac{1}{2}} e^{-iqx} e^{-\varepsilon|q|L/2\pi} b_{q\sigma} \quad ,$$

$$A_{p\sigma} = \exp(\tfrac{1}{2}\pi\, i \sum_{p'\sigma'} \varepsilon_{p\sigma,p'\sigma'} N_{p'\sigma'}) \quad . \tag{8}$$

The factor $A_{p\sigma}$ ensures the anticommutation of fields with different labels p: $\varepsilon_{p\sigma,p'\sigma'}$ is a matrix whose off-diagonal elements are antisymmetric and equal to ± 1. A suitable choice is

$$A_{p\sigma} = \exp[\tfrac{1}{2}\pi i (p \sum_{\sigma'} N_{-p\sigma'} + \sigma \sum_{\sigma'} N_{p\sigma'})] \quad . \tag{9}$$

It is next convenient to form (i) number (N) and current (J) combinations, then (ii) form charge (c) and spin (s) combinations:

$$N^c = \sum_{p\sigma} N_{p\sigma} \;,\quad \bar{\theta}^c_N = \tfrac{1}{4}\sum_{p\sigma} \bar{\theta}_{p\sigma} \;,\quad \phi^c_N(x) = \tfrac{1}{4}\sum_{p\sigma} \phi_{p\sigma}(x) \quad ;$$

$$J^c = \sum_{p\sigma} pN_{p\sigma} \;;\quad N^s = \sum_{p\sigma} \sigma N_{p\sigma} \;;\quad J^s = \sum_{p\sigma} p\sigma N_{p\sigma} \quad ; \quad \text{etc.} \tag{10}$$

$$b_{qc} = 2^{-\frac{1}{2}} \sum_{\sigma} b_{q\sigma} \;,\quad b_{qs} = 2^{-\frac{1}{2}} \sum_{\sigma} \sigma b_{q\sigma} \;. \tag{11}$$

Then

$$\phi^c_N(x) = \tfrac{1}{2}\pi(x/L)J^c + 2^{-\frac{1}{2}} i \sum_{q\neq 0} \left(\frac{2\pi}{L|q|}\right)^{\frac{1}{2}} e^{-iqx} b_{qc} \quad ; \tag{12}$$

$$\phi^c_J(x) = \tfrac{1}{2}\pi(x/L)N^c + 2^{-\frac{1}{2}} i \sum_{q\neq 0} \operatorname{sgn}(q)\left(\frac{2\pi}{L|q|}\right)^{\frac{1}{2}} e^{-iqx} b_{qc} \quad ; \tag{13}$$

there are equivalent expressions for ϕ^s_N and ϕ^s_J. When the net charge current is calculated, it is found to be $v_F\,(J^c/L)$, while (N^c/L) is the mean charge density; likewise N^s and J^s control the (azimuthal) spin density and current.

In the new basis, the Hamiltonian and momentum operators separate into charge and spin components:

$$H^o = H^o_c + H^o_s \;;\quad P = P_c + P_s \quad ; \tag{14}$$

$$H^o_c = v_F\Big[\sum_q |q|\, b^\dagger_{qc} b_{qc} + \tfrac{1}{4}(\pi/L)(N^c - N_o)^2 + (J^c)^2\Big] \;; \tag{15}$$

$$P_c = \tfrac{1}{2}(\pi/L)N^c J^c + \sum_q q\, b^\dagger_{qc} b_{qc} \quad . \tag{16}$$

The expressions for H^o_s and P_s are similar (but without N_o). The quantum numbers N^c, N^s, J^c, J^s are all integers, but they are not independent: the constraint that the various combinations giving $N_{p\sigma}$ are themselves integral imposes certain "selection rules" on physical states, that couple the charge and spin degrees of freedom:

$$(-1)^{N^s} = (-1)^{J^s} = (-1)^{N^c} = (-1)^{J^c} = \pm 1 \quad ; \tag{17}$$

$$\eta = \exp(\tfrac{1}{2}\pi i(N^c + J^c)) = \exp(\tfrac{1}{2}\pi i(N^s + J^s)) = \pm 1 \quad . \tag{18}$$

N^s is twice the azimuthal spin component. When $(-1)^{N^s} = +1$ (integral spin), $\eta = +1$, but when the spin is half-integral, there are two possible values for η, which are interchanged by spatial inversion ($p \to -p$).

The spin density operators must now be constructed. The primary spin density is

$$\vec{\sigma}_p(x) = \tfrac{1}{2} \sum_{p\sigma\sigma'} \vec{\sigma}_{\sigma\sigma'} \psi^\dagger_{p\sigma}(x) \psi_{p\sigma'}(x) \quad . \tag{19}$$

This operator yields the global spin

$$\vec{S} = {}_0\int^L dx \sum_p \vec{\sigma}_p(x) \ . \tag{20}$$

It is composed only of spin-labelled operators:

$$\sigma_p^z(x) = \tfrac{1}{4}L^{-1}(N+pJ) + 2^{-\frac{1}{2}} i \sum_q \left[\frac{|q|}{2\pi L}\right]^{\frac{1}{2}} e^{iqx}[\theta(pq)b_q^\dagger + \theta(-qp)b_{-q}] \ ,$$

$$\sigma_p^+(x) = \eta(p)^{N}L^{-1} e^{2i(\phi_N^\dagger + p\phi_J^\dagger)} e^{2i(\bar{\theta}_N + p\bar{\theta}_J)} e^{2i(\phi_N + p\phi_J)} \ . \tag{21}$$

At first glance, it might be expected that the local electronic spin density operator $\vec{\sigma}(x)$ is just $\Sigma_p \vec{\sigma}_p(x)$; in fact this is not so because of the non-local relation between p-labelled fermion fields and the electrons and positrons. The true electronic spin density operation is a complicated non-local object when expressed using the boson representation, but an Ansatz, valid at low energies $E << v_F k_F$, is

$$\vec{\sigma}(x) \sim \tfrac{1}{2} \sum_{pp'\sigma\sigma'} \vec{\sigma}_{\sigma\sigma'} \psi_{p\sigma}^\dagger(x)\psi_{p'\sigma'}(x) \ . \tag{22}$$

In the boson representation,

$$\psi_{p\sigma}^\dagger(x)\psi_{-p\sigma}(x) = -\eta\, p\sigma(-\sigma)^{N_s}(\chi_J^c)^p(\chi_J^s)^{p\sigma}$$

$$\psi_{p\sigma}^\dagger(x)\ \psi_{-p-\sigma}(x) = p\sigma(-p\sigma)^{N_s}(\chi_J^c)^p(\chi_N^s)^{\sigma} \tag{23}$$

where χ are suitably labelled combinations of the form

$$\chi = L^{-\frac{1}{2}} e^{2i\phi^\dagger} e^{2i\bar{\theta}} e^{2i\phi} \ . \tag{24}$$

Then the total spin density field has the form

$$\vec{\sigma}(x) \sim \tfrac{1}{2} \sum_p [\vec{\sigma}_p(x) + (-p)^{(N_s+1)} \chi_J^c\, \vec{\tau}(x)] \ . \tag{25}$$

$\vec{\tau}(x)$ is also composed only of spin operators:

$$\tau^z(x) = \tfrac{1}{2}\eta \sum_p (p)^N (\chi_J)^{-p}$$

$$\tau^\pm(x) = (\pm)^N(\chi_N)^{\pm 1} \ . \tag{26}$$

The field $\vec{\tau}(x)$ does not obey "continuum spin"-type commutation relations, as it is in a form valid only in a low-energy subspace; however, as required by spin-rotation invariance, it does have the correct commutation relations with the global spin operator $\vec{S}$.

If the free-fermion spin correlation functions are evaluated using (25), the long wavelength $q \sim 0$ component comes from $\vec{\sigma}_p(x)$, and falls off as $|x-x'|^{-2}$, reflecting the linear construction (21) of $\sigma_p^z(x)$ in terms of bosons. The $2k_F$ component comes both from the charge and spin degrees of freedom:

$$< \chi_J^c(x)\ (\chi_J^c(x'))^{-1} > \sim e^{2ik_F x}\ |x-x'|^{-1}$$

$$<\vec{\tau}(x)\ \cdot\ \vec{\tau}(x')> \sim |x-x'|^{-1} \tag{27}$$

Both terms contribute a factor decaying as $|x-x'|^{-1}$.

We are now ready to examine the effects of interactions. Provided $2k_F$ is not near a reciprocal lattice vector, there are three types of interactions generated by the linearisation of (5) : long wavelength ($p+p' \to p+p'$), backscattering ($p+(-p) \to (-p)+p$), and Umklapp scattering ($p+p \to (-p)+(-p)$), which is allowed if a reciprocal lattice vector $G \sim 4K_F$. The Pauli-matrix identity $2\delta_{\alpha\alpha'}\delta_{\alpha\sigma'} = (\delta_{\alpha\alpha'}\delta_{\sigma\sigma'} + \vec{\sigma}_{\alpha\alpha'}\cdot\vec{\sigma}_{\sigma\sigma'})$ can be used to rearrange the backscattering term into an explicitly spin-rotationally invariant form. Then one obtains [11,17] terms

$$H^1 = \frac{U}{2\pi} \sum_q \rho_q^c\ \rho_{-q}^c \tag{28}$$

$$H^2 = U \int_0^L dx \sum_p e^{\frac{1}{2}ip(4k_F-G)x} (\chi_J^c)^{2p} \tag{29}$$

$$H^3 = -U \int_0^L dx \sum_p \vec{\sigma}_p(x) \cdot \vec{\sigma}_{-p}(x)\ . \tag{30}$$

The charge density operators ρ_q^c are just the charge subspace analogues of σ^z. The key feature to note is that in this linearised approximation, the interaction separates into terms H^1 and H^2 acting only on the charge subspace, and H^3 acting only in the spin subspace. This is the key simplification in the development of the continuum theory of the spin-$\frac{1}{2}$ Heisenberg chain. Unfortunately, this simple factorisation property does not survive in analogous treatments of spin $>\frac{1}{2}$ systems. The effect of the Umklapp term H^2 in the presence of H^1 has been studied by Emery et al. [11]: for positive U, it induces an instability against a $4k_F$ pinned charge density wave if $4k_F$ is equal to the reciprocal lattice vector G of the Umklapp potential. This state is a charge insulator with a gap $\sim v_F/R_s$ for excitations which can be characterised as spinless "solitons" or mobile topological defects where the phase of the CDW slips by 2π, and which carry charge ± 1. R_s is the characteristic lengthscale for phase slippage: It is very large in the weak coupling limit, but becomes of the order of the lattice spacing in the strong coupling limit.

The order parameter of this CDW is essentially given by χ_J^c: this develops long-range order as $|x-x'| \to \infty$:

$$<\chi_J^c(x)(\chi_J^c(x'))^{-1}> \sim R_c^{-1}\, e^{iG(x-x')} \quad . \tag{31}$$

When the charge subspace state is analysed in more detail, it emerges that it is apparently doubly degenerate: The Umklapp interaction (29) allows the current J_c to change in units of 4: since the allowed values of J_c differ by units of 2, there are two distinct states characterised by different values of $(-1)^{\frac{1}{2}J_c}$ which remains a good quantum number. However the selection rules (17), (18) coupling spin and charge quantum numbers remove this additional degeneracy, and given the spin quantum numbers N_s and J_s, the charge ground state is uniquely determined. Since a $4k_F$-density wave corresponds to one electron per unit cell, $N_c = N_a$, the number of atoms on the chain, then

$$(-1)^{N_s} = (-1)^{N_a} \quad ; \quad \exp \tfrac{1}{2}\pi i(N_s+J_s-N_a) = (-1)^{\frac{1}{2}J_c}$$

$$P^c = \tfrac{1}{4}G\, J_c \quad . \tag{32}$$

For N_a even, J_c takes values 0, 2 (modulo 4), for N_a odd, $J_c = \pm 1$. The two possible values of P^c differ by $\frac{1}{2}G$.

We have now essentially obtained the effective Hamiltonian for the Heisenberg chain, which is described by the remaining low-energy spin degrees of freedom:

$$H^o = v_s[\sum_q |q| b_q^\dagger b_q + \tfrac{1}{4}(\tfrac{\pi}{L})(N^2+J^2)]$$

$$H = H^o - U_o \int_o^L dx \int_o^L dy\, f\,(\tfrac{x-y}{R})\vec{\sigma}_p(x)\cdot\vec{\sigma}_{-p}(y)$$

$$P = \tfrac{1}{4}G(2-N_a+N) + [\tfrac{1}{4}G+\tfrac{1}{2}(\tfrac{\pi}{L})N]\, J + \sum_q q b_q^\dagger b_q$$

$$S^z = \tfrac{1}{2}N \; ; \quad (-1)^N = (-1)^J = (-1)^{N_a} \quad . \tag{33}$$

An explicit cutoff length R (derived from the lattice lengthscale) has been included in the interaction term. The kinetic term H^o can be written in an explicitly rotationally invariant form using (19):

$$H^o = \lim_{a\to 0} \int_o^L dx \sum_p (\tfrac{1}{3}\vec{\sigma}_p(x)\cdot\vec{\sigma}_p(x+a) - 1/2\pi a^2)$$

$$P = \tfrac{1}{3}\int_o^L dx \sum_p p\; \vec{\sigma}_p(x)\cdot\vec{\sigma}_p(x) \quad . \tag{34}$$

Except for the explicit rotational symmetry, all the details in (33) can be independently obtained from a linearisation of the Jordan-Wigner form (3) of the model, essentially following the procedure of Luther and Peschel [4]. From this point of view, the interaction term is generated by 2-particle Umklapp scattering in the equivalent spinless Fermi gas. Finally, the expression (25) for the spin density operator must be modified to take the charge ordering into account:

$$\vec{\sigma}(x) \sim \tfrac{1}{2} \sum_p (\vec{\sigma}_p(x) + R^{-\frac{1}{2}} e^{\frac{1}{2}iGpx}(-p)^{N_a+1}\vec{\tau}(x)) \quad . \tag{35}$$

The lengthscale R here is the scale generated by the charge solitons, but since the strong coupling limit has notionally been taken, R must again be identified with the lattice scale.

4. Properties of the Continuum S = ½ Antiferromagnet

If it were not for the non-linear interaction term in (33) the model describing the low-energy states of the antiferromagnet would be a trivial linear boson model. In fact this term can be treated by a scaling theory in the cutoff length R [18]; if a dimensionless coupling $\lambda = UR/\pi v_F$ is defined, it is found to undergo marginal scaling as the lengthscale R is increased:

$$\frac{d\,\lambda}{d\,\ln R^{-1}} = \lambda^2 + \tfrac{1}{2}\lambda^3 \;\ldots . \tag{36}$$

Since λ is positive, it always scales to weak coupling provided the magnetisation N_s is zero: a bare value $\lambda \sim 1$ can be assumed, so the interaction is characterised solely by the cutoff length R. In the presence of finite magnetisation, a preferred axis is introduced and the scaling equation (36) became more complicated, and involves an additional anisotropy parameter. However, in zero field, the non-linear term is scaled right away: the correlation functions will then be given by the spin factor of the free Fermi gas: the oscillatory part dominates:

$$\langle\vec{\sigma}(x) \cdot \vec{\sigma}(x')\rangle \sim R^{-1} \cos[\tfrac{1}{2}G(x-x')]/|x-x'| \quad . \tag{37}$$

The fall-off with exponent 1 emerges naturally for the isotropic Heisenberg chain, in this rotationally invariant formulation, though it was deduced less directly in the previous work based on the Jordan-Wigner form (3).

In finite field, the renormalisation of the interaction term leads to an effective low energy form [2]

$$H = v_s[\sum_q |q|\, a_q^\dagger a_q + \tfrac{1}{4}(\tfrac{\pi}{L})(e^{-2\varphi}N^2 + e^{2\varphi}J^2)] \tag{38}$$

where the bosons $a_q^\dagger$ are related to the $b_q^\dagger$ by a Bogoliubov transformation:

$$a_q^\dagger \sim \cosh\varphi\, b_q^\dagger + \sinh\varphi\, b_{-q} \quad . \tag{39}$$

The correlation functions are controlled by the dimensionless coupling parameter φ, which parametrises the residual quadratic boson terms of the interaction left after the renormalisation process.

Another feature of the model (33) is the fermion-like momentum $2k_F = \tfrac{1}{2}G+(\pi/L)N_s$, characterising spin current excitations $J\to J+2$. In the spin $-\tfrac{1}{2}$ fermion view-point, this is the momentum $\tfrac{1}{2}(k_{F\uparrow}+k_{F\downarrow})$ associated with the excitation of a spin-↑ particle from the left to the right Fermi point, simultaneous with the scattering of a spin-↓ particle in the opposite direction. This is a process that leaves the charge state of the system unchanged, and is hence unaffected by the presence of the charge density wave.

The basic derivation of the Heisenberg chain from an "itinerant" fermion system provides a tool for further study. In particular, the complex behaviour of the system in weak fields and weak uniaxial exchange anisotropy [19] can be identified with the renormalisation treatment of the interaction term [18]. Here I will go on to mention a few other possible developments.

5. The Spin-Peierls Instability

In the absence of an underlying periodicity other than G, and with full spin-rotation symmetry, the only possible couplings are of the form $\vec{\sigma}_p \cdot \vec{\sigma}_p$, $\vec{\sigma}_p \cdot \vec{\sigma}_{-p}$ and $\vec{\tau} \cdot \vec{\tau}$. This last turns out to vanish identically when evaluated and normal ordered. On the other hand, if the exchange is modulated with a period g, there can be terms like

$$H' = W \int dx \sum_p e^{i(g-\frac{1}{2}G)px} \vec{\sigma}_p(x) \cdot \vec{\tau}(x) \ . \tag{40}$$

If $g = \frac{1}{2}G + (\pi/L)N_s$ this term becomes active in a way analogous to the interaction in (33) in the uniform system. In the Jordan-Wigner fermion language, such a term translates into one-particle Umklapp processes. The scaling theory of this rotationally invariant form confines the earlier results of Cross and Fisher [20].

6. The Isotropic Antiferromagnet with $S > \frac{1}{2}$

Though the Jordan-Wigner method is particular to spin-$\frac{1}{2}$ systems, the approach sketched above is not so obviously limited. Indeed, one expects that the physics is essentially independent of the microscopic spin length S, except in as much as it affects parameters like v_s, k_F, , etc. Though neither has been fully carried through, I sketch two possible generalisations to $S > \frac{1}{2}$ systems.

The first consists of coupling 2S isotropic spin-$\frac{1}{2}$ chains to form a spin S chain. The Hamiltonian would have the form (with cutoffs implicit):

$$H = \sum_{\alpha=1}^{2S} H_\alpha^o - g_o \int^L dx \sum_{\alpha \neq \alpha} \vec{\sigma}_\alpha \cdot \vec{\sigma}_{\alpha'} \tag{41}$$

where $\alpha = 1, \ldots, 2S$ is an index distinguishing the 2S "replica" spin-$\frac{1}{2}$ systems. Quite complex possible couplings are generated, but the "isotropic" mode should remain gapless, while gaps for excitation of "internal" modes should develop.

A second approach would return to the spin-$\frac{1}{2}$ fermion system, this time generalising the Hubbard chain to have an SU(2S) "orbital" symmetry, and 2S degenerate bands. With just a Hubbard-type SU(2S)-isotropic coupling, this model does not have a gap until above a critical coupling for instability of the 4S-particle Umklapp process, and then only to an SU(2S) generalisation of the SU(2)/spin-$\frac{1}{2}$ system, rather than a spin-S chain. However, if a Hund's rule coupling to maximize the local spin is added, the instability that generates a gap for non-SU(2S)-singlet states and yields a spin-S-chain-like spectrum again occurs at arbitrarily weak U. However the simplifying feature of the decoupling between charge and spin degrees of freedom is no longer present, and technical problems again remain for the development of this route to the spin-S system.

7. Summary

The Heisenberg model is usually discussed as an idealisation of a highly localised electronic spin system. In this talk I have tried to convey how, at least in one-dimensional physics, the Heisenberg model can be best understood by "putting back" the itinerant character into the model. The success of this approach of course reflects some very special physics peculiar to one-dimensional systems, but provided only information about long-wavelength / low-energy behaviour is required, an analogous technique could be devised to study higher dimensional systems based on Hubbard models with "perfect nesting" of the Fermi surface in the half-filled band limit, such as the square and simple cubic tight binding lattices. In these systems too, the instability against the magnetic insulating state sets in at arbitrarily weak positive coupling.

References

1. H.A. Bethe: J. Phys. 71, 205 (1930)
2. F.D.M. Haldane: Phys. Rev. Lett. 20, 1445 (1980)
3. P. Jordan and E. Wigner: Z. Phys. 47, 631 (1928)
4. A. Luther and I. Peschel: Phys. Rev. B12, 3908 (1975)
5. J.M. Luttinger: J. Math. Phys. 4, 1154 (1963)
6. D.C. Mattis and E.H. Lieb: J. Math. Phys. 6, 304 (1965)
7. R.J. Baxter: Ann. Phys. (N.Y.) 70, 193 (1972)
8. T. Matsubara and H. Matsuda: Prog. Theor. Phys. (Kyoto) 16, 569 (1956)
9. M. Cieplak and L.A. Turski: J. Phys. C 13, 5741 (1980)
10. P.W. Anderson: in "Solid State Physics", ed. F. Seitz and D. Turnbull (Academic, New York 1963), Vol.14, page 99
11. V.J. Emery, A. Luther, and I. Peschel: Phys. Rev. B13, 1272 (1976)
12. E.H. Lieb and F.Y. Wu: Phys. Rev. Lett. 20, 1445 (1968)
13. A. Luther and I. Peschel: Phys. Rev. B9, 2911 (1974)
14. D.C. Mattis: J. Math. Phys. 15, 609 (1974)
15. R. Heidenreich, B. Schroer, R. Seiler, and D. Uhlenbrock: Phys. Lett. 54A, 119 (1975)

16. F.D.M. Haldane: J. Phys. C12, 4971 (1979)

17. A. Luther and V.J. Emery: Phys. Rev. Lett. 33, 389 (1974)

18. S.T. Chui and P.A. Lee: Phys. Rev. Lett. 35, 315 (1975)

19. F.D.M. Haldane: to be published

20. M.C. Cross and D. Fisher: Phys. Rev. B19, 402 (1979)

Magnetic Susceptibility of the Strongly Correlated Hubbard Model

Kenn Kubo

Institute of Physics, University of Tsukuba, Ibaraki 305, Japan

1. Introduction

The problem of metallic ferromagnetism has been an outstanding problem for solid-state theorists [1]. In recent years large progress has been made in understanding of the problem from the viewpoint of spin fluctuations [2]. It still seems to us, however, no definite answer has been given to the question what is essential for the appearance of ferromagnetism. For instance, although the single-band Hubbard model has been employed by many investigators, it is not known whether the model can be ferromagnetic or not. Therefore it still has some meaning to study this simplest model in detail and to examine under which conditions it simulates a ferromagnet.

The main difficulty of the problem stems from the evaluation of the correlation effect which reduces the interaction energy of the paramagnetic state. Really in the strong correlation limit, the energy difference between para- and ferromagnetic state comes only from the kinetic part. In addition, one must take into account of possibilities of other types of magnetic states, i.e. anti-ferromagnetism or spin density waves. The most stable state must be chosen from many possibilities for a given set of parameters. The parameters coming into our discussions are electron number density n ($n = N_e/N_L$), the strength of the intraatomic Coulomb interaction U, the energy band parameters and the temperature T. Following discussions are restricted to the energy bands with only nearest-neighbor hopping t, which is assumed to be positive without loss of generality.

We are only concerned below with the strongly correlated case ($U/t \gg T$). This is the typical case of itinerant magnetism in the sense that one can hardly imagine a spin density fixed at a lattice site unless n is unity. In the strong correlation limit ($U = \infty$) the ground states of the systems with $N_e = N_L - 1$ were rigorously proved to be totally ferromagnetic [3]. The problem in the case of the thermodynamic limit seems to be still open. The uniform susceptibility of sc lattice was calculated perturbationally with respect to t by several authors [4,5]. Their results showed that the fourth-order term favors ferromagnetism at any density n for $U = \infty$. However the results were erroneous. The high-temperature expansion of the uniform susceptibility of fcc lattice with $U = \infty$ was calculated by PLISCHKE [6]. His result was also erroneous perhaps due to erroneous treatment of fermion commutation relations.

We have obtained the perturbational corrections to the uniform and the staggered susceptibilities up to t^4 for strongly correlated system ($U/t \gg 1$) [7]. The wave-vector dependence of the susceptibility for arbitrary $\vec{Q}$ has been also studied in the strong correlation limit [8]. From analyses of the susceptibility the ferromagnetism is concluded to be hard to appear in sc lattice. The possibility remains in nearly half-filled bcc and fcc lattices.

2. Magnetic Susceptibility

The Hamiltonian considered is

$$H = H_0 + H', \tag{1}$$

where H_0 is the site-diagonal part

$$H_0 = U\sum_i n_{i\uparrow}n_{i\downarrow} - \sum_{i,\sigma} \mu_{i\sigma}n_{i\sigma} \tag{2}$$

and H' is the kinetic term which is treated as a perturbation

$$H' = t \sum_{\langle ij\rangle,\sigma} (c^+_{i\sigma}c_{j\sigma} + \text{c.c.}). \tag{3}$$

Here $c^+_{i\sigma}$ ($c_{i\sigma}$) and $n_{i\sigma}$ represent, respectively, creation (annihilation) and number oberators for electrons with spin σ at the i-th site. The summation in H' is done over nearest-neighbor pairs. The site-dependent chemical potential $\mu_{i\sigma}$ includes the external magnetic field

$$\mu_{i\sigma} = \mu + \sigma h_0 \cos(\vec{Q}\cdot\vec{R}_i), \tag{4}$$

where $\vec{R}_i$ is the position vector of the i-th site. The grand partition function of the system is expanded in term of H' as

$$\Xi = \Xi_0[1+ \sum_{n=1}^{\infty} \frac{(-)^n}{n!} \int_0^\beta d\tau_1 \cdots \int_0^\beta d\tau_n < T_\tau[H'(\tau_1)\cdots\cdot H'(\tau_n)] >_0]. \tag{5}$$

Ξ_0 is the unperturbed grand partition function and $\beta = (k_BT)^{-1}$. The bracket $<\ \ >_0$ represents an average over the grand canonical ensemble of the unperturbed sytem and T_τ denotes the time ordering. The n-th order terms in the series (5) correspond to diagrams with n bonds connecting lattice sites. Diagrams appearing up to the fourth order are shown below.

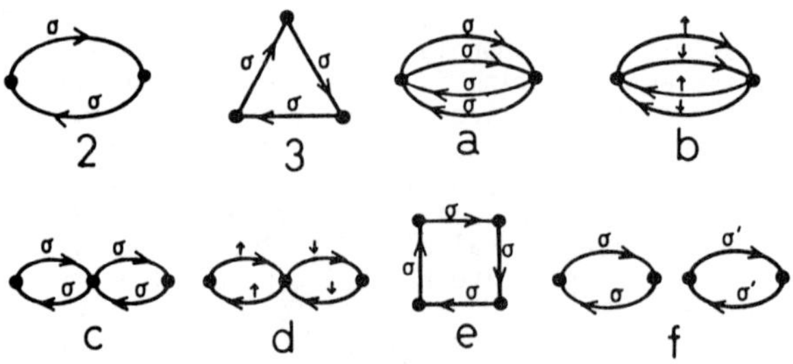

Fig.1 Diagrams Corresponding to the terms in (5) up to t^4.

The thermodynamic potential Ω is expanded in terms corresponding to these diagrams as

$$\Omega \cong \Omega_0 + \Omega_2 + \Omega_3 + \Omega_a + \Omega_b + \cdots + \Omega_f, \tag{6}$$

with for example

$$\Omega_d = k_B T t^4 \sum_{i=j=k} \iiint d\tau_1 \cdots d\tau_4 G_{i\uparrow}(\tau_3,\tau_1) G_{k\downarrow}(\tau_4,\tau_2) G_{j\uparrow\downarrow}(\tau_1,\tau_2,\tau_3,\tau_4) \quad (7)$$

and so on. Unperturbed Green functions are defined as usual. Although Wick's theorem does not apply as the unperturbed states are interacting ones, we can make use of a similar rule to decouple Green functions with parallel spins. They reduce to Green functions of non-interacting electrons feeling random potentials caused by electrons with antiparallel spins. Evaluations of Ω_2, Ω_3, Ω_a, Ω_c and Ω_e are largely simplified by making use of this rule. Anyhow we can calculate all the terms appearing up to t^4 straightforwardly. The susceptibility is derived from the thermodynamic potential as

$$\chi(\vec{Q}) = -2\frac{\partial^2}{\partial h_0^2}\Omega\Big|_{h_0=0}, \quad (8)$$

which is a function of μ and include the perturbational corrections on n. The susceptibility as a function of n is obtained by making use of the relation $n = -\frac{\partial\Omega}{\partial\mu}$. The obtained expression of the uniform susceptibility for $n \leq 1$ reads

$$\begin{aligned}
\chi_0 &= \beta n,\\
\chi_2 &= -zt^2\frac{\beta^2}{U}n^2,\\
\chi_3 &= p_3 t^3(\beta^4 + 6\frac{\beta^3}{U})n^2(1-n),\\
\chi_{a+b} &= -2zt^4(\frac{\beta^3}{U^2} - 2\frac{\beta^2}{U^3})n^2,\\
\chi_{c+d} &= -\frac{1}{2}z(z-1)t^4[\frac{\beta^4}{U}n^2(1-n) + 2\frac{\beta^3}{U^2}n^3 - 2\frac{\beta^2}{U^3}n^2], \qquad (9)\\
\chi_e &= p_4 t^4[-\frac{1}{6}\beta^5 n^2(1-n)(8-11n) - 2\frac{\beta^4}{U}n^2(1-n)(2-3n),\\
&\quad + 2\frac{\beta^3}{U^2}n^2(1-n)(4-5n) - 4\frac{\beta^2}{U^3}n^2(2-5n)],\\
\chi_f &= z(2z-1)t^4[\frac{\beta^4}{U}n^3(1-n) + \frac{\beta^3}{U^2}n^4] \quad \text{and}\\
\chi_g &= z^2 t^4[\frac{\beta^4}{U}n^2(1-n)(1-2n) + 2\frac{\beta^3}{U^2}n^3(1-n)].
\end{aligned}$$

Terms proportional to $e^{-\beta U}$ are assumed to be negligibly small and omitted in above expressions. The number of nearest neighbors is denoted by z and $p_3(p_4)$ is the number of triangles (quadrilaterals) per site composed of nearest-neighbor bonds. The last term χ_g comes from the correction on the chemical potential in χ_2. As was expected the second-order term represents antiferromagnetic kinetic exchange interaction. Noticeable is that corrections vanish in the strong correlation limit except for χ_3 and χ_e. This results from that the energy spectrum of the system with infinite repulsions is independent of spin configurations in one dimension [9.10]. The first correction in fcc lattice comes from the third-order term which is always positive (negative) for $n<1$ ($n>1$). In sc or bcc lattices the correction starts from χ_e, which changes its sign at $n=8/11$ and acts in favor of ferromagnetism only for $0<|n-1|<3/11$. Above result is consistent with that by NAGAOKA and also with that in the low density limit by KANAMORI [11]. Former authors gave an erroneous result which always favors ferromagnetism [4,5]. The staggered susceptibility χ_s in lattices which can be devided into AB sublattices has been also obtained. As will be shown in (10) all the corrections remain finite for $U=\infty$. The response of a one-dimensional system to spatial fluctuations of magnetic field is different from that of free spins. The correction comes from the spatial dependence of the wave functions.

2.1 Analysis of χ and χ_s in the strongly Correlated Case

In order to clarify physical implications of obtained results we practice a qualitative analysis for the magnitude of ferromagnetic and antiferromagnetic couplings in sc and bcc lattices. The inverse susceptibility correct up to t^4 are extrapolated to the lower temperatures and the Curie temperature T_c (the Neel temperature T_N) is determined as the temperature where χ^{-1} (χ_s^{-1}) vanishes. The phase diagrams in n-t/U plane obtained in this way for sc are shown in Fig.2.

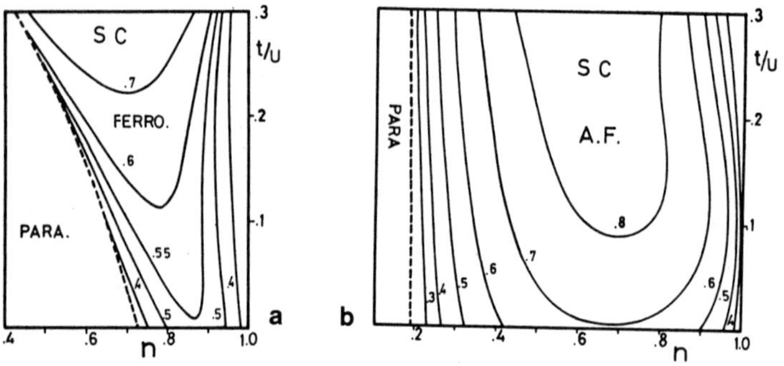

Fig.2 The phase diagrams for sc lattice determined from χ(a) and χ_s(b). The solid curves are T_c and T_N contours. Numbers are denoting T_c and T_N scaled by t.

Ferromagnetic coupling works for $.7\lesssim n\lesssim 1$ and is most effective at $n\simeq.85$. As t/U increases the coupling increases gradually except for $n\simeq 1$ and the ferromagnetic region extends to the lower electron densities. This feature reflects the fact that stronger interactions lead to larger correlation effects and thus hinders the occurrence of ferromagnetism contrary to the weak interaction case. It is rather surprising that the antiferromagnetic region is as large as to extend out to $n\simeq.2$. The coupling is strongest around $n\simeq.7$ and gradually increases with t/U. The effect of the kinetic exchange is quite small compared to that of $(\beta t)^4$-terms except for small region at $n\simeq 1$. In the half-filled case T_N is given by $4t^2/U$ for small t/U. It vanishes at $t/U\simeq.045$ as the other corrections than the kinetic exchange prevail. T_N is always higher than T_c in sc for $t/U\lesssim.3$. As are shown in Fig.3 results are similar in bcc lattice except for the relative magnitude of the

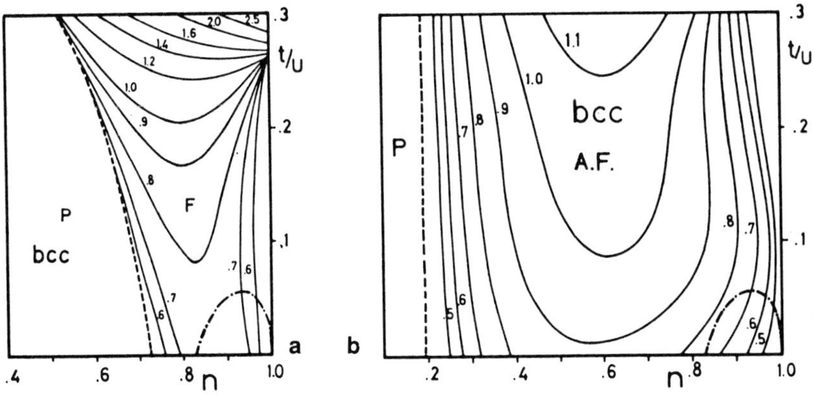

Fig.3 The phase diagrams for bcc lattice. The dash-dotted line encircles the region where $T_c > T_N$.

critical temperatures. In this case there is a small region at $.83 \lesssim n < 1$ and $t/U \lesssim .05$ where T_c is greater than T_N. The difference from sc comes from a larger contribution of the diagram 1e. The higher connectivity of bcc structure gives a large value of p_4 as twelve compared to the value three in sc.

Results shown above suggest that ferromagnetism is difficult to appear in strongly correlated sc lattice but may be possible in bcc.

2.2 Wave-Vector Dependence

In fact we must examine the possibility of ferromagnetism compared with all states with wave-vector $\vec{Q}$. For that purpose we examine the wave-vector dependence of the susceptibility $\chi(\vec{Q})$. We restrict ourselves in the following to the case $U=\infty$. Main features in the strongly correlated case may be guessed from the results for $U=\infty$, as we have seen above that T_c and T_N are only weakly dependent on t/U. In this limit $\chi(\vec{Q})$ reads

$$\chi_2 = -\frac{1}{12} z\beta(\beta t)^2 \, n(1 - n)a_2,$$

$$\chi_3 = p_3\beta(\beta t)^3 n(1 - n)[n + \frac{1}{4}(1 - \frac{7}{2}n)a_2],$$

$$\chi_{a+b} = -\frac{1}{120} z\beta(\beta t)^4 n(1 - n)a_2,$$

$$\chi_{c+d} = -\frac{1}{60} z(z - 1)\beta(\beta t)^4 n(1 - n)[(1 + \frac{n}{2})a_2 + \frac{1}{4}(1 - n)a_3], \tag{10}$$

$$\chi_e = -\frac{4}{3} p_4\beta(\beta t)^4 n(1 - n)[n(1 - \frac{11}{8}n) + \frac{1}{20}(1 - 11n + \frac{51}{4}n^2)(a_2 + \frac{1}{4}a_4)],$$

$$\chi_f = \frac{1}{12} z(2z - 1)\beta(\beta t)^4 n^2(1 - n)^2 a_2,$$

and

$$\chi_g = \frac{1}{24} z^2 \beta(\beta t)^4 n(1-n)(1-2n)^2 a_2.$$

Wave-vector dependence is involved in a_2, a_3 and a_4, which are defined as

$$a_2(\vec{Q}) = \frac{2}{z} \sum_{j \in N(i)} (1 - \cos\vec{Q}\cdot\vec{R}_{ji}),$$

$$a_3(\vec{Q}) = \frac{2}{z(z-1)} \sum_{j \neq k \in N(i)} (1 - \cos\vec{Q}\cdot\vec{R}_{jk}) \tag{11}$$

and

$$a_4(\vec{Q}) = \frac{1}{p^4} \sum_{\square(i)} (1 - \cos\vec{Q}\cdot\vec{D}_i).$$

Here $N(i)$ denotes the set of the nearest neighbors of the i-th site, $\vec{R}_{ji} = \vec{R}_j - \vec{R}_i$, $\square(i)$ the set of quadrilaterals which includes the i-th site as their vertex and $\vec{D}_i$ is the diagonal vector through the i-th site. The coefficients a_i's are all non-negative and vanish at $\vec{Q}=0$. For the sublattic antiferromagnetism $a_2=4$ and $a_3=a_4=0$. Explicit formulas are given in [8]. As was done above the critical temperature $T_c(\vec{Q})$ is obtained by extrapolating $\chi(\vec{Q})^{-1}$ to the temperature where it vanishes. The wave vector $\vec{Q}$max giving the maximal T_c is determined for a given value of n. The state with $\vec{Q}$max might be the most probable to appear. In fact only wave-vectors on the symmetry lines are examined to give $\vec{Q}$max(n) and T_c(n). The result for sc lattice is shown in Fig.4. Five density regions are distinguished according to which symmetry line $\vec{Q}$max lies on. As has been already shown the sublattice antiferromag-

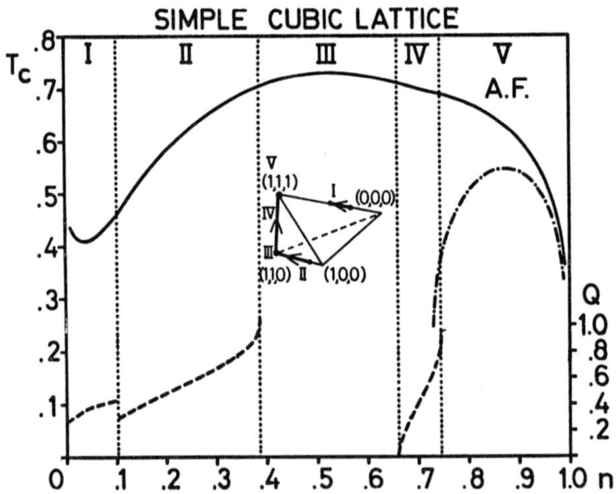

Fig.4 T_c and $\vec{Q}$max in sc lattice. The dashed lines show the values of Q_x(I), Q_y(II) and Q_z(III) of $\vec{Q}$max along corresponding lines depicted in the inset. Dash-dotted line shows the ferromagnetic temperature. The unit of $\vec{Q}$ is π/(lattice const.) as well as in the following figures.

netism always prevails over ferromagnetism. Other types of antiferromagnetism appear for intermediate densities. For low densities states with spin density waves appear.

The result for bcc is shown in Fig.5. The ferromagnetic state appears in the same density region (V) as in Fig.3. In bcc the sublattice antiferromagnetism appears only in a small density region. It is interesting that there appears a phase with quadruple period for intermediate densities. No transition appears for low densities.

A fairly large region n 1 is occupied by ferromagnetism in fcc. This results from the fact that both the third-order and the fourth-order corrections favor ferromagnetism. An antiferromagnetism appears for intermediate densities (V). There appears a small ferromagnetic region also for n>1. A state with quadruple period occupies a large region for intermediate densities (III).

We have seen that the conclusion in the preceding subsection has not been altered by analyses of the wave-vector dependence of $\chi(\vec{Q})$. Ferromagnetism does not appear for low densities in any lattice structures.

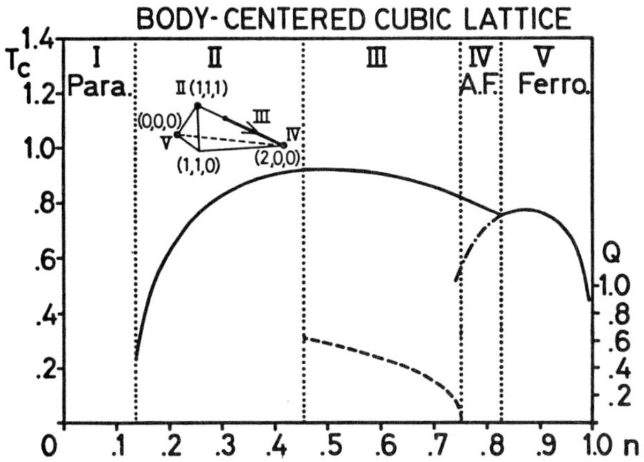

Fig.5 T_c and $\vec{Q}$max in bcc lattice. The dashed lines show the values of Q_y on the line III depicted in the inset.

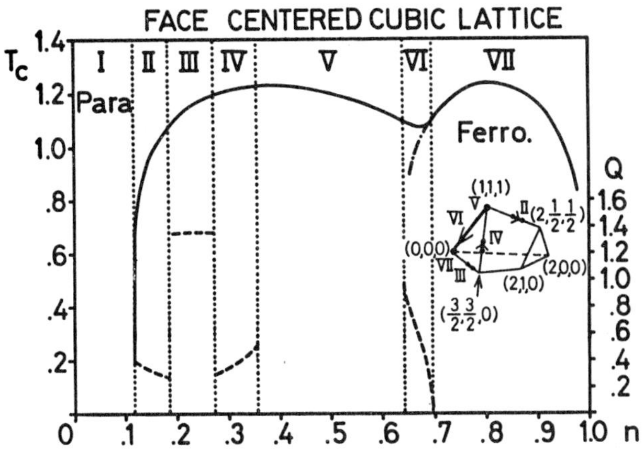

Fg.6 T_c and $\vec{Q}$max in fcc lattice for n<1. Dashed lines show $2-Q_x$(II), Q_x(III) and Q_x(IV). In the region IV the dashed line shows the value of $2-Q_x$ on the same line as in II.

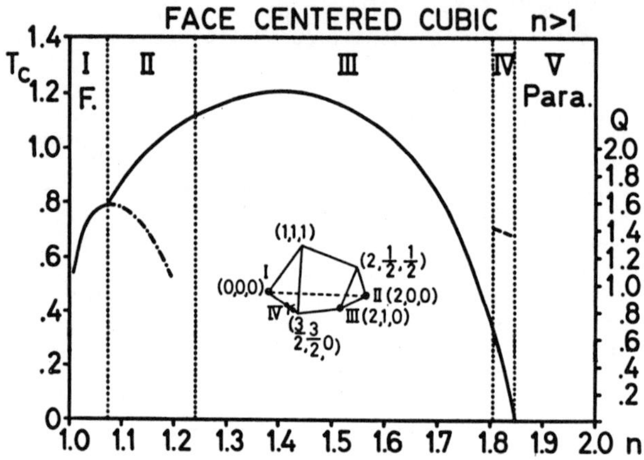

Fig.7 T_c and $\vec{Q}$max in fcc lattice for n>1. The dashed line shows the value of $\vec{Q}$max on the line depicted in the inset.

2.4 Higher Order Corrections

The values of the parameter βt at the transition temperatures obtained above are around unity and by no means small. Therefore contributions of higher order terms may be important. In order to go up to higher orders it is convenient to write down the susceptiblility in terms of the spin correlation function. The uniform susceptibility is expressed as

$$\chi = \beta n + \beta \sum_{r=3}^{\infty} (\beta t)^r c_r \tag{12}$$

where

$$c_r = \frac{(-)^r}{r!} \frac{2}{N_L} \sum_{i=j} \langle s_i^+ s_j^- \left(\frac{H'}{t}\right)^r \rangle_c . \tag{13}$$

Here the bracket $\langle \; \rangle_c$ denotes a cummulant average and $s_i^+ = c_{i\uparrow}^+ c_{i\downarrow}$. The problem is quite similar to that of obtaining high-temperature series in XY model [1]. The difference is that we are dealing with true fermions with

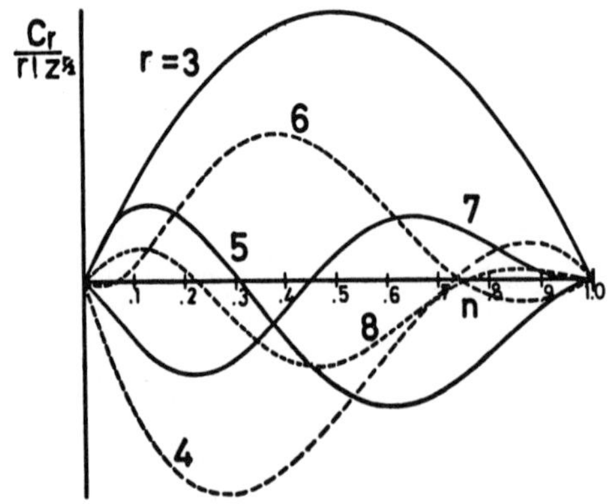

Fig.8 The coefficients c_3~c_8 for fcc lattice. They are scaled by factors $z^{r/2}$.

spin 1/2 and not the spinless quasi fermions. In addition we need to obtain the series of Ω to renormalize μ. Terms are expressed in terms of diagrams as usual. It is easily shown that no one-dimensional diagram contributes to the uniform susceptibility. Neglecting the difference in the positions of the spin vertices we have 3, 7, 12 and 31 diagrams for $c_5 \sim c_8$ respectively. The obtained coefficients are shown for fcc lattice in Fig.8. The higher order coefficients oscillate the more rapidly as n varies. The coefficients for sc and bcc lattice behave like corresponding ones for fcc.

We have made preliminary analyses by making use of Pade approximant [13]. So far obtained results are rather scattered. More terms seem necessary for obtaining a convergence. You can imagine from Fig.8. that the convergence is really slow. Oscillatory behavior of c_r's results from the fermion commutation relation which gives a factor -1 by a permutation of operators.

3. Discussion

We have seen above that the appearance of ferromagnetism is by no means obvious in the strongly correlated single-band Hubbard model. Our lowest order analyses have given no ferromagnetic state in sc lattice contrary to the result of Hartree-Fock method [14]. Our result seem to contradict the result by NAGAOKA. It is however not proved that his result corresponds to the lowest order of the density expansion in the thermodynamic limit. Therefore the system might not be ferromagnetic even for low densities of holes. The band-narrowing effect was also studied earlier for a system with one hole [15]. Ferromagnetism was shown to reduce the energy of the hole. The result however cannot be applied directly to many hole cases because holes we are treating are not usual fermions and have necessarily many-body character. As has been seen in 2.3 our result is also not conclusive. Higher order corrections may change the result drastically. Anyhow we have seen the competition between ferromagnetism and other magnetic states is more tight than has been expected. The old problem whether metallic ferromagnetism is possible or not without band degeneracies still remains to be answered.

References

1. C. Herring, in Magnetism, Vol.IV, ed. by G. T. Rado and H. Suhl (Academic Press, New York 1966) and references cited therein.
2. For example lectures by T. Moriya, H. Hasegawa and J. Hubbard in these Proceedings and references cited therein.
3. Y. Nagaoka, Phys. Rev. 147 (1966) 392.
4. G. Beni, P. Pincus and D. Hone, Phys. Rev. B8 (1973) 3389.
5. C. Lyon-Caen and M. Cyrot, J. Phys. C 8 (1975) 2091.
6. M. Plischke, J. Stat. Phys. 11 (1974) 159.
7. K. Kubo, Prog. Theor. Phys. 64 (1980) 758.
8. K. Kubo, to appear in Prog. Theor. Phys. Suppl.
9. J. B. Sokoloff, Phys. Rev. B2 (1970) 779.
10. G. Beni, T. Holstein and P. Pincus, Phys. Rev. B8 (1973) 312.
11. J. Kanamori, Prog. Theor. Phys. 30 (1968) 1445.
12. D. D. Betts, in Phase transitions and Critical Phenomena, Vol.3, ed. by C. Domb and M. S. Green (Academic Press, New York 1974), p. 570.
13. K. Kubo and M. Tada, in preparation.
14. D. Penn, Phys. Rev. 142 (1966) 350.
15. W. F. Brinkman and T. M. Rice: Phys. Rev. B2 (1970) 1324.

Electron Correlation in the Small Hubbard Model

Arisato Kawabata

Department of Physics, Gakushuin University,
Mejiro, Toshima-ku, Tokyo, Japan

Abstract

Hubbard model on 2×2×2 latice is investigated with the use of the computor diagonalization of the Hamiltonian. The eigenenergy, the magnitude of the total spin and the spin correlation function of the ground state are calculated for various values of the total electron number and the Coulomb repulsive energy.

1. Introduction

In recent years great progress has been made on the understanding of the magnetism of the metals[1]. There are, however, some problems which are left unsolved. For instance, it is not known whether the degeneracy of the atomic orbital is essential to the occurrence of ferromagnetism in metals, and the study of the problem of the strongly interacting electrons is still very important. Hubbard model is the simplest model of strongly correlated electron system. Several methods of treating this model have been proposed[2, 3, 4]. Qualitatively they could account for some features of strongly correlated electron systems, but it is not clear how those methods are quantitatively reliable. Generally, very small differences in energies of the various spin states are involved in the magnetism of the metals, and hence we need a quantitative method of treating electron correlation in order to completely understand the magnetism of the metals.

Here we will investigate Hubbard model composed of eight atomic sites on the corners of a cube. The same model was studied by TAKAHASHI[5] in the case of infinitely large Coulomb repulsive energy. We will calculate the ground-state energy and the spin correlation function for finite values of the Coulomb energy using computor diagonalization of the Hamiltonian.

2. Formulations

The Hamiltonian is of the form

$$H = \sum_{i,j,\sigma} t_{ij} c_{i\sigma}^{\dagger} c_{j\sigma} + I \sum_{i} n_{i\uparrow} n_{i\downarrow}, \tag{2.1}$$

where the notations are standard. The ground-state energy and the eigenfunction was calculated with the use of so-called power method. We start with a many-body state vector u_0 which is expected not to be orthogonal to the correct ground-state eigenvector v_0. We operate the Hamiltonian H on u_0 n times , and we obtain

$$u_n = H^n u_0 = \sum_i b_i E_i{}^n v_i \;, \tag{2.2}$$

where $b_i = (v_i, u_0)$, v_i being the correct eigenstate of H belonging to the eigenenergy E_i. Since the freedom of the system is finite the values of E_i are bounded , and we can choose the zero of energy so that $|E_0|$ is the largest among $|E_i|$. Then, for n large enough, we obtain

$$u_n = b_0 E_0{}^n [\; v_0 + O(E_1/E_0)^n \;] \;, \tag{2.3}$$

where E_1 is the eigenenergy with the second largest absolute value. Since $b_0 \neq 0$ by assumption, it follows that

$$E_0 = \lim_{n\to\infty} u_{n+1}/u_n \;, \tag{2.4}$$

and at the same time we obtain v_0 by normalizing u_n. The Hamiltonian H commutes with the total spin. Therefore, if we start with u_0 which is an eigenstate of S, the magnitude of the total spin, and of S_z, its z component, the state vector u_n is also an eigenstate of S and S_z belonging to the same eigenvalue, and we obtain the state with the lowest energy in the subspace spanned by those eigenstates of S and S_z. This energy will be called $E_n(S)$. In the course of the numerical calculation there is a possibility that the components of the different eigenvalues of S and S_z are mixed in u_n, and hence the magnitude of the mixing has to be chequed by calculating the expectation value of S of v_0 obtained by the calculation. As u_0 we take the state of the lowest energy in non-interacting case. For the details of the choice of u_0, the reader is referred to [6]. Because of the rotational symmetry we consider only the case $S_z = S$. The calculation was done in the representation in which the hopping energy is diagonal. Therefore the procedure of the calculation is little affected by the dependence of the hopping integral t_{ij} on i and j. However, here we consider the case of the nearest–neighbor hopping. We put $t_{ij} = -1$ for nearest-neighbor pairs, and the energies will be measured in this unit.

In Fig.1 we show the results of the calculation of $E_N(S)$ for N = 7. The zero of the energy is that of the complete ferromagnetic state $E_N(N/2)$, which is independent of the value of I (this choice of the zero of the energy will be applied to any value of N). As is seen from the figure, S of the ground state is 1/2 for I < 39.0 and is 7/2 for I > 39.0. In Fig.2 we show $E_N(S)$ for

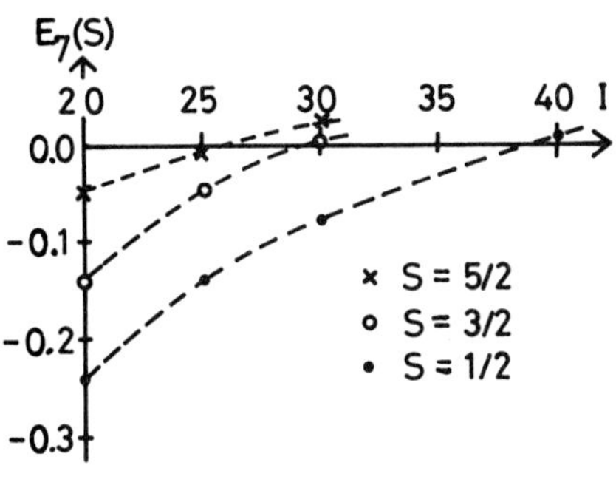

Fig.1 $E_7(S)$ in the region $20 \leqq I \leqq 40$. The zero of the energy is $E_7(7/2)$, i.e. that of the complete ferromagnetic state.

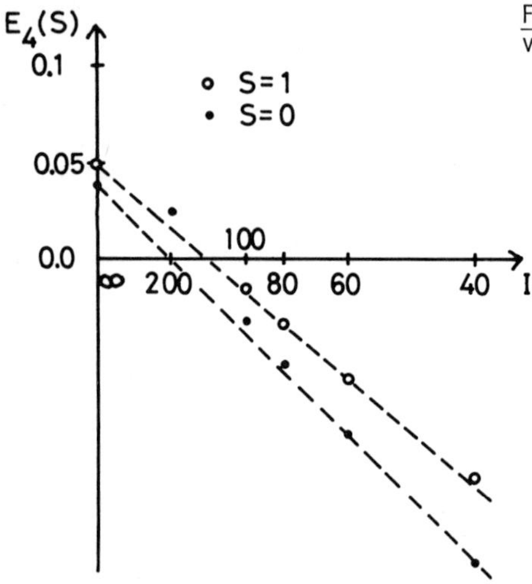

Fig.2 $E_4(S)$ in the region $I \leqq 40$. The value at $I = \infty$ was taken from [5].

$N = 4$. We find that the total spin of the ground state is zero when $I \lesssim 150$. When I is very large , however, we must be careful in analysing the numerical results, for a small error in the average number of the doubly occupied sites gives rise to a large error in the energy. In fact, for very large I, $E_N(S)$ should depend on I like

$$E_N(S) = E_N^{\infty}(S) + C/I, \tag{2.5}$$

where $E_N^{\infty}(S)$ is $E_N(S)$ for $I = \infty$ and C is a constant dependent on N and S. In the limit $I = \infty$, we have to diagonalize the Hamiltonian without the interaction term excluding the configrations which contains doubly occupied sites, and we obtain $E_N(S)$ in high accuracy[5]. As is seen in Fig.2, the calculated value of $E_N(s)$ does not tend to $E_N^{\infty}(S)$ in the limit $I = \infty$, and by means of interporation, as is shown in the figure with the dotted lines, we find that the ground state becomes completely ferromagnetic for $I > 200$. When $N = 5$, S of the ground state is 3/2 irrespective of the value of I. When N = 2, 3, 6, S of the ground state assumes the minimum value, i.e. 0 or 1/2, for any value of I. The above resulst are summmalized in Table 1.

N	I: 0. – 39.	39. – 200.	200. – ∞
2	0	0	0
3	1/2	1/2	1/2
4	0	0	2
5	3/2	3/2	3/2
6	0	0	0
7	1/2	7/2	7/2

Table 1 The magnitude of the total spin S of the ground state for various values of the total electron number N.

An important feature of the case of N = 4 and 7 is that the complete ferromagnetic state appears at a critical value of I and that the states of intermediate values of S never become the ground state. Namely, $E_N(S)$ behaves as is shown in Fig.3 schematically, and it shows that generally the appearance of

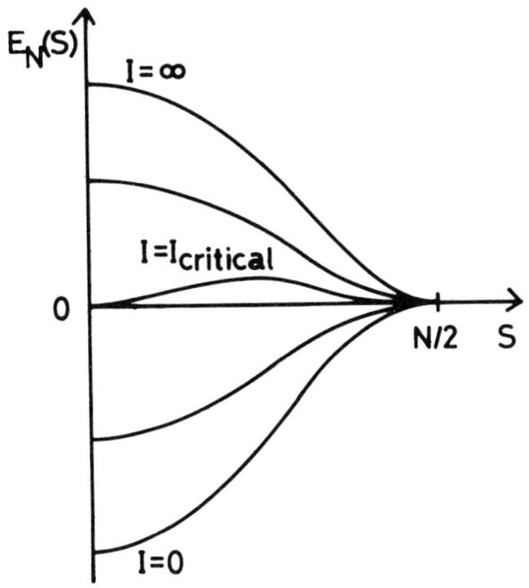

Fig.3 Schematic behavior of $E_N(S)$ as the function of S in the case when the transition from the state of S = 0 to the complete ferromagnetic state occurs.

the ferromagnetic ground state is not necessarily accompanied with the divergence of the magnetic susceptibility , i.e. $1/d^2E_N(S)/dS^2$ at S = 0. Such a behavior of $E_N(S)$ can be expected if the transition is that from antiferromagnetic to ferromagnetic state under the existence of the localized monent. It is an interesting problem whether the localized moments exist (strictly speaking, whether the local moment picture is a good approximation) in non-half-filled case. Although its existence is indicated in the above examples, we can not draw a clear conclusion from them.

3. Spin Correlation Functions

The spin correlation function is defined by

$$F_{ij} = 4(v_0, \vec{S}_i \cdot \vec{S}_j v_0), \tag{3.1}$$

where $\vec{S}_i$ is the spin operator at the site i. From v_0 obtained in the way described in 1, we can calculate F_{ij}. Here we will consider the case S = 0. Firstly, from the relation

$$\sum_{i,j} F_{ij} = 4(v_0, S^2 v_0) = 4S(S + 1), \tag{3.2}$$

we can check whether the components of the state vector of other value of S are mixed in v_0. It has been found that (3.1) holds up to 10^{-4} in all the cases shown below. In Table 2 we show the results of the calculation of F_{ij} for N = 4. It is easy to see that the ground state has threefold degeneracy when I = 0, and we obtain three different v_0, which are degenerate, if we start from those u_0. The values of F_{ij} shown in Table 2 are the averages of

I	i=j	1st	2nd	3rd
0.	1.125	-0.250	-0.125	0.
20.	1.488	-0.292	-0.237	0.102
40.	1.497	-0.283	-0.244	0.085
100.	1.499	-0.274	-0.235	0.028

Table 2 Spin correlation function F_{ij}. 1st means F_{ij} between the first-neighbor sites, and so forth.

F_{ij} obtained from those three v_Q. As is seen in the table, when I is not very large, F_{ij} between the first-neighbor sites decreases as I increases, and then it increases when I becomes very large. On the contrary, F_{ij} between the third-neighbor sites increases as I increases and then decreases for larger I. Such behavior of the spin correlation function may be attributed to the tendency of the ferromagnetic short-range order in the case of very large I.

References

1. For example, the reveiw by T. Moriya in this proceedings.
2. J. Kanamori, Prog. Theor. Phys. 30 (1963) 275.
3. J. Hubbard, Proc. Roy. Soc. A281 (1964) 401.
4. M. Gutzwiller, Phys. Rev. 137A (1965) 1725.
5. M. Takahashi, J. Phys.Soc. Jpn. 49 (1979) 49.
6. A. Kawabata, Solid State Commun. 32 (1979) 895.

Part IV

Mixed Valence in 4f Compounds

Dynamics of 4*f* Electrons in Mixed Valence Compounds

E. Müller-Hartmann

Institut für Theoretische Physik, Universität Köln, Zülpicher Str. 77,
D-5000 Köln, Fed. Rep. of Germany

1. Introduction

The 4f shells of rare-earth atoms have a radius which is extraordinarily small for a partially filled shell [1]. As is indicated in Fig.1, the 4f orbitals are localized within the Xenon core of the atoms and have a radius that is 4 to 5 times smaller than the atomic radius.

Nowhere else in the periodic table does one find such an extreme localization of a partially filled shell, and only the 5f shell of the heavy actinides may come close to this [2]. In view of their small orbital radius it is not surprising that 4f electrons in rare-earth compounds are coupled rather weakly to the other degrees of freedom of the solid. First of all, there is no direct overlap between 4f orbitals on different rare-earth atoms. The most important coupling is in general the hybridization of 4f orbitals with states in the conduction or valence bands. This coupling is considerably weaker than the interactions which are responsible for the intraionic correlations within a 4f shell. For this reason there is no doubt that any serious description of rare-earth compounds has to take appropriate account of intraionic correlations.

In most rare-earth compounds the 4f shells are found in a specific, unique configuration, let us say $4f^n$. In this case we say that the rare-earth ions

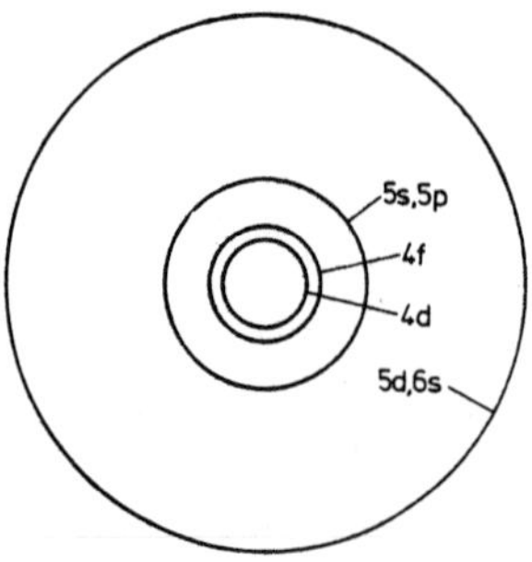

Fig. 1. Radii of rare-earth orbitals

have a stable valence. It may, however, happen that the 4f shells coexist in two neighboring configurations, say $4f^{n-1}$ and $4f^{n}$. In this case we speak of a mixed valence compound. At present some thirty compounds with mixed valence are known, an additional dozen compounds go into a mixed valence state under high pressure, and numerous alloys appear to show mixed valence, too. There are also compounds where a second configuration is not assumed with any appreciable probability, but where it is energetically close by. Refering to those compounds which are often called KONDO compounds we prefer to speak of an unstable valence rather than of mixed valence.

In case of a stable valence the 4f shells have a rotational degree of freedom which exhibits itself in the well-known magnetic properties of such compounds. Quite differently, the 4f electrons of mixed valence compounds may jump out of a $4f^{n}$ comfiguration into the conduction bands and back into a $4f^{n-1}$ configuration. Via these valence fluctuation processes the 4f electrons acquire a translational degree of freedom and a certain itinerant character. This suggests considering mixed valence compounds as very similar to transition metals. The situation is, however, quite different from that in transition metals where d-band widths and intraionic correlation energies are of comparable size. As already pointed out above, the f-band width in rare-earth compounds is much smaller than correlation energies. For this reason mixed valence compounds provide us with the unique opportunity to study systematically the motion of electrons in the extreme limit of high correlation.

Experimentally it is well established that all rare-earth ions in a mixed valence compound are equivalent [3]. One speaks of a homogeneously mixed valence. It seems to be extremely difficult to obtain a serious understanding of what stabilizes homogeneous valence mixing and why there is no inhomogeneous separation of the 4f configurations. Since rare-earth radii depend sensitively on the configuration – atomic volumes of neighboring configurations differ by some 15 percent – the stability problem requires a proper treatment of screening and elastic energies as well as hybridization. At this point it is certainly advisable to accept homogeneity as an experimental fact in much the same way as one generally accepts the stability of a crystal structure.

Homogeneous valence mixing implies either temporal fluctuations of the valence or coherent superposition of two configurations on each rare-earth atom. At temperatures above the 4f-band width (room temperature) one can visualize mixed valence compounds as performing real temporal fluctuations of the valence. At low temperatures, however, they approach a unique coherent state which in general is nonmagnetic. It is one of the fundamental problems of the theory of mixed valence compounds to find a proper description of the quantum correlations of their low-temperature state and a description of their low-lying electronic excitations.

Other interesting questions arise if one realizes that valence fluctuations have a similar time scale as lattice vibrations. Together with the large atomic volume fluctuations already mentioned this suggests resonant coupling of valence fluctuations with certain lattice modes which should be visible via phonon anomalies. It was always surprising to me that so few phonon anomalies were found in mixed valence compounds. The quantitative analysis of relaxation rates which will be given below will shed some new light onto this question.

Most theoretical work on mixed valence compounds has been done with models which neglect the orbital degeneracy of the 4f shell [4]. Such a simplification is certainly justified if one aims at the basic clarification of fundamental many-body aspects like the structure of quantum correlations at low temperatures. Too little attention has, however, been paid to the impact of orbital degeneracy onto the behavior of real mixed valence compounds. It is a major goal of the present talk to demonstrate that intraionic correlations of the various rare-earth ions may cause almost order of magnitude differences in certain properties of mixed valence compounds. This can lead to qualitatively different behavior of compounds containing different rare-earth ions.

We have calculated dynamical susceptibilities of mixed valence compounds. It has been recently shown by KURAMOTO [5] that using MORI's formalism one can understand relaxation rates of the 4f shell by a straightforward lowest order evaluation of the memory function. We have generalized KURAMOTO's work to properly include orbital degeneracy and intraionic exchange and spin-orbit interaction. Not only magnetic, but also other response functions of the 4f shell have been calculated. Some results of this calculation are presented in Sect.4 of this talk. Their implications for mixed valence compounds are discussed in Sect.5. The model used for the calculation and the application of the MORI method to the present problem are outlined in Sects.2 and 3. The talk will end with a short outlook.

The magnetic dynamical susceptibility of mixed valence compounds has been considered by other authors before. BALSEIRO and LOPEZ [25] have investigated the Anderson model with a method by which they missed the dominating relaxation process. FOGLIO [26] aimed at the compound $CePd_3$; he employed a decoupling procedure which seems more ambiguous than the Mori method and his results do not agree with ours. GONÇALVES DA SILVA [27] has presented an interesting analysis of sum rules.

2. Model with Orbital Degeneracy

The most popular model for mixed valence compounds is the periodic ANDERSON model [6] which neglects orbital degeneracy of the 4f shell. What we want to emphasize here is that one of the most urgent improvements of this model is to include orbital degeneracy. There is rather little chance to make reasonable contact with experiment without using a realistic model of the 4f shell. It may be important to refine the model into other directions also, e.g. by including screening effects, but this should not be done without including orbital degeneracy.

Since the Hamiltonian is written in second quantization, one should be aware of the one-particle basis in which it is formulated. It is indispensable to have quasi-atomic 4f orbitals in this basis, because one needs them to include intraionic correlations. In view of the small atomic 4f radii we can easily make orbitals on different ions orthogonal to each other. In addition to 4f orbitals the basis should contain appropriate valence (conduction) band states. They should be Bloch states which are orthogonal to the 4f orbitals and can be chosen, e.g., as discussed by HIRST [7]. It has to be emphasized that a very different basis is obtained in a band structure calculation which includes 4f states. The eigenstates one obtains this way are all hybridized Bloch states between the 4f orbitals and valence band states described above.

The periodic Anderson model refined to include realistic ionic correlations is given by the Hamiltonian

$$H = H_{4f} + H_{cond} + H_{hybr} = H_0 + H_{hybr} \quad . \tag{1}$$

The ionic Hamiltonian H_{4f} contains the intraionic Coulomb repulsion between 4f orbitals and exchange and spin-orbit interaction. It looks quite complicated, but fortunately we will never need it explicitly. Coulomb repulsion which is experimentally known to be about 7 eV [8] only has the effect of excluding the coexistence of more than two neighboring 4f configurations. Exchange and spin-orbit interaction exclude all excited states of the two configurations from low-energy processes. We thus will be able to take full account of H_{4f} by simply restricting initial, final, and intermediate states of the 4f shell to Hund's rule ground states of the two coexisting configurations.

Conduction (or valence) electrons are considered as noninteracting and are described by the Hamiltonian

$$H_{cond} = \sum_{\underline{k}\nu\sigma} \varepsilon_{\underline{k}\nu} c^{+}_{\underline{k}\nu\sigma} c_{\underline{k}\nu\sigma} \quad . \tag{2}$$

They are classified as usual by wave number $\underline{k}$, band index ν and spin direction σ.

Conduction band states and 4f orbitals are mixed by the hybridizing term of the Hamiltonian (1),

$$H_{hybr} = \frac{1}{\sqrt{N}} \sum_{\underline{k}\nu\underline{\ell}m\sigma} \left[V_{\underline{k}\nu m} e^{i\underline{k}\underline{\ell}} f^{+}_{\underline{\ell}m\sigma} c_{\underline{k}\nu\sigma} + V^{*}_{\underline{k}\nu m} e^{-i\underline{k}\underline{\ell}} c^{+}_{\underline{k}\nu\sigma} f_{\underline{\ell}m\sigma} \right] \quad . \tag{3}$$

Rare-earth ions are numbered by the lattice vector $\underline{\ell}$ which counts the N elementary cells of the lattice. The magnetic quantum number of the 4f orbitals is denoted by m ($-3 \leq m \leq 3$). Translational symmetry of the lattice is explicitly built in into (3). Additional rotational symmetries of a compound are reflected by the hybridization matrix $V_{\underline{k}\nu m}$. For all proper or improper symmetry rotations R which leave a rare-earth atom fixed we can assume

$$V_{R^{-1}\underline{k}\nu m} = \sum_{m'=-3}^{3} V_{\underline{k}\nu m} R^{(3)}_{m'm}(R) \quad , \tag{4}$$

where $R^{(3)}_{m'm}$ is a rotation matrix.

Crystal fields will be supposed to be included in the Hamiltonian also. As we shall see hybridization is the source of covalent contributions to crystal field splitting. Any crystal field of electrostatic origin will be assumed to be part of H_{4f} and is easily diagonalized in each configuration.

3. Application of Projection Methods

We want to calculate dynamical susceptibilities of the 4f shell of mixed valence compounds by applying MORI's formalism [9] to the model described above. Our notation will be close to the one used by FORSTER [10]. We will first outline the method in general.

Since we wish to include multivariate processes, e.g. as caused by crystal field transitions, we will use a matrix formulation. If A_α ($\alpha=1,\ldots,r$) are the "slow" operators of the physical system considered, one should look at the whole matrix susceptibility

$$\chi_{\alpha\beta}(z) = i \int_0^\infty \left\langle \left[A^{+}_{\alpha}(t), A_\beta \right] \right\rangle e^{izt}\, dt \quad . \tag{5}$$

Denoting matrices by a hat we write Mori's identity in the form

$$\hat{\chi}(z) = [-\hat{\omega}+i\hat{\sigma}(z)] \cdot [z\hat{\chi}(0)-\hat{\omega}+i\hat{\sigma}(z)]^{-1}\hat{\chi}(0) \quad . \tag{6}$$

In order to explain the meaning of the various symbols one has to introduce the positive Hermitian scalar product

$$<A_\alpha|A_\beta> = \int_0^\beta d\tau \left\langle e^{\tau H} A_\alpha^+ e^{-\tau H} A_\beta \right\rangle \tag{7}$$

on the linear space of operators $A_\alpha, \ldots$. The Liouville operator

$$LA_\alpha = [H, A_\alpha] \tag{8}$$

acts on the same space and will also be used.

The isothermal susceptibility which appears in (6) is given by

$$\chi_{\alpha\beta}(0) = <A_\alpha|A_\beta> \quad . \tag{9}$$

The $(z=\infty)$ part of the relaxation kernel,

$$\omega_{\alpha\beta} = <A_\alpha|L|A_\beta> \quad , \tag{10}$$

is reactive and describes crystal field splittings in our application. The memory matrix

$$\sigma_{\alpha\beta}(z) = \left\langle A_\alpha|LQ \frac{i}{z-LQ} L|A_\beta \right\rangle \tag{11}$$

contains the absorptive part of the relaxation and will give us information about line widths. It is of crucial importance for the method to be meaningful that the projection operator

$$Q = 1 - P \quad , \quad P = \sum_{\alpha,\beta} |A_\alpha>[\hat{\chi}(0)^{-1}]_{\alpha\beta}<A_\beta| \tag{12}$$

removes all slow parts of the dynamics from the memory matrix.

In the dynamics of 4f shells of mixed valence compounds all operators acting within the $4f^{n-1}$ and $4f^n$ configurations are slow. If the Hund's rule ground states of an ion are $|J_{n-1}J^z_{n-1}>$ and $|J_nJ^z_n>$, the set of slow operators consists of all operators

$$|J_{n-1}J^z_{n-1}><J_{n-1}J^{z'}_{n-1}| \quad \text{and} \quad |J_nJ^z_n><J_nJ^{z'}_n| \tag{13}$$

for each rare-earth ion. Spacial Fourier transform partially diagonalizes the susceptibility matrix such that one is left with a matrix of dimension

$$r = (2J_{n-1}+1)^2 + (2J_n+1)^2 \tag{14}$$

for each momentum. The dimension is still rather high in some cases ($r=13^2+8^2=233$ for Tm). But as we shall see point symmetries allow to break this matrix down into rather small pieces.

For the calculation of crystal fields we split $\hat{\omega}$ into two parts,

$$\omega^0_{\alpha\beta} = <A_\alpha|L_0|A_\beta> \quad , \quad \omega^{hybr}_{\alpha\beta} = <A_\alpha|L_{hybr}|A_\beta> \quad . \tag{15}$$

According to (6) crystal field splittings are given by the eigenvalues of the matrix $[\hat{\chi}(0)]^{-1}\hat{\omega}$. Since the slow operators form an invariant subspace of L_0, $L_0A_\beta = \sum_\gamma A_\gamma E^0_{\gamma\beta}$, we obtain $[\hat{\chi}(0)]^{-1}\hat{\omega}^0 = \hat{E}^0$ which does not depend on hybridization whatsoever and is completely determined by the electrostatic crystal field Hamiltonian. All covalent contributions to crystal field splitting are contained in $\hat{\omega}^{hybr}$.

4. Lowest Order Results

As KURAMOTO [5] has shown lowest order evaluation of the memory function leads to magnetic relaxation rates which naturally explain several characteristic features of the quasi-elastic line widths of mixed valence compounds measured by neutron scattering: (a) line widths are much larger than those corresponding to Korringa relaxation of stable 4f shells; (b) line widths are more or less independent of temperature; (c) line widths are not observed to depend on momentum. In the present context we are satisfied with a lowest order calculation, too.

We first consider the memory matrix (11). Since Q projects all slow operators out of LA_β and LA_α in (11), it follows that $QL|A_{\alpha,\beta}> = L_{hybr}|A_{\alpha,\beta}>$, such that the memory matrix is proportional to the square of the hybridization. In lowest order in hybridization it turns out that Q drops out of (11), since L_0 conserves the number of 4f electrons but L_{hybr} does not. Therefore the memory matrix to lowest order is given by the simple expression

$$\sigma_{\alpha\beta}(z) = \left\langle A_\alpha \middle| L_{hybr} \frac{i}{z-L_0} L_{hybr} \middle| A_\beta \right\rangle_0 \quad . \tag{16}$$

It is straightforward to calculate $\hat{\omega}^{hybr}$ of (15) to lowest, i.e., second, order in the hybridization by a thermodynamic perturbation expansion. For the static susceptibility $\hat{\chi}(0)$ it is sufficient to use the unperturbed (zeroth-order) expression. This way we obtain covalent crystal field splittings as well as line widths to second order in hybridization.

Obviously operators (13) on different rare-earth ions are not coupled by lowest order charge fluctuation processes which leaves the susceptibilities independent of momentum.

In lowest order calculations the hybridization appears in terms of hybridization integrals

$$v_{mm'} = \frac{1}{N} \sum_{\underline{k}\nu} V^*_{\underline{k}\nu m} V_{\underline{k}\nu m'} \delta(\varepsilon_{\underline{k}\nu} - \mu) \tag{17}$$

which, with a density of conduction band states weight factor, average the square of hybridization matrix elements over momenta whose energy is equal to the chemical potential μ. There are $7 \times 7 = 49$ such integrals to begin with. But due to point symmetries (4) their number is greatly reduced. In particular we shall consider the case of full rotational symmetry which implies

$$v_{mm'} = W_0 \cdot \delta_{mm'} \quad . \tag{18}$$

In the important case where the rare-earth ions are centers of octahedral symmetry there are only two further invariants which are

$$W_4 = (v_{00} + v_{11} - v_{22} - v_{33})/7 \quad , \quad W_6 = (2v_{11} + v_{33} - 2v_{00} - v_{22})/7 \quad , \tag{19}$$

if the quantization axis of 4f orbitals coincides with the fourfold axis. All $v_{mm'}$ can be expressed in terms of W_0, W_4 and W_6.

Since results depend on the average valence of a compound we need a convenient way of parametrizing this dependence. If δ $(0<\delta<1)$ is the probability of finding the configuration $4f^n$, the average number of 4f electrons on a rare-earth atom is

$$n_{4f} = (n-1) \cdot (1-\delta) + n \cdot \delta = (n-1) + \delta \quad . \tag{20}$$

We will use the parameter δ to identify the average valence.

Only six of the rare earths have been observed in a mixed valence state: $Ce(4f^0\text{-}4f^1)$, $Pr(4f^1\text{-}4f^2)$, $Sm(4f^5\text{-}4f^6)$, $Eu(4f^6\text{-}4f^7)$, $Tm(4f^{12}\text{-}4f^{13})$ and $Yb(4f^{13}\text{-}4f^{14})$. We therefore will present results for these six rare earths.

4.1 Full Rotational Invariance

We first discuss the somewhat idealized case (18), because some important features connected with intraionic correlations can be made particularly clear in this case. In order to reduce the dimension (14) of matrices (6) as much as possible we take linear combinations of the operators (13) which are irreducible tensor operators of the rotation group:

$$|JJ_z\rangle\langle JJ_z'| \rightarrow T_{\Lambda\lambda} \qquad (\Lambda=0,\ldots,2J;\ \lambda=-\Lambda,\ldots,\Lambda) \quad . \tag{21}$$

The operators $T_{\Lambda\lambda}$ are the well-known multipole operators. It is easily shown that all matrices in (6) are diagonal in the representation indices Λ and λ and are independent of the value of λ. Since each representation appears at most once in each configuration and since there are exactly two relevant configurations, rotational symmetry leads us to matrices whose dimensions are not larger than two. We obtain two by two matrices which couple the operators $T^{(n-1)}_{\Lambda\lambda}$ and $T^{(n)}_{\Lambda\lambda}$ for $\Lambda \leqq 2 \min (J_{n-1}, J_n)$ and one-dimensional matrices for larger values of Λ.

In the present case of rotational symmetry $\hat{\omega}$ turns out to vanish, because there is no crystal field splitting in this case. The problem therefore reduces to a calculation of the trivial static susceptibilities $\hat{\chi}(0)$ and of the memory functions (16). For the quasi-elastic line widths we actually only need the zero frequency limit $\hat{\sigma}(i\varepsilon)$.

The monopole operators $T^{(n-1)}_0$ and $T^{(n)}_0$ are given by the projection operators onto the corresponding configuration. What do these two operators represent physically? Their difference measures the total charge of the 4f shell which is an important observable. Their sum, however, is the identity operator in the Hilbert space of our model. We therefore expect the relaxation matrix of the monopole operators to have one eigenvalue zero. The nonvanishing eigenvalue is the relaxation rate Γ_0 of the 4f charge (D for "density") and the charge susceptibility is

$$\chi_D(z) = \chi_D(0)\, \frac{i\Gamma_D}{z+i\cdot\Gamma_D} \quad . \tag{22}$$

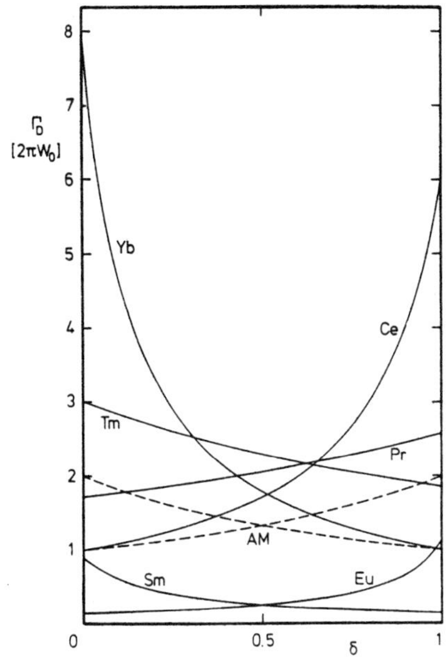

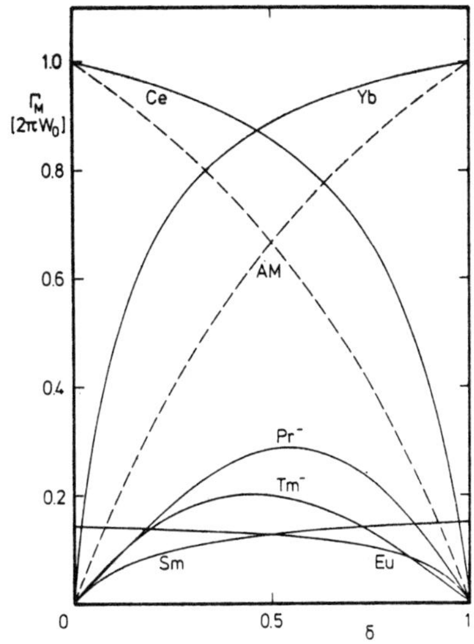

Fig. 2. Charge relaxation rates

Fig. 3. Magnetic relaxation rates

Charge relaxation rates are shown in Fig.2 for the six relevant rare earths together with those for the Anderson model (AM). There are two curves shown for the Anderson model corresponding to the average occupation of the localized nondegenerate orbital being smaller or larger than 1, respectively.

For all higher multipole operators we have to distinguish two cases: there can be either one or two such operators in the two configurations and corresponding to this there will be either one or two relaxation rates. Four of the relevant rare earths, namely Ce, Sm, Eu and Yb, are particularly simple with this respect, since one of their configurations has a nonmagnetic ground state (J=0). They have just one relaxation rate for all higher multipole operators and it turns out that this rate does not even depend on multipolarity. The physical relevance of the higher operators is obvious: the vector operator is proportional to the magnetic moment of the 4f shell, the second-rank tensor is proportional to the electric quadrupole moment, etc. We thus obtain one common relaxation rate Γ_M of magnetic moment, electric quadrupole moment, etc., for the four simple rare earths listed above. These rates are shown in Fig.3, again together with the magnetic relaxation rates of the Anderson model.

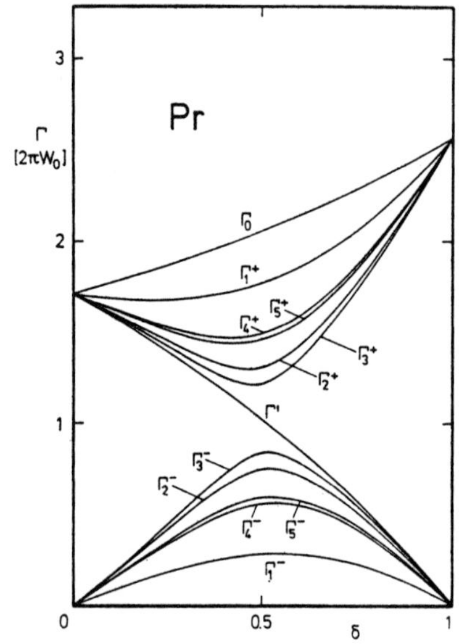

Fig. 4. Relaxation rates for Pr

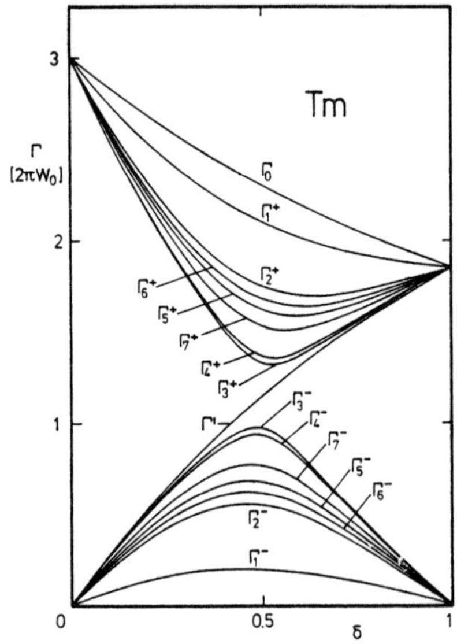

Fig. 5. Relaxation rates for Tm

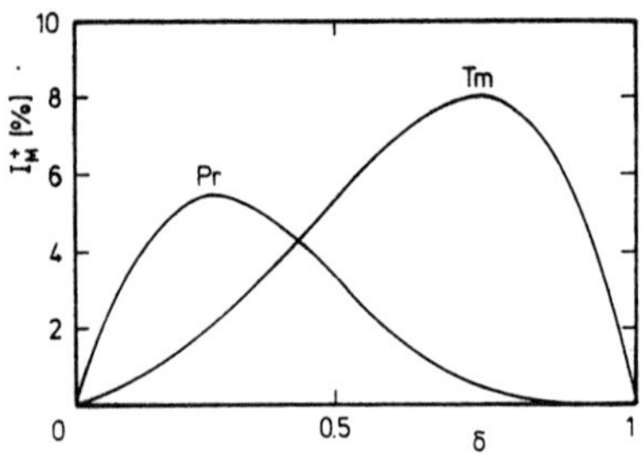

Fig. 6. Weight of the fast vector relaxation mode in the magnetic susceptibility

The situation is more complicated for Pr and Tm. Angular momenta of the Hund's rule ground states of these ions are $J_{n-1} = 5/2$, $J_n = 4$ and $J_{n-1} = 6$, $J_n = 7/2$, respectively. Therefore we find two relaxation rates for tensor operators of rank $\Lambda \leq 5$ (Pr) and $\Lambda \leq 7$ (Tm), respectively, which we call Γ^+_Λ and Γ^-_Λ. All operators of higher rank have one common relaxation rate Γ'. Figures 4 and 5 show all these rates for Pr and Tm.

Of course, relaxation of operators of high multipolarity is not really very relevant to measurable properties of compounds. Most important to know is the dynamics of vector operators which determines the magnetic dynamical susceptibility. From what was said above it is not surprising that the magnetic susceptibility of Pr and Tm compounds has two quasi-elastic lines and has the form

$$\chi_M(z) = \chi_M(0)\left[\frac{i\Gamma_1^+}{z+i\Gamma_1^+}\cdot I_M^+ + \frac{i\Gamma_1^-}{z+i\Gamma_1^-}\cdot I_M^-\right] \quad . \tag{23}$$

The intensities $I_M^\pm$ $(I_M^+ + I_M^- = 1)$ describe the weight with which the two eigenoperators of the relaxation matrix couple to the magnetic moment of the 4f shell. As shown in Fig.6, the fast vector relaxation mode contributes rather weakly to the magnetic susceptibility, never more than 8 per cent for Tm and less than 6 per cent for Pr.

4.2 Realistic Crystalline Symmetry

In a realistic crystalline environment the quasi-elastic poles of the dynamic susceptibilities described above are split into quasi-elastic and inelastic poles. At temperatures which are larger than crystal field splittings susceptibilities can again be written as a sum of Lorentzians:

$$\chi(z) = \chi(0) \sum_{\kappa \doteq -k}^{k} \frac{-\Delta_\kappa + i\Gamma_\kappa}{z-\Delta_\kappa + i\Gamma_\kappa}\cdot I_\kappa \quad . \tag{24}$$

(For Tm and Pr there are two such sums.) As usual, inelastic lines appear in pairs corresponding to excitation and deexcitation of crystal field states: $\Delta_{-\kappa} = -\Delta_\kappa, \Gamma_{-\kappa} = \Gamma_\kappa, I_{-\kappa} = I_\kappa$.

Without going into too many quantitative details for the various rare earths we shall briefly describe the main features using, if necessary, the simple example of Ce in an octahedral environment which has only a single crystal field transition $(\Gamma_7 \leftrightarrow \Gamma_8)$.

(a) There are two causes for crystal field splitting Δ which are contained in our theory: electrostatic fields and hybridization. In terms of (15) the first mechanism is represented by $\hat{\omega}^0$, the second one by $\hat{\omega}^{hybr}$. The first mechanism is in fact the conventional crystalline electric field, whereas the second gives a covalent contribution to the splitting. It is proportional to the anisotropic parts of the hybridization integrals, as given by (19) for octahedral symmetry. What is striking about this covalent crystal field is that in mixed valence compounds it turns out to have an intrinsic dependence on temperature. In the case of Ce our lowest order result is

$$\Delta = 6W_4 \ln[D(\delta)]/T \quad . \tag{25}$$

(D denotes an energy related to the effective valence band width.) The logarithmic variation with temperature arises because hybridization removes the configuration degeneracy. The infrared divergence has the same origin as in the one-particle self-energy [11].

(b) Line widths of the various lines are modified due to anisotropic hybridization integrals. The modification is such that the average line width is unaltered:

$$\sum_{\kappa=-k}^{k} \Gamma_\kappa \cdot I_\kappa = \Gamma_{isotr} \quad . \tag{26}$$

Only the overall factor W_0, not the variation with δ is altered. In the case of Ce, e.g., we obtain

$$\Gamma_{qu.-el.} \propto W_0 + \frac{32}{31} W_4 \quad , \quad \Gamma_{inel.} \propto W_0 - W_4 \quad . \tag{27}$$

(c) The Hund's rule ground state of Eu^{2+} has vanishing orbital angular momentum (s-state ion) and no crystal field splitting. (Eu^{3+} is nonmagnetic.) In fact, only tiny cubic crystal fields ($\lesssim 100$ mK) are observed in stable Eu^{2+} ions which have their origin in relativistic corrections to Hund's rules and in hybridization effects. The latter are tiny because the excitation of neighboring configurations provides large energy denominators (several eV). The dependence on these energy denominators is very strong since the splitting is of high order in spin-orbit interaction [12].

In mixed valence compounds there are no large energy denominators. We therefore obtain covalent crystal field splittings for Eu compounds which have the same size as for Sm compounds. In an octahedral environment the transitions $\Gamma_6 \rightarrow \Gamma_8$ and $\Gamma_7 \rightarrow \Gamma_8$ couple to the magnetic susceptibility. For these transitions we obtain

$$\Delta_{68} = -\frac{36}{77}(W_4 - W_6) \ln\left[\underset{68}{D(\delta)}\right]/T \quad , \quad \Delta_{78} = \frac{4}{77}(15W_4 + 7W_6) \ln\left[\underset{78}{D(\delta)}\right]/T \quad . \tag{28}$$

5. Discussion of Results

One has to be aware that the expansion with respect to hybridization used in the last section is a high-temperature expansion in the sense that, a priori, its validity is limited by the condition $T \gg W_0$. Despite the limitation the results of this simple calculation are surprisingly rich and useful. In this section we wish to point out some aspects of the results which we believe to be important and discuss some simple implications.

(a) Relaxation of 4f shells in mixed valence compounds is dominated by charge fluctuation processes. One might have thought that these processes are the source of a single relaxation rate. Our results show that this is an incorrect expectation. In particular, relaxation rates for charge and magnetic moment may differ by large numerical factors. Charge relaxation is always faster than magnetization relaxation.

(b) The influence of Hund's rule correlations on relaxation rates is very strong. For Ce and Yb (for which there is no exchange) deviations from the Anderson model are due to the large angular momenta of the real ions in the magnetic configuration. This explains why in the stable magnetic limit rates are larger by factors of 6 and 8, respectively, than in the Anderson model. For the other four ions exchange effects lead to a pronounced reduction of relaxation rates. This happens because two-body correlations in the 4f shell reduce the effectiveness of the relaxation mechanism.

(c) Magnetic relaxation rates are directly measured by neutron scattering. In mixed valence compounds they are almost independent of temperature (except for Tm Se) [13]. We take this as an indication that higher order corrections to our calculation are not extremely important; such corrections would lead to variation of the rates with temperature since the perturbation expansion is in terms of W_0/T (leave aside factors of $\ln(D)/T$). Comparing Γ_M with experiments we can deduce quantitative information about hybridization integrals W_0. It seems that for the understanding of relaxation rates the variation of Γ_M/W_0 which we have calculated is more important than the variation of W_0. If this is generally true, the line widths of Sm and Eu compounds which have not yet been measured will come out conspicuously small.

(d) Knowing W_0 from experiment we are able to calculate charge relaxation rates Γ_D quantitatively. These rates cannot be measured directly, but are important for understanding some other dynamical properties of mixed valence compounds. Most obvious is the relevance of charge relaxation for phonon anomalies. Quantitative comparison of charge relaxation rates with phonon frequencies explains why mixed valence compounds do not always show phonon anomalies and helps to select compounds which do. Let us look at two compounds, $Sm_{1-x}Y_xS$ and $CeSn_3$, which do [14] and do not show [15] phonon anomalies, respectively. The valence of Sm in $Sm_{1-x}Y_xS$ is around 2.5-2.7 ($\delta_{Sm} \approx 0.3$-0.5), and the valence of Ce in $CeSn_3$ is close to 3 ($\delta_{Ce} = 0.9$). From Fig.7 it is obvious that charge relaxation rates of these two compounds can be expected to differ by an order of magnitude. For $CeSn_3$ we know from neutron scattering [16] that $\Gamma_M = 25$ meV, thus Γ_D is much faster than phonons and there should be no phonon anomalies. We do not know Γ_M for $Sm_{1-x}Y_xS$ precisely, but as dis-

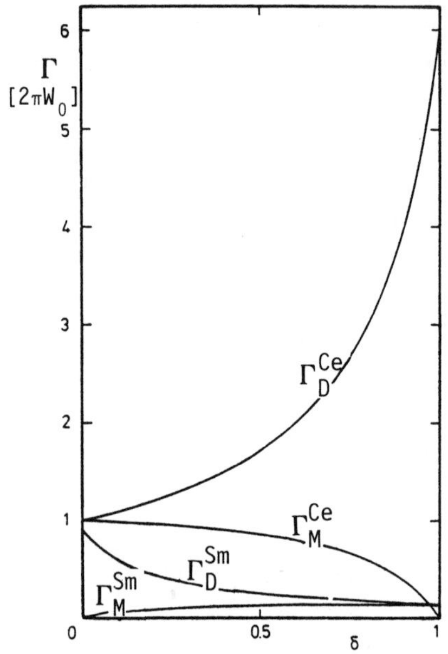

Fig. 7. Relaxation rates for the discussion of phonon anomalies

cussed in (c), Fig.7 is probably showing the right proportions. Due to the small relaxation rate lifetime effects which have not been included into the theory [17] can be expected to be unimportant for $Sm_{1-x}Y_xS$.

(e) In the limit of stable valence ($\delta \to 0$ or $\delta \to 1$) our lowest order relaxation rates drop to zero in all cases in which the static susceptibility in this limit remains finite (see, in particular, Fig.3). This only indicates that lowest order processes no longer dominate in such cases. There will be a cross-over to a Korringa- or Kondo-like relaxation which keeps rates finite in the stable valence limit, but which cannot be represented in a one-parameter plot like Fig.3.

(f) The compound TmSe is the only available mixed valence compound in which two quasi-elastic lines are expected to exist in the magnetic susceptibility. One such line (Γ_M=7 meV) has been found in neutron scattering at temperatures above 100 K [18]. According to our calculation the second line should be about 60 meV broad and should contain only about 6 per cent of the quasi-elastic intensity. It is not at all surprising that the second line has not been seen.

(g) At low temperatures TmSe has exceptional properties which are not found in any other mixed valence compounds investigated: (i) It orders magnetically at a few Kelvin [19] and (ii) its quasi-elastic magnetic line width is given by $\Gamma_M \approx k_BT$ below 100 K [18]. One can think of several reasons for this unusual behavior. One might get the idea both (i) and (ii) are due to interaction effects between different Tm ions, as is suggested by the philosophy of [20]. This possibility is convincingly excluded by recent experiments [21] which show the same line width for alloys $La_{1-x}Tm_xSe$ in the dilute limit (x=0.05). This experiment shows conclusively that the Korringa-like low-temperature line width (ii) is a single ion effect. Another origin of (ii) might lie in higher order corrections to the simple calculation discussed here. Although this possibility cannot be easily excluded, it does not seem particularly probable that higher order corrections (in terms of hybridization) should be so much different for Tm than for the other rare earth. A third way of approaching (ii) is to consider the influence of screening on the relaxation rates [22]. The idea is that excitonic correlations (e.g. 4f-hole-5d-electron attraction) may create a bottleneck for vector relaxation. This implies that some other mixed valence compounds, in particular Sm compounds, may show property (ii), too.

(h) Crystal field splitting has been observed in two mixed valence compounds only [23,24]. This is, in fact, not surprising. In most mixed valence compounds crystal field splittings appear to be smaller than the large intrinsic line widths caused by charge fluctuation processes. Crystal fields can be resolved, if the line width happens to be small enough. Our calculation should help to find favorable situations for observing a splitting.

(i) The origin of crystal field splittings in metals is not well understood. In particular, it is not at all clear what the relative importance of electrostatic and covalent contributions is. Electrostatic contributions are difficult to calculate, because they are quite sensitive to screening. Covalent contributions are also difficult to calculate, because they depend sensitively on the anisotropy (19) of the hybridization. In mixed valence compounds covalent contributions to the splitting according to (25) show a striking variation with temperature. This coincides with the observations in TmSe [23] and $Ce_{0.9-x}La_xTh_{0.1}$ [24]. A unique opportunity opens to distinguish experimentally between covalent and electrostatic contributions: From the dependence on temperature one can directly read off anisotropic hybridization integrals, e.g. W_4 in (25).

(j) The sizable reduction of the line width of $Ce_{0.9-x}La_xTh_{0.1}$ [24] as compared to $Ce_{0.9}Th_{0.1}$ cannot be due to a decrease of hybridization. If it was, the temperature-dependent splitting (25) would have been reduced in proportion to the line width and would never have been observed. The small line width may have been caused by a change of valence (see Fig.3).

(k) The apparent, though small, variation of quasi-elastic line widths with temperature [13] may, at least in part, be due to the variation of the splitting of unresolved crystal field levels.

(l) As can be seen from comparing (28) and (25) covalent crystal field splitting in Eu compounds in comparison to Ce compounds is reduced by the same correlation factor as the line widths (Fig.3). The same is true for Sm compounds. Thus, it is not easier to resolve covalent crystal field splitting in Eu and Sm compounds than in Ce compounds. But Sm compounds also have electrostatic crystal field splitting which is not reduced by the correlation factor. Therefore, mixed valence compounds of Sm provide the best opportunity to identify electrostatic crystal fields.

6. Outlook

We have tried to demonstrate that a proper description of correlations within the 4f shell leads to a remarkable degree of insight into the physics of mixed valence compounds. In the context of what has been discussed in this talk the most outstanding unsolved problem is the low-temperature behavior of TmSe.

The approach applied here to dynamical susceptibilities can be applied to other physical properties as well. We are presently in the process of calculating one-particle propagators with realistic intraionic correlations. In this calculation we are aiming at extending the lowest order treatment of the self-energy self-consistently to low temperatures.

References

1. M. Goeppert-Mayer: Phys. Rev. *60*, 184 (1941);
   K.M. Saxena, G. Malli: Phys. Rev. A*3*, 1278 (1971)
2. A.J. Freeman, D.D. Koelling: Physica B-C *86-88*, 16 (1977)
3. See, e.g., J.M.D. Coey, O. Massenet: *Valence Instabilities and Related Narrow-Band Phenomena*, ed. by R.D. Parks (Plenum, New York 1977) p.211
4. For recent reviews see J.M. Jefferson, K.W. Stevens: J. Phys. *C11*, 3919 (1978);
   J.M. Robinson: Phys. Rep. *51*, 1 (1979);
   N. Grewe, H.J. Leder, P. Entel: Festkörperprobleme *XX*, 413 (1980)
5. Y. Kuramoto: Z. Phys. B *37*, 299 (1980)
6. P.W. Anderson: Phys. Rev. *124*, 41 (1961);
   H.J. Leder, B. Mühlschlegel: Z. Phys. B *29*, 341 (1978)

7. L.L. Hirst: Phys. kond. Mat. *11*, 255 (1970)
8. J.-N. Chazalviel, M. Campagna, G.K. Wertheim, P.H. Schmidt, L.D. Longinotti: Physica *86-88* B+C, 237 (1977)
9. R. Zwanzig: In *Lectures in Theoretical Physics*, Vol.3 (Interscience, New York 1961);
H. Mori: Progr. Theoret. Phys. *34*, 423 (1965)
10. See, e.g., D. Forster: *Hydrodynamic Fluctuations, Broken Symmetry, and Correlation Functions* (Benjamin, London, Amsterdam 1975) Chap.5
11. A.F. Barabanov, K.A. Kikoin, L.A. Maksimov: Solid State Commun. *15*, 977 (1974);
E. Müller-Hartmann: Solid State Commun. *31*, 113 (1979)
12. S.E. Barnes, K. Baberschke, M. Hardiman: Phys. Rev. B *18*, 2409 (1978)
13. E. Holland-Moritz, M. Loewenhaupt, W. Schmatz, D. Wohlleben: Phys. Rev. Lett. *38*, 983 (1977);
S.M. Shapiro, J.D. Axe, R.J. Birgeneau, J.M. Lawrence, R.D. Parks: Phys. Rev. B *16*, 2225 (1977);
E. Holland-Moritz, D. Wohlleben, M. Loewenhaupt: J. Phys. Coll. *39*, C6-835 (1978);
E. Holland-Moritz, H. Loewenhaupt: J. Phys. Coll. *40*, C5-359 (1979);
M. Loewenhaupt, E. Holland-Moritz: J. Appl. Phys. *50*, 7456 (1979);
E. Holland-Moritz: Jül-Spez-14 (1978);
W.C.M. Mattens, F.R. de Boer, A.P. Murani, G.H. Lander: J. Magn. and Magn. Mat. *15-18*, 973 (1980)
14. H.A. Mook, R.M. Nicklow, T. Penney, F. Holtzberg, M.W. Shafer: Phys. Rev. B *18*, 2925 (1978)
15. L. Pintschovius, E. Holland-Moritz, D. Wohlleben, S. Stähr, J. Liebertz: Solid State Commun. *34*, 953 (1980)
16. E. Holland-Moritz: Jül-Spez-14 (1978)
17. P. Entel, N. Grewe, M. Sietz, K. Kowalski: Phys. Rev. Lett. *43*, 2002 (1979);
H. Bilz, G. Güntherodt, W. Kleppmann, W. Kress: Phys. Rev. Lett. *43*, 1998 (1979)
18. M. Loewenhaupt, E. Holland-Moritz: J. Magn. Magn. Mat. *9*, 50 (1978)
19. H. Bjerrum-Møller, S.M. Shapiro, R.J. Birgeneau: Phys. Rev. Lett. *39*, 1021 (1977);
M. Loewenhaupt, E. Holland-Moritz: J. Appl. Phys. *49*, 2124 (1978)
20. C.M. Varma: Solid State Commun. *30*, 537 (1979)
21. E. Holland-Moritz, H. Scheuer: private communication
22. Y. Kuramoto: private communication
23. E. Holland-Moritz, M. Loewenhaupt: J. Phys. Coll. *40*, C5-359 (1979)
24. B.H. Grier, S.M. Shapiro, C.F. Majkrzak, R.D. Parks: Phys. Rev. Lett. *45*, 666 (1980)
25. C.A. Balseiro, A. Lopez: Solid State Commun. *17*, 1241 (1975)
26. M.E. Foglio: J. Phys. C*11*, 4171 (1978)
27. C.E.T. Gonçalves da Silva: Phys. Rev. Lett. *42*, 1305 (1979)

Effect of Strong Correlation on a Mixed Valence System

Kazuo Ueda
Department of Applied Physics, University of Tokyo, Tokyo, 113, Japan

1. Introduction

In recent years a great deal of research has been done on some rare-erath metals and their compounds which show the mixed valence or valence fluctuating phenomena [1]. Properties of this kind of materials are full of variety from material to material. Among them, in this paper we consider the case where one configuration is magnetic, e.g. $Sm^{3+}(^6H_{5/2})$ and $Ce^{3+}(^2F_{5/2})$, and the other is nonmagnetic, e.g. $Sm^{2+}(^7F_0)$ and $Ce^{4+}(^1S_0)$. Magnetic property of this class of materials has a characteristic feature. At high temperatures susceptibility is expressed by the formula

$$\chi = C_I\chi_I + C_{II}\chi_{II}, \tag{1}$$

where $\chi_I(\chi_{II})$ is the susceptibility expected for one valence state and $C_I(C_{II})$ is its fraction. In a word, high-temperature property is characterized by a local-moment-type behavior. On the other hand, at low temperatures the susceptibility tends to a constant and no magnetic order is observed.

1.1 Hamiltonian

To consider above-mentioned magnetic property and related thermodynamical ones, we make use of the periodic array of the Anderson model (Anderson lattice) [2]:

$$H = H_0 + H'(U),$$
$$H_0 = \sum_{k\sigma}\varepsilon_k d^+_{k\sigma}d_{k\sigma} + E_0\sum_{i\sigma} f^+_{i\sigma}f_{i\sigma} + \frac{V}{\sqrt{N_0}}\sum_{ik\sigma}(e^{-ikR_i}f^+_{i\sigma}d_{k\sigma} + e^{ikR_i}d^+_{k\sigma}f_{i\sigma}), \tag{2}$$
$$H'(U) = U\sum_i f^+_{i\uparrow}f_{i\uparrow}f^+_{i\downarrow}f_{i\downarrow},$$

where N_0 is the number of atoms, the orbital degeneracy of d and f states is neglected for simplicity and the Coulomb repulsion U and the hybridization V are assumed to be local.

As a typical example we assume two electrons per atom. If E_0 lies below the center of the d band and E_0+U above, Fig.1a, the system lies in an integral valence regime. If E_0 is raised through the center, Fig.1b, or E_0+U is lowered, Fig.1c, a mixed valence state is realized. These two mixed valence states are equivalent because of an electron-hole symmetry. In the most qualitative sense, we may consider that the mixed valence states of Ce and its compounds can be simulated by Fig.1b and those of SmS and SmB_6 by Fig.1c. For SmS the integral valence state can be regarded as the doubly occupied state, instead of the singly occupied integral valence state shown in Fig.1a.

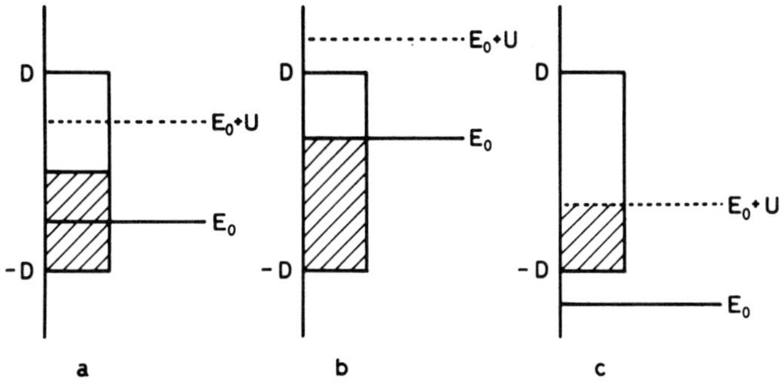

Fig.1 Integral versus mixed valence regimes

1.2 Hartree-Fock, RPA Theory

One of standard methods to deal with a many-body Hamiltonian like the Anderson lattice is a mean-field theory. LEDER and MÜHLSCHLEGEL [3] studied susceptibility and specific heat by the Hartree-Fock (HF) approximation and RISEBOROUGH and MILLS [4] investigated dynamical susceptibility by the random phase approximation (RPA)[1]. Here we review the HF-RPA theory. The dynamical susceptibility of the total system consists of

$$\chi(q,\omega) = \chi_{dd}(q,\omega) + \chi_{df}(q,\omega) + \chi_{fd}(q,\omega) + \chi_{ff}(q,\omega), \tag{3}$$

where $\chi_{df}(q,\omega)$, for example, is the response function defined by the ratio of induced moment of the d band to applied field on the f band. In the mean-field approximation, the following relations are obtained,

$$\chi_{dd} = \chi^{o}_{dd} + \chi^{o}_{df}\frac{U}{2N_o}\chi_{fd}, \tag{4.a}$$

$$\chi_{df} = \chi^{o}_{df} + \chi^{o}_{df}\frac{U}{2N_o}\chi_{ff}, \tag{4.b}$$

$$\chi_{fd} = \chi^{o}_{fd} / [1 - \frac{U}{2N_o}\chi^{o}_{ff}], \tag{4.c}$$

$$\chi_{ff} = \chi^{o}_{ff} / [1 - \frac{U}{2N_o}\chi^{o}_{ff}], \tag{4.d}$$

1 The expressions for the dynamical susceptibility in [4] are somewhat different from and more complicated than (3)-(5). The discrepancy arises from that in [4] the mean-field term is not included in the non-interacting part of the Hamiltonian.

where the arguments (q,ω) are suppressed and χ^o_{dd}, χ^o_{df}, χ^o_{fd}, and χ^o_{ff} are the response functions of the non-interaction system except that E_o is replaced by $E_o+U\langle f^{\dagger}_{i-\sigma}f_{i-\sigma}\rangle$. Inserting (4.a)-(4.d) into (3), we get

$$\chi = \chi^o_{dd} + \chi^o_{df} + \chi^o_{fd} + \chi^o_{ff} + \frac{U}{2N_o}(\chi^o_{df} + \chi^o_{ff})^2 / [1 - \frac{U}{2N_o}\chi^o_{ff}]. \tag{5}$$

The HF-RPA theory is justified when V, more exactly the average kinetic energy of the f electrons, is larger than U. For the present problem the position in the parameter space is in the opposite limit: U is much greater than V. The drawbacks of the HF-RPA theory are typically seen in the following fact. The HF-RPA theory predicts a magnetic state with very high transition temperature for almost all reasonable range of parameters and cannot reproduce the correct paramagnetic Curie susceptibility.

1.3 Organization of This Paper

Owing to the strong Coulomb repulsion among f electrons and the small hybridization between f and conduction (d) electrons, f electrons can move keeping away from each other with small loss of kinetic energy. This effect of correlation should naturally lead to the correct Curie susceptibility and a significant reduction of the effective Coulomb repulsion. If the system deviates from the integral valence the loss of the kinetic energy becomes smaller and the paramagnetic state will be stabilized for a reasonable range of parameters. Very recently the present author has developed a simple theory which takes account of this effect [5,6]. We deal with the free energy F(M,N;T) as a function of total magnetization and electron number. In the next section we summarize the method of approach together with some discussions on the nature of the approximation. On the basis of the free energy obtained, we discuss the thermodynamical properties such as spin susceptibility and specific heat in section 3. Then we apply the method to an extended Anderson lattice model which takes into account a direct f-f transfer. Final section is devoted to summary of conclusions and several discussions. A possible improvement on the simple theory is also discussed there.

2. Free Energy

2.1 Approximation Method

To calculate the free energy F(M,N;T) as a function of the total magnetization M and the total electron number N, we introduce the magnetization M_f and the electron number N_f of the f electrons as explicit variables:

$$F(M,N;T) = \mathcal{F}(M,N;M_f,N_f;T), \tag{6}$$

with M_f and N_f to be determined from

$$\frac{\partial}{\partial M_f}\mathcal{F}(M,N;M_f,N_f:T) = 0, \tag{7}$$

$$\frac{\partial}{\partial N_f}\mathcal{F}(M,N;M_f,N_f;T) = 0. \tag{8}$$

Throughout the present paper we assume g=2 for both d and f states and measure M and M_f in units of μ_B. Although N_f and M_f are not constants of motion, the above procedure is exact because they are macroscopic variables. By making use of the coupling constant integration, we have

$$F(M,N;M_f,N_f;T) = F_0 + F_U, \tag{9}$$

$$F_U = \int_0^U dU \left\langle \frac{1}{U} H'(U) \right\rangle_{CU}. \tag{10}$$

where F_0 is the free energy for U=0 and $\langle \cdots \rangle_{CU}$ denotes the conditional statistical average under fixed values of M, N, M_f and N_f and U. If the integrand of (10) is approximated by the value of the non-interacting system (U=0), the HF free energy is obtained

$$F_{HF} = F_0 + \frac{U}{4N_0}(N_f^2 - M_f^2). \tag{11}$$

As we have pointed out, the HF theory is justified for weak coupling.

Our main idea to develop a theory suitable for the present strong correlation case is to evaluate (10) in the limit V=0 first. Because M_f and N_f are specified, we have the well-defined limit of V=0 and F_U may be divided as

$$F_U = \int_0^U dU \left\langle \frac{1}{U} H'(U) \right\rangle_{CU,V=0} + \int_0^U dU \left\{ \left\langle \frac{1}{U} H'(U) \right\rangle_{CU} - \left\langle \frac{1}{U} H'(U) \right\rangle_{CU,V=0} \right\}. \tag{12}$$

The first term is exactly integrable and is given by

$$F_U^{(0)} = F_{fU}(M_f,N_f;T) - F_{f0}(M_f,N_f;T), \tag{13}$$

where F_{fU} is the atomic limit free energy for the f band with Coulomb repulsion U, U=0 for F_{f0}. It is a very difficult task to evaluate the second term faithfully. To estimate it the following observations are helpful. The integrand, which is given by the difference between the probabilities of the doubly occupied state with and without V, becomes smaller as the temperature is raised. At a given temperature it is the largest for the paramagnetic

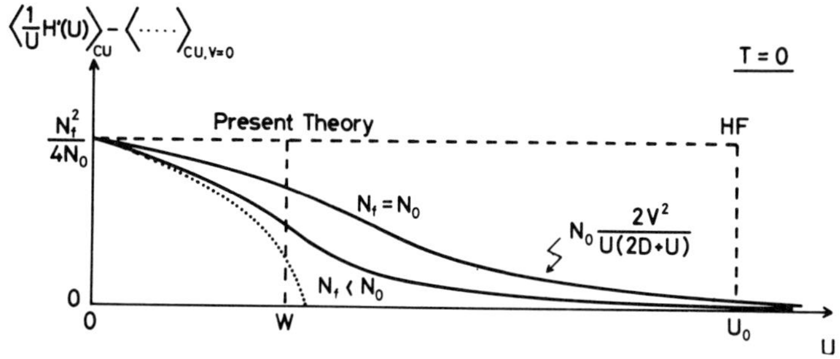

Fig.2 Schematic behavior of the integrand as a function of U in the paramagnetic state at T=0. The bare interaction is denoted by U_0 in this figure. The HF theory and the present one use the approximated integrands shown by the dashed lines. The dotted line is the integrand evaluated by using the RPA dynamical susceptibility (section 5).

state. So we examine the behavior of the integrand in the paramagnetic state at T=0, Fig.2. First we consider the integral valence case ($N_f=N_o$). We can show that it starts from $N_f^2/4N_o$ and decreases linearly with U. For U greater than V^2/D, 2D being the width of the d band, a perturbation theory with respect to the mixing term shows that the integrand falls off according to

$$2N_o V^2 / [U(2D+U)]. \tag{14}$$

In the mixed valence regime the probability of the double occupation should fall much faster with U. Thus we may expect that the second term can be estimated by introducing a certain cutoff W of the order of the effective f-band width. After a further simplification we have

$$\begin{aligned} F_U^{(1)} &\cong \int_0^W dU\{<\frac{1}{U}H'(U)>_{CU} - <\frac{1}{U}H'(U)>_{CU,V=0}\} \\ &\cong \int_0^W dU\{<\frac{1}{U}H'(U)>_{CO} - <\frac{1}{U}H'(U)>_{CU,V=0}\} \\ &= \frac{W}{4N_o}(N_f^2 - M_f^2) - F_{fW}(N_f,M_f;T) + F_{fo}(N_f,M_f;T). \end{aligned} \tag{15}$$

By putting together (9), (13) and (15), the following expression is obtained

$$F = F_o + F_{fU} - F_{fW} + \frac{W}{4N_o}(N_f^2 - M_f^2). \tag{16}$$

It should be mentioned that our approximation procedure, which introduce the cutoff W, is not applicable for all cases. However, it is important to note that the integral valence case lies in the very limit where the cutoff is not justified: Eq.(14) shows that the integrand falls off as 1/U for $V^2/D<U<D$. On the other hand, for low-density cases ($N_f \ll N_o$ or $2N_o - N_f \ll N_o$) the above procedure is justified. Our method can be considered as an approach from the low-density side. Although the present author has not been able to show to what extent the present method is applicable, he expects that it holds in a considerably wide range.

To evaluate W quantitatively, the T-matrix (Kanamori) approximation [7] can be used, which is justified in the case of the low density. The explicit expression for W is given in [5].

2.2 Free Energies in the Restricted Subspace

To calculate (16) explicitly, it is necessary to evaluate the free energies in the restricted subspace where M, N, M_f and N_f are given. This can be done by using grand canonical ensembles. The non-interacting part can be calculated as

$$F_o(M,N;M_f,N_f;T) = \Omega_o(B,\mu;B_f,\mu_f;T) + BM + \mu N + B_f M_f + \mu_f N_f, \tag{17}$$

where B, μ, B_f and μ_f are determined by

$$M = \frac{\partial\Omega_0}{\partial B}\,, \quad N = -\frac{\partial\Omega_0}{\partial\mu}\,, \quad M_f = -\frac{\partial\Omega_0}{\partial B_f}\,, \quad N_f = -\frac{\partial\Omega_0}{\partial\mu_f}\,. \tag{18}$$

In the same way we have

$$\mathcal{F}_{fU}(M_f,N_f;T) = \Omega_{1U}(B_{1U},\mu_{1U};T) + B_{1U}M_f + M_{1U}N_f, \tag{19}$$

with

$$M_f = -\frac{\partial\Omega_{1U}}{\partial B_{1U}}\,, \quad N_f = -\frac{\partial\Omega_{1U}}{\partial\mu_{1U}}\,. \tag{20}$$

The same expression with W instead of U is obtained for $\mathcal{F}_{fW}$. The conditions to determine M_f (7) and N_f (8) now read as

$$B_f + B_{1U} - B_{1W} - \frac{W}{2N_0} M_f = 0, \tag{21}$$

$$\mu_f + \mu_{1U} - \mu_{1W} + \frac{W}{2N_0} N_f = 0. \tag{22}$$

The same conditions for the HF theory are

$$B_f - \frac{U}{2N_0} M_f = 0, \tag{23}$$

$$\mu_f + \frac{U}{2N_0} N_f = 0, \tag{24}$$

which are nothing but the HF self-consistent equations.

2.3 Ground State

To see some aspects of the present approximation, we discuss ground-state properties predicted by it. To be explicit we assume a constant density of states $\rho=1/2D$ for the d band and consider the case of two electrons per atom, $N=2N_0$. The results are summarized in Fig.3. A semiconducting paramagnetic solution always exists irrespective of E_0 as a locally stable state. For $-U+W/2<E_0<-W/2$, the f electron number is fixed to N_0 and the chemical poten-

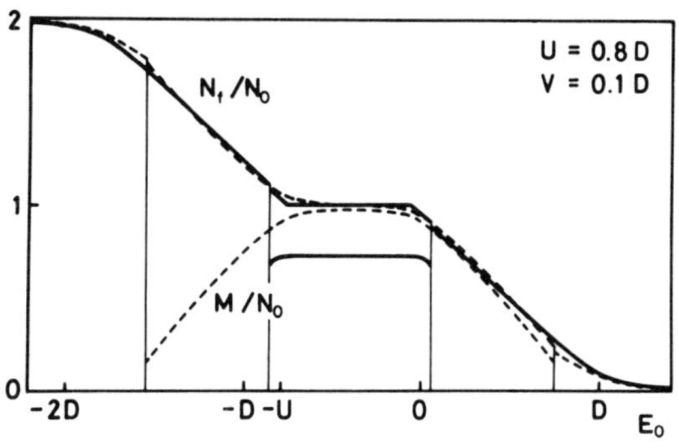

Fig.3 The f electron number N_f and the magnetization M of the ground state. The results obtained by the present theory are shown by the solid lines and the HF results by the dashed lines. The vertical lines denotes the boundary between the ferro- and paramagnetic states

tial is pinned at the center of the conduction band. If E_0 is raised or lowered from the range, a mixed valence state is realized. If $E_0>-W/2$ the chemical potential is near E_0 and if $E_0<-U+W/2$ it is near E_0+U. Near the range of E_0, where the valence is integral, a ferromagnetic solution also exists. Of course the true magnetic state will, in general, be a helical or antiferromagnetic state. Energy of the paramagnetic and the ferromagnetic states is calculated in [5]. In Fig.3 we have shown the state which has lower energy. As the paramagnetic state given by the present theory is semiconductive the transition between the para- and ferromagnetic states is first order in character. The HF results are also shown in the figure. We may see that the paramagnetic state is stabilized in a reasonable range of parameters in the present theory in contrast with the HF theory.

It may be of some interests to examine the pseudo-particle density of states, which is given by the one-particle energy level defined by

$$\varepsilon_N = F(M=0,N;T=0) - F(M=0,N-1;T=0), \tag{25}$$

where we have assumed paramagnetic state. Fig.4 shows schematically the pseudo-density of states. In all cases, there are two large peaks near E_0 and E_0+U whose character is f like. Each peak can receive 1/2 f electron per spin. In the mixed valence regime the hybridization gap splits one of the two peaks into two parts. The chemical potential lies in the hybridization gap.

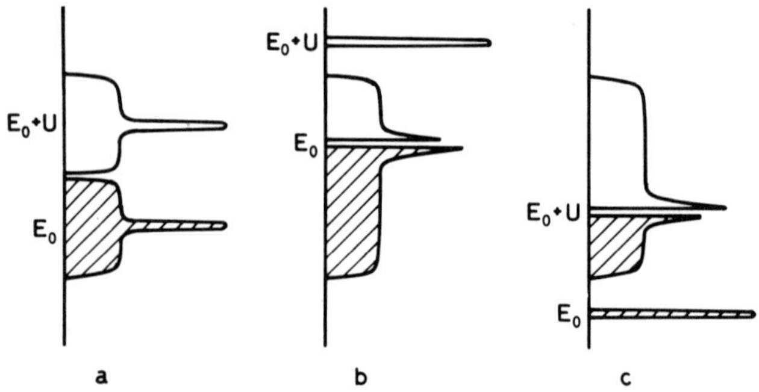

Fig.4 Pseudo-particle density of states in the paramagnetic state at T=0 (schematic)

Some remarks are necessary for the paramagnetic states. The reason why the present theory predicts always the semiconducting state for the paramagnetic one is that it treats the self-energy as k independent. As we have pointed out, the present theory is justified for the low-density case. When N_f is near N_0, whether the paramagnetic state is semiconductive or not is still under debate. We note that the present theory is not sufficiently accurate to draw any definite conclusion. More elaborate theoretical works are necessary. One of such studies is reported in this conference by YAMADA and YOSIDA [8]. In any case the semiconducting state is rather special. If we consider the case $N=N_0$ or take into account degeneracy of the conduction band, the ground state will, in general, be a highly correlated Fermi liquid. Furthermore, even if $N=2N_0$ and if the conduction band is non-degenerate, the ground state can be metallic when we consider a small direct f-f transfer, because the hybridization gap is very narrow. The latter case is discussed in section 4.

3. Thermodynamical Properties

3.1 Spin and Charge Susceptibilities

Now we discuss thermodynamical properties in the paramagnetic mixed valence state on the basis of the free energy obtained in the preceding section. The spin susceptibility per atom is defined by

$$\frac{1}{\chi} = N_o \left.\frac{\partial^2 F(M,N;T)}{\partial M^2}\right|_{M=0} . \tag{26}$$

The spin susceptibility is a measure of spin fluctuations. In the mixed valence state charge fluctuations also are expected to be considerable. The charge susceptibility which is a measure of the charge fluctuations is defined by

$$\frac{1}{\eta} = N_o \frac{\partial^2 F(M=0,N;T)}{\partial N^2} . \tag{27}$$

After some calculations we have

$$\chi = L_o + \left(\frac{W}{2} + \chi_{fW}^{-1} - \chi_{fU}^{-1}\right) \frac{L_1^2}{1 - \left(\frac{W}{2} + \chi_{fW}^{-1} - \chi_{fU}^{-1}\right)(L_2 + L_3)} , \tag{28}$$

$$\eta = L_o - \left(\frac{W}{2} - \eta_{fW}^{-1} + \eta_{fU}^{-1}\right) \frac{L_1^2}{1 + \left(\frac{W}{2} - \eta_{fW}^{-1} + \eta_{fU}^{-1}\right)(L_2 + L_3)} , \tag{29}$$

The expressions for L_o, L_1, L_2 and L_3 are given in [5]. Here we note that they are related to the susceptibilities of the non-interacting system:

$$L_o = \chi_{dd}^o + \chi_{fd}^o + \chi_{df}^o + \chi_{ff}^0 , \tag{30.a}$$

$$L_1 = \chi_{df}^o + \chi_{ff}^o , \tag{30.b}$$

$$L_2 + L_3 = \chi_{ff}^0 . \tag{30.c}$$

On the other hand χ_{fU} and χ_{fW} are the spin susceptibilities of the f electrons in the atomic limit with Coulomb repulsion U and W, respectively. In the same way $\eta_{fU}(\eta_{fW})$ is the charge susceptibility in the atomic limit. For comparison we show the HF results,

$$\chi_{HF} = L_o + \frac{U}{2} \frac{L_1^2}{1 - \frac{U}{2}(L_2 + L_3)} , \tag{5}$$

$$\eta_{HF} = L_o - \frac{U}{2} \frac{L_1^2}{1 + \frac{U}{2}(L_2 + L_3)} . \tag{31}$$

We may see that in the present theory the spin susceptibility is expressed by the effective exchange given by

$$\frac{W}{2} + \chi_{fW}^{-1} - \chi_{fU}^{-1},$$

while η is expressed by the effective Coulomb given by

$$\frac{W}{2} - \eta_{fW}^{-1} + \eta_{fU}^{-1},$$

which is different from the effective exchange. It should be noted that the difference between the present theory and the HF one is not limited to this. Although the expressions for L_0, L_1, L_2 and L_3 are the same for both theories, they contain different self-energy μ_f, which is given by (22) for the present theory and by (24) for the HF one.

3.2 Temperature Dependence

A characteristic temperature of the system is given by the effective f-band width. The width of virtual level $\Delta = \pi\rho V^2$ may be considered as the effective f-band width. As we have pointed out, W is of the same order as Δ. At low temperatures ($T<\Delta$) the susceptibilities obtained in the preceding subsection reduce to the HF type results with the effective interaction W. At high temperatures (28) and (29) reduce to

$$\chi \sim \chi_{fU} \cong \beta n_f, \tag{32}$$

$$\eta \sim \eta_{fU} \cong \beta n_f(1 - n_f), \tag{33}$$

where we have assumed $N_f < N_0$. The spin susceptibility at high temperatures is given by the average of the susceptibilities expected for two configurations, cf.(1). β is the Curie law expected for the f^1 configuration and $n_f = N_f/N_0$ is its fraction. Apparently f^0 configuration has no contribution. The charge susceptibility at high temperatures is also given by that of the atomic limit. It takes the largest value when the f band is quarter-filled. It is generally smaller than the spin susceptibility and the difference decreases as the system deviates from the half-filled case.

Specific heat also can be calculated from the free energy. To investigate its temperature dependence, it is convenient to discuss the entropy given by

$$S(N;T) = -\frac{\partial}{\partial T} F(N;T) = -\frac{\partial \Omega_0}{\partial T} - \frac{\partial \Omega_{1U}}{\partial T} + \frac{\partial \Omega_{1W}}{\partial T}. \tag{34}$$

The limiting values of each term is easily calculated [6] and the entropy of the system is

$$S(N;T) = \begin{cases} 0 & (T<<\Delta,W) \\ K_B \log({}_{N_0}C_{N_f} \cdot 2^{N_f}) & (\Delta,W<<T<<U) \\ K_B \log{}_{2N_0}C_{N_f} & (U<<T) \end{cases}. \tag{35}$$

As Δ and W are of the same order and small, we may expect a large Schottky type maximum at low temperatures. This specific heat is due to the appearance of N_f "local moments" which slowly change their position from site to site.

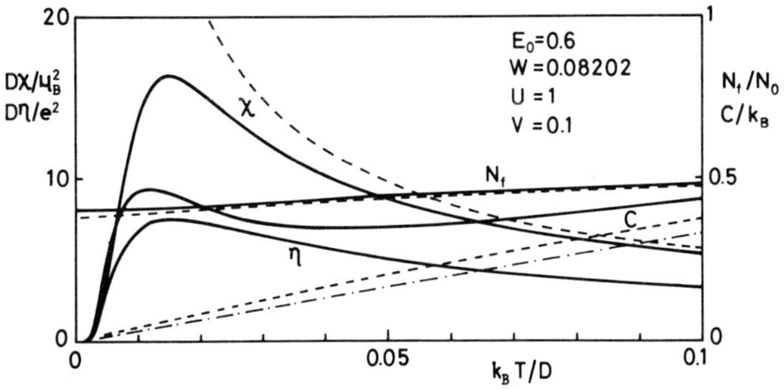

Fig.5 Temperature dependence of the physical quantities. The results obtained by the perturbation theory within the most dominant approximation are also shown by the dashed lines. The dot-dashed line is the T-linear specific heat of the d band.

Now we show a numerical example. Fig.5 shows N_f, χ, η and C for E_0=0.6D. In this example the constant density of states is used for the d band, U=D is assumed and W evaluated by the T-matrix approximation is used.

3.3 Perturbation Theory ($U\to\infty$)

To discuss high-temperature properties we may use the perturbation theory which treats the f-d mixing as the perturbation [9-13]. Within the most dominant approximation [13] the total electron number is given by

$$\frac{N}{N_0} = 1 + \frac{\mu}{D} + n_f, \tag{36}$$

where

$$n_f = 2/\{2 + \exp[\beta(E_0 + \lambda - \mu)]\}, \tag{37}$$

with

$$\lambda = \frac{V^2}{2D}\begin{cases} \ln |D/E_0-\mu| & (|E_0-\mu|>k_BT) \\ \ln \beta D & (|E_0-\mu|<k_BT) \end{cases}, \tag{38}$$

is the f electron number per atom. Within the same approximation we have

$$\chi = \beta n_f, \tag{39}$$

$$C/k_B = \frac{\pi^2}{3}\frac{k_BT}{D} + \{\ln[2(1-n_f)/n_f]\}^2 \frac{n_f(1-n_f)}{1+\beta D n_f(1-n_f)}, \tag{40}$$

where the second term of the specific heat originates from the T dependence of N_f and is almostly linear in T. The detailed discussions on the present results are found in [13]. Here we only note the following two points.

Firstly, the logarithmic shift of the self-energy λ, which is dominant in the impurity problem, plays only a minor role because the total electron number should be conserved according to (36). Secondly, the present results on χ and C are valid only at very high temperatures. Numerical results on N_f, χ and C are also plotted in Fig.5. For comparison the T-linear specific heat of the conduction band is also shown by the dot-dashed line.

4. Application to a Periodic Anderson Model with a Direct f-f Transfer

The paramagnetic state obtained by the theory developed in section 2 is semiconductive for $N=2N_0$. As we have stated there, whether the true ground state is semiconductive or not is a difficult problem except for the low-density case. The conclusion will depend on $E_0(N_f)$ and also on the structure of the d band. Leaving this problem for future study, here we consider, as an example which can be metallic, an extended periodic Anderson model with a direct f-f transfer term;

$$H = (2) + \sum_{i\neq j,\sigma} t_{ij} f^+_{i\sigma} f_{j\sigma} = (2) + \sum_{k\sigma} \varepsilon_{fk} f^+_{k\sigma} f_{k\sigma}. \tag{41}$$

The approximation method developed in section 2 is easily applied to this Hamiltonian. The difference is that V=0 in eq.(12) is replaced by V=0 and $t_{ij}=0$. The effect of the direct transfer appears in the non-interacting term F_0 and the value of W. To avoide mathematical complexity we assume

$$\varepsilon_{fk} = \alpha\varepsilon_k, \tag{42}$$

and a semielliptic density of states for the d band

$$\rho(\varepsilon) = \frac{2}{\pi D} [1 - (\frac{\varepsilon}{D})^2]^{1/2}. \tag{43}$$

Thus the density of states of the f band is also the semielliptic one with the width $2|\alpha|D$. In the following example we take $\alpha=0.02$ and measure energy in units of D.

The ground-state properties are summarized in Fig.6. In this figure all quantities are calculated by assuming the paramagnetic state and plotted as functions of an implicit variable $E_0-\mu_f$. As E_0 is lowered towards the center of the d band, the magnitude of the hybridization gap decreases. Region III is semiconductive and regions I and II are metallic. In the latter cases the susceptibility of the non-interacting system χ_0 is, of course, finite. At each E_0 or $E_0-\mu_f$, the effective interaction W is evaluated by the T-matrix approximation. The spin and the charge susceptibilities are calculated by using it. At the boundary between I and II the spin susceptibility diverges. Thus the paramagnetic metallic state is obtained in region II.

Some numerical examples of temperature dependence of the physical quantities are shown in Fig.7 and Fig.8. We take $E_0=0.6967$ ($E_0-\mu_f=0.7$) in Fig.7 and $E_0=0.5942$ ($E_0-\mu_f=0.6$) in Fig.8. The former parameter is close to the boundary of the metal-semiconductor transition. For comparison the local moment type susceptibility χ_{fU} is plotted by dot-dashed lines and the T-linear specific heat of the d band is also shown by dashed lines. At T=0 χ and n have finite values and C is proportional to T at very low temperatures.

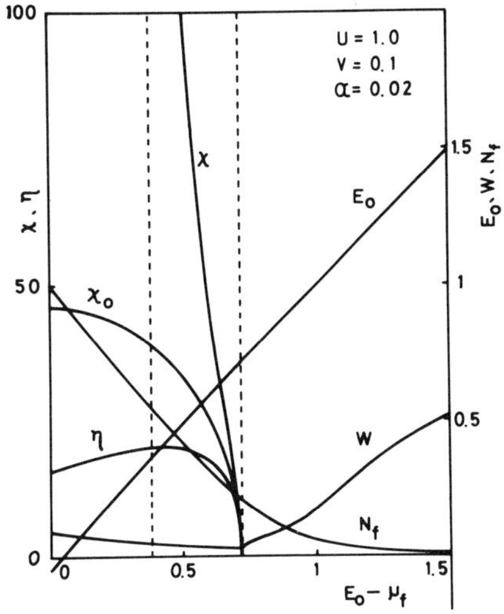

Fig.6 Ground state properties of the periodic Anderson model with the direct f-f transfer

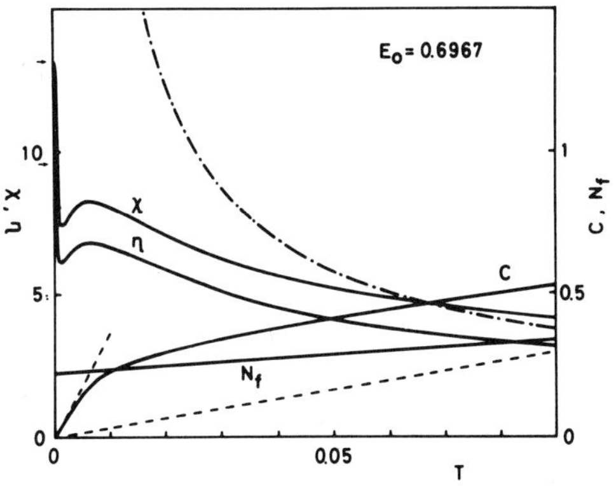

Fig.7 Temperature dependence of the physical quantities for E_0=0.6967

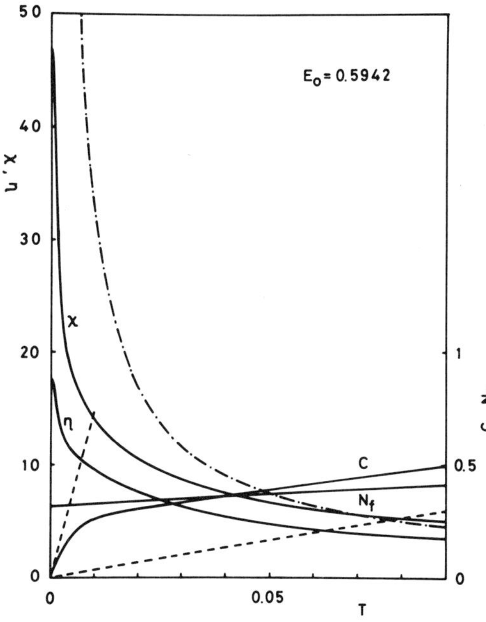

Fig.8 Temperature dependence of the physical quantities for E_0=0.5942

Within the present approximation the coefficient of the T-linear term is given by the total density of states of the non-interacting system, which is nothing but χ_0:

$$C = k_B^2 \frac{\pi^2}{3} T\chi_0 . \tag{44}$$

This low-temperature limiting behavior is also shown by dashed lines.

5. Summary and Discussions

The main features of the present theory may be summarized as follows.
(i) The paramagnetic mixed valence state is stabilized for reasonable range of parameters.
(ii) The thermodynamical properties at low temperatures are described as if the system has a reduced effective interaction.
(iii) The high-temperature properties are characterized by the local moment type behaviors. In particular χ and η obtained by the present theory can show Curie law with the correct Curie constants.
(iv) The present theory smoothly interpolates the above low- and high-temperature region. As a result it predicts a large maximum of the specific heat.

Possible Improvement of the Simple Theory

Although the above mentioned main features seem to be reasonable in a qualitative sense, the simple theory hitherto discussed has several points to be improved. One of the most significant points is that at low temperatures the simple theory predicts the HF type results with the reduced interaction. The low-temperature specific heat in the metallic case (44) is a typical example. To describe the low-temperature properties more accurately, it is necessary to take account of low-energy processes more faithfully. It means that we need to evaluate the first line of (15) without employing the simplification made in the second line. For that purpose we may use a method developed by MORIYA and KAWABATA [14]. The integrand may be rewritten by using the transversal dynamical susceptibility of the f band. Then the correction to the HF type term is given in terms of the RPA dynamical susceptibility. Thus we obtain

$$F = F_0(M,N;M_f,N_f;T) + F_{fU}(M_f,N_f;T) - F_{fW}(M_f,N_f;T) + \frac{W}{4N_0}(N_f^2 - M_f^2)$$
$$+ \sum_q \frac{1}{2\pi}\int_{-\infty}^{\infty} d\omega \coth\left(\frac{\beta\omega}{2}\right)\mathrm{Im}\{\log[1 - W\chi_{fo}^{+-}(q,\omega)] + W\chi_{fo}^{+-}(q,\omega)\}, \qquad (45)$$

where χ_{fo}^{+-} is the transversal dynamical susceptibility of the non-interacting system under fixed values of M, N, M_f and N_f and in the paramagnetic phase it is related to the dynamical susceptibility discussed in subsection 1.2,

$$\chi_{fo}^{+-}(q,\omega) = \frac{1}{2}\chi_{ff}^{o}(q,\omega). \qquad (46)$$

The nature of the present approximation is also shown in Fig.2. One of the notable points is that the results of this approximation will not depend on the choice of W so seriously in so far as we choose W so that the integrand at the cutoff is small. This condition at T=0 can be rewritten as

$$\frac{N_f}{2} \cong \sum_q \frac{1}{2\pi}\int_{-\infty}^{\infty} d\omega\, \mathrm{sgn}(\omega)\chi_{fo}^{+-}(q,\omega)/[1 - W\chi_{fo}^{+-}(q,\omega)]. \qquad (47)$$

Conversely, we may use (47) as another method to determine W. Although the present program is feasible in principle, a huge scale of numerical calculation is necessary and the author has not yet performed it. If it is carried out for such a metallic case as we discussed in section 4, we obtain, for example, a larger T-linear specific heat.

In the present paper we have considered the periodic Anderson model and slightly extended one. To discuss experimental results and to compare them with theoretical predictions, the Hamiltonians considered here are apparently oversimplified: They neglect orbital degeneracy of the d and f states and also neglect a coupling between electronic states and lattice etc. Nevertheless, the author expects that the simple theory discussed in this paper will serve as an initial step to our understanding of some aspects of the heart of the mixed valence phenomena.

References

1 See for example, Valence Instabilities and Related Narrow-Band Phenomena, ed. R.D. Parks (Plenum, New York, 1977).
2 P.W. Anderson: Phys. Rev. 123(1961)41
3 H.J. Leder and B. Mühlschlegel: Z. Physik B29(1978)341
4 P.S. Riseborough and D.L. Mills: Phys. Rev. B21(1980)5338
5 K. Ueda: To be published in J. Phys. Soc. Jpn
6 K. Ueda: To be published in J. Phys. Soc. Jpn
7 J. Kanamori: Prog. Theor. Phys. 30(1963)275
8 K. Yamada and K. Yosida: in this book
9 A.C. Hewson: J. Phys. C10(1977)4973
10 A. Bringer and H. Lustfeld: Z. Physik B28(1977)213
11 E. Müller-Hartmann: Solid State Commun. 31(1979)113
12 T.K. Lee and S. Chakravarty: Phys. Rev. B22(1980)3609
13 K. Ueda and Y. Ishii: To be published
14 T. Moriya and A. Kawabata: J. Phys. Soc. Jpn 34(1973)639, 35(1973)669 and Proc. Int. Conf. on Magnetism, 1973, Moscow, IV, 5

Ground State of the Periodic Anderson Hamiltonian

Kōsaku Yamada and Kei Yosida*

College of Engineering, Shizuoka University, Hamamatsu 432

*Institute for Solid State Physics, University of Tokyo,
Roppongi, Minato-ku, Tokyo 106, Japan

The ground-state properties of the periodic Anderson Hamiltonian are investigated. This Hamiltonian describes the system consisting of conduction electrons and localized electrons which are connected with the mixing term. The localized electrons have strong Coulomb repulsion on the same atom.

Starting with the unperturbed system without the Coulomb repulsion, we consider the effect of the Coulomb repulsion on the unperturbed ground state. This way of approach is in contrast with that by the s-d model.

Our investigations consist of the following four parts :
1. General consideration for the unperturbed ground state.
2. Discussion on the basis of the Hartree-Fock approximation for Ne/N=1, where Ne and N are, respectively, the total number of electrons and atoms.
3. Consideration for the low-density case of the localized electrons on the basis of the t-matrix approximation for Ne/N=2.
4. Consideration on the basis of the perturbation method with respect to Coulomb repulsion U for symmetric bands and Ne/N=2.

1. Introduction

In connection with the mixed valence or valence fluctuation phenomena in rare-earth metals and compounds, we investigate the ground-state properties of the periodic Anderson Hamiltonian which describes the system consisting of the conduction s (or d)-electrons and the localized d (or f) electrons with strong Coulomb repulsion between them.

There are two approaches for this system. One is the approach in which the s-d mixing is treated as a perturbation for the conduction electrons and the localized electrons. This method is usually called the s-d model and is considered to be valid for the strong limit of Coulomb repulsion U and for the case in which there exist an integral number of localized electrons per atom. In the other approach, the unperturbed Hamiltonian, in which Coulomb repulsion U is ignored, is first diagonalized and then the effect of the Coulomb repulsion on the ground state of the unperturbed Hamiltonian is taken into account. In this paper, we take the second approach bearing in mind the valence mixing.

The initial ground state of the unperturbed Hamiltonian is metallic for total number of electrons per atom, Ne/N equal to one but it is non-metallic for two electrons per atom. For these two cases, the effects of the Coulomb repulsion between localized d electrons on the same site are studied on three approximate methods: Hartree-Fock, low-density approximation and perturbation method.

In the next section we describe the physical properties of the unperturbed system, which are needed for later calculations. In Sec.3 consideration based on the Hartree-Fock approximation is given for Ne/N=1, and in Sec.4 the low-density case is considered on the basis of the t-matrix approximation mainly for Ne/N=2. The ground state is investigated on the basis of the perturbation method with respect to U in Sec.5, where, for simplicity, the symmetric conduction bands and Ne/N=2 are assumed. Discussion is given in the final section.

2. Physical Properties of the Unperturbed Ground State

In this paper, we discuss the ground state of the following periodic Anderson Hamiltonian:

$$H = \sum_{k,\sigma} \varepsilon_k n_{k\sigma} + \sum_{i,\sigma} E_0 \, n_{i\sigma} + U \sum_i n_{i\uparrow} n_{i\downarrow}$$

$$+ \frac{V}{\sqrt{N}} \sum_{i,k,\sigma} (e^{-ikR_i} a^+_{i\sigma} c_{k\sigma} + e^{ikR_i} c^+_{k\sigma} a_{i\sigma}) \quad . \qquad (2.1)$$

$c^+_{k\sigma}$ and $c_{k\sigma}$ are, respectively, the creation and annihilation operators for conduction electron with energy ε_k and $n_{k\sigma}$ is given by $c^+_{k\sigma}c_{k\sigma}$. Similarly, $a^+_{i\sigma}$ and $a_{i\sigma}$ are those operators for d or f electron localized at i-th site and $n_{i\sigma}$ is given by $a^+_{i\sigma}a_{i\sigma}$. E_0 denotes the energy level of localized electrons and U the Coulomb repulsion between two d electrons on the same site. The hybridization between the conduction electron state and the localized electron state are described by the last term of the Hamiltonian. N is the total number of lattice sites.

In this section we describe some expressions for physical quantities in the unperturbed state, which will be necessary for later calculations. In one-body approximation we can also use these expressions by replacing E_0 with $E_{0\sigma}=E_0+U\langle n_{-\sigma}\rangle$.

The Green's function of the localized electron, $g^0_{k\sigma}(\omega)$ and that of the conduction electron, $G^0_{k\sigma}(\omega)$, are, respectively, given by

$$g^0_{k\sigma}(\omega) = [i\omega - E_0 - \frac{V^2}{i\omega-\varepsilon_k}]^{-1} \quad , \qquad (2.2)$$

and

$$G^0_{k\sigma}(\omega) = [i\omega - \varepsilon_k - \frac{V^2}{i\omega-E_0}]^{-1} \quad . \qquad (2.3)$$

Eqs.(2.2) and (2.3) can be rewritten as

$$g^{o}_{k\sigma}(\omega) = \frac{\alpha_k^+}{i\omega - E^+_{k\sigma}} + \frac{\alpha_k^-}{i\omega - E^-_{k\sigma}} \quad , \tag{2.4}$$

and

$$G^{o}_{k\sigma}(\omega) = \frac{\alpha_k^-}{i\omega - E^+_{k\sigma}} + \frac{\alpha_k^+}{i\omega - E^-_{k\sigma}} \quad , \tag{2.5}$$

where

$$E^{\pm}_{k\sigma} = \frac{1}{2}\,[E_o + \varepsilon_k \pm \sqrt{(E_o-\varepsilon_k)^2 + 4|V|^2}] \quad , \tag{2.6}$$

and

$$\alpha^{\pm}_{k\sigma} = \frac{1}{2}\,[1 \pm \frac{E_o - \varepsilon_k}{\sqrt{(E_o-\varepsilon_k)^2+4|V|^2}}] \quad . \tag{2.7}$$

The density of states per atom for the localized state $\rho_\ell(\omega)=-\frac{1}{\pi}\frac{1}{N}\sum_k \mathrm{Im}\, g^{o}_{k\sigma}(\omega)$ and that for the conduction electron state $\rho_c(\omega)=-\frac{1}{\pi}\frac{1}{N}\sum_k \mathrm{Im}\, G^{o}_{k\sigma}(\omega)$ are, respectively, given by

$$\rho_\ell(\omega) = \rho_o(\omega + \frac{V^2}{E_o-\omega})\,\frac{V^2}{(E_o-\omega)^2} \quad , \tag{2.8}$$

and

$$\rho_c(\omega) = \rho_o(\,\omega + \frac{V^2}{E_o-\,\omega}) \quad , \tag{2.9}$$

where $\rho_o(\varepsilon)$ is the density of states for the free conduction electrons.

In this expression, as ω tends to E_o the argument of $\rho_o(\varepsilon)$ is out of the conduction band. Therefore, a band gap exists around E_o as far as the conduction band has a finite width. The gap width Δ_G is given by

$$\Delta_G = E^+_{k\sigma}(\varepsilon_k=D_B) - E^-_{k\sigma}\,(\varepsilon_k=D_T) \quad , \tag{2.10}$$

where D_B and D_T are the bottom and the top of the conduction band, respectively. If we put $D_T=-D_B=D$ and $E_o=0$, we have

$$\Delta_G = -D + \sqrt{D^2+4V^2} \sim \frac{2V^2}{D} \quad , \quad \text{for} \quad V<<D \quad . \tag{2.11}$$

Hereafter, we assume the following conduction band, for simplicity,

$$\rho_o = \frac{1}{2D} \quad , \quad \text{for} \quad -D<\varepsilon<D \quad ,$$
$$\rho_o = 0 \quad , \quad \text{for} \quad |\varepsilon|>D \quad . \tag{2.12}$$

Then, $\rho_\ell(\omega)$ and $\rho_c(\omega)$ are, respectively, given by

$$\rho_\ell(\omega) = \rho_o \frac{V^2}{(E_o-\omega)^2} \, , \quad \text{for} \quad \omega_B^-<\omega<\omega_T^- \quad \text{and} \quad \omega_B^+<\omega<\omega_T^+ \quad ,$$
$$\rho_c(\omega) = \rho_o \quad , \tag{2.13}$$

where

$$\omega_B^\pm = \frac{1}{2} \{-D + E_o \pm \sqrt{(D+E_o)^2 + 4V^2}\} \quad ,$$

and

$$\omega_T^\pm = \frac{1}{2} \{D + E_o \pm \sqrt{(D-E_o)^2 + 4V^2}\} \quad .$$

The number of localized electrons for this case is given by

$$n_{\ell\sigma} = \rho_o \{ \frac{V^2}{E_o-\mu} + \frac{1}{2}(E_o+D) - \frac{1}{2} \sqrt{(D+E_o)^2+4V^2}\} \quad , \quad \text{for} \quad \mu<E_o \quad , \tag{2.14}$$

or

$$n_{\ell\sigma} = \frac{1}{2} + \rho_o \{ \frac{V^2}{E_o-\mu} + \frac{1}{2}(D+E_o) + \frac{1}{2} \sqrt{(D-E_o)^2+4V^2}\} \, , \quad \text{for} \quad \mu>E_o \quad , \tag{2.15}$$

where μ is the chemical potential. The number of conduction electrons is also given by

$$n_{c\sigma} = \rho_o \{\mu - \frac{1}{2} (E_o - D - \sqrt{(D+E_o)^2+4V^2})\} \, , \qquad \text{for} \quad \mu<E_o \quad , \tag{2.16}$$

or

$$n_{c\sigma} = \frac{1}{2} + \rho_o \{\mu + \frac{1}{2} \, (D - E_o - \sqrt{(D-E_o)^2+4V^4})\} \, , \quad \text{for} \quad \mu>E_o \quad . \tag{2.17}$$

The total electron number per spin is given as

$$n_{t\sigma} = n_{\ell\sigma} + n_{c\sigma} = \rho_o \{\mu + \frac{V^2}{E_o-\mu} + D\}$$

$$= \frac{1}{2} + \rho_0 \{\mu + \frac{V^2}{E_0-\mu}\} \quad , \quad \text{for} \quad \mu < E_0 \quad , \tag{2.18}$$

or

$$n_{t\sigma} = \frac{3}{2} + \rho_0 \{\mu + \frac{V^2}{E_0-\mu}\} \quad . \quad \text{for} \quad \mu > E_0 \quad . \tag{2.19}$$

Therefore, if the total electron number per atom, Ne/N, is given, μ is determined as

$$\mu + \frac{V^2}{E_0-\mu} = 0 \quad , \quad \text{for} \quad Ne/N = 1 \text{ or } 3 \quad , \tag{2.20}$$

$$\mu + \frac{V^2}{E_0-\mu} = \pm D \quad , \quad \text{for} \quad Ne/N = 2 \quad . \tag{2.21}$$

The susceptibility of the localized electrons as an external field is only applied to localized electrons is given by

$$\chi_0(0) = \rho_0 \left(\mu + \frac{V^2}{E_0-\mu}\right) \frac{V^4}{(E_0-\mu)^2} \frac{1}{[(E_0-\mu)^2+V^2]}$$

$$+ \frac{\rho_0}{2} \left\{ \frac{\mu + \frac{V^2}{E_0-\mu} - E_0}{\sqrt{(\mu+\frac{V^2}{E_0-\mu} - E_0)^2+4V^2}} + \frac{D + E_0}{\sqrt{(D+E_0)^2+4V^2}} \right\} \quad . \tag{2.22}$$

If we assume the relation, $\varepsilon_{k+G}=-\varepsilon_k$ for the conduction band, the antiferromagnetic susceptibility is given by

$$\chi_G^0(0) = 2\rho_0 \left\{ \int_{-D}^{\mu+\frac{V^2}{E_0-\mu}} d\varepsilon \frac{\alpha_k^- \cdot \beta_k^+}{\varepsilon_k^+ - E_k^-} + \int_{-D}^{-|\mu+\frac{V^2}{E_0-\mu}|} d\varepsilon \frac{\alpha_k^- \beta_k^-}{\varepsilon_k^- - E_k^-} \right\} \quad ,$$

$$\text{for} \quad 0 \leq Ne/N \leq 2 \quad , \tag{2.23}$$

where $\varepsilon_k^{\pm}=E_k^{\pm}(-\varepsilon_k)$, $\beta_k^{\pm}=\alpha_k^{\pm}(-\varepsilon_k)$, and $\vec{G}$ is K/2, K being the reciprocal vector.

Using (2.20) for Ne/N=1, we obtain

$$\chi_G^0(0) = 2\rho_0 \left\{ \int_{-D}^{0} d\varepsilon \frac{\alpha_k^- \beta_k^+}{\varepsilon_k^+ - E_k^-} + \int_{-D}^{0} d\varepsilon \frac{\alpha_k^- \beta_k^-}{\varepsilon_k^- - E_k^-} \right\} \quad , \tag{2.24}$$

where

$$\varepsilon_k^- - E_k^- = \frac{1}{2}\left(-2\varepsilon_k + \sqrt{(E_o-\varepsilon_k)^2+4V^2} - \sqrt{(E_o+\varepsilon_k)^2+4V^2}\right) \quad . \tag{2.25}$$

This term tends to zero as $\varepsilon_k \to 0$. Thus, $\chi_G^o(0)$ diverges logarithmically for Ne/N=1. This divergence of $\chi_G^o(0)$ for Ne/N=1 occurs for such conduction bands as simple cubic and body-centered-cubic lattices with nearest-neighbor transfer satisfying the relation $\varepsilon_{k+G}=-\varepsilon_k$. Thus, we conclude that the system with Ne/N=1 and $\varepsilon_{k+G}=-\varepsilon_k$ shows the antiferromagnetic instability for infinitesimal Coulomb repulsion.

In the case of Ne/N=2, we obtain $\mu+V^2/(E_o-\mu)=D$ from (2.21). If we put $E_o=0$, $\chi_G^o(0)$ is given by

$$\chi_G^o(0) = \frac{\rho_o}{2V^2}\left[D\sqrt{D^2+4V^2} - 2V^2 \log\frac{D+\sqrt{D^2+4V^2}}{2|V|}\right]$$

$$\simeq \rho_o\left[\frac{D^2}{2V^2} - \log\frac{D}{|V|}\right] \quad , \quad \text{for } \frac{|V|}{D} \ll 1 \quad . \tag{2.26}$$

On the other hand, $\chi_o^o(0)$ is given by

$$\chi_o^o(0) = \frac{\rho_o D}{\sqrt{D^2+4|V|^2}} \simeq \rho_o \quad . \tag{2.27}$$

From (2.26) and (2.27), we conclude for Ne/N=2

$$\chi_G^o(0) > \chi_o^o(0) \quad . \tag{2.28}$$

On the basis of ferromagnetic and antiferromagnetic susceptibilities in the unperturbed state we can discuss the instability which is expected with increasing U. We show the phase diagram in Fig.1 which indicates the larger

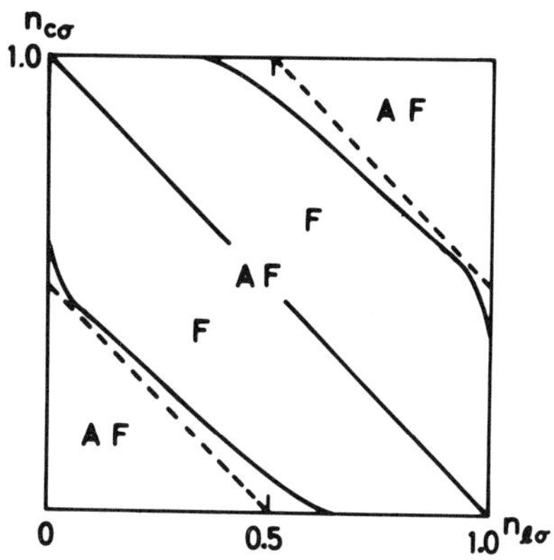

Fig.1 Phase diagram which indicates the larger susceptibility of χ_o^o and χ_G^o. χ_o^o and χ_G^o denote, respectively, the uniform or ferromagnetic susceptibility and the staggered or antiferromagnetic susceptibility in the unperturbed state. The conduction band is assumed to have a constant density of states and satisfy the relation $\varepsilon_{k+G}=-\varepsilon_k$. V=0.2.

susceptibility between χ_G^0 and χ_0^0 in the unperturbed state. On the line of $n_{c\sigma}+n_{\ell\sigma}=1$, antiferromagnetic instability occurs at a finite value of U. On the other hand, on the line of $n_{c\sigma}+n_{\ell\sigma}=1/2$, antiferromagnetic instability occurs for infinitesimal value of U.

In the case of the actual conduction bands for the simple cubic lattice $\chi_G^0(0)$ for Ne/N=2 is shown in Fig.2. In this figure, $\chi_q^0(0)$ has always a maximum at $\vec{q}=\vec{G}=(\frac{\pi}{a},\frac{\pi}{a},\frac{\pi}{a})$. Therefore, the antiferromagnetic instability occurs with increasing U [1].

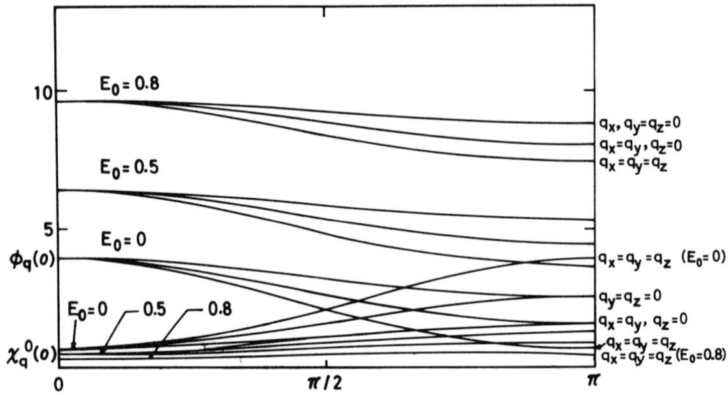

Fig.2 The wave-vector dependence of $\chi_q^0(0)$ and $\phi_q(0)$ in the unperturbed state for the simple cubic lattice. $\varepsilon_k=-\frac{1}{3}(\cos k_x+\cos k_y+\cos k_z)$. Ne/N=2, V=0.2.

3. Hartree-Fock Approximation for Ne/N=1

In the previous section, we obtain the conclusion that the system with Ne/N=1 and $\varepsilon_{k+G}=-\varepsilon_k$ shows the antiferromagnetic instability for infinitesimal Coulomb repulsion. In this section we study the stability of this antiferromagnetic ground state for Ne/N=1 on the basis of the Hartree-Fock approximation.

We show $n_{\ell\uparrow}$,$n_{\ell\downarrow}$ and $2S_{\ell z}=n_{\ell\uparrow}-n_{\ell\downarrow}$ in the sublattice for several values of E_0 in Fig.3. There, we have assumed that the conduction band has a constant density of states and satisfies the relation $\varepsilon_{k+G}=-\varepsilon_k$. In the Hartree-Fock approximation we obtain also paramagnetic or ferromagnetic solution for large values of U. Actually we compared these states with antiferromagnetic state in energy, and the antiferromagnetic state is confirmed to be always lowest in its energy. Therefore, the ground state is always antiferromagnetic except for the case of U=0.

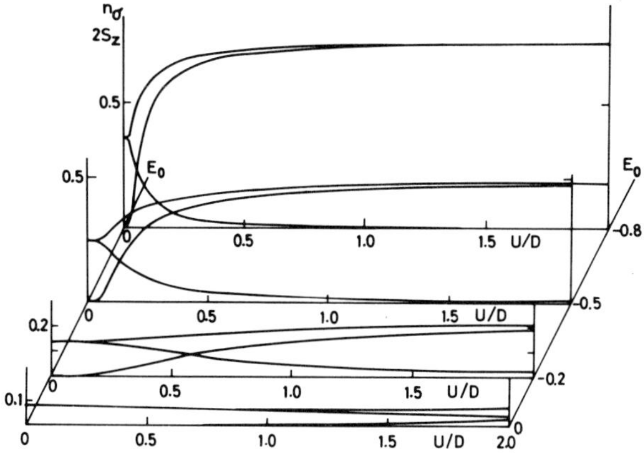

<u>Fig.3</u> $n_{\ell\uparrow}$, $n_{\ell\downarrow}$ and $2S_{\ell z}=n_{\ell\uparrow}-n_{\ell\downarrow}$ of the sublattice for several values of E_0. Ne/N is taken as 1. The conduction band is assumed to have a constant density of states and to satisfy the relation, $\varepsilon_{k+G}=-\varepsilon_k$. V=0.2.

In Fig.4, we show the density of states for antiferromagnetic ground state, assuming V=0.2 and U=1.0. Among four parts devided by three gaps, only the lowest part is occupied, as we assume Ne/N=1.

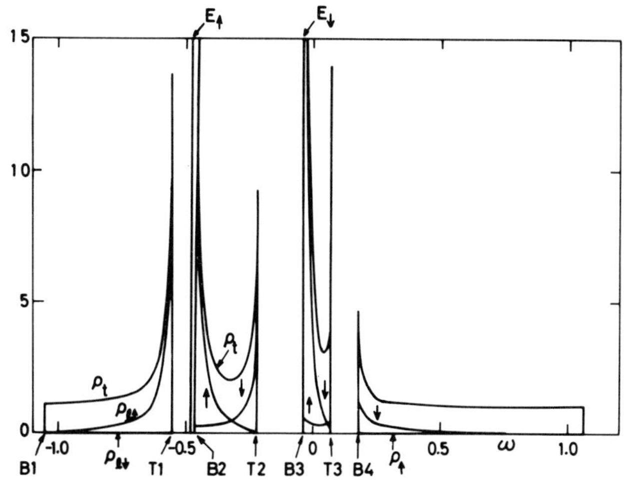

<u>Fig.4</u> The density of states for the antiferromagnetic ground state of the Hartree-Fock approximation. V=0.2, U=1.0, E_0=0.5 and Ne/N=1. $n_{\ell\uparrow}$=0.461 ($E_\uparrow$=-0.484), $n_{\ell\downarrow}$=0.016 ($E_\downarrow$=-0.039).

4. Low-Density Approximation for Ne/N=2

First, we show localized electron number $n_{\ell\sigma}$ as a function of E_0 for Ne/N=2 in Fig.5, where we assume $\varepsilon_k=-\frac{1}{3}(\cos k_x+\cos k_y+\cos k_z)$ for the simple cubic lattice and $\varepsilon_k=-\cos k$ for one-dimensional case.

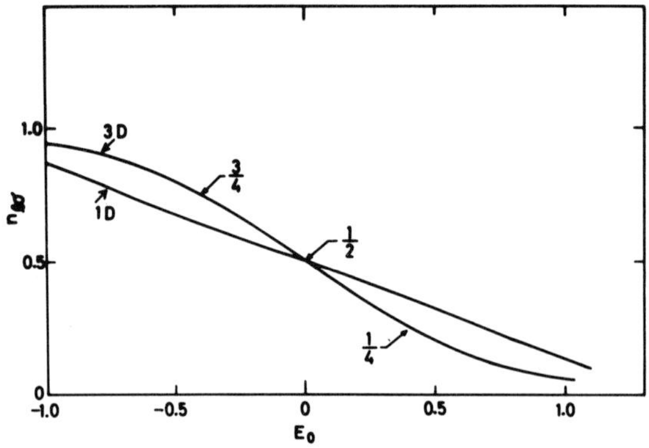

Fig.5 The relation between localized electron number, $n_{\ell\sigma}$, and energy level of the localized electron, E_0, in the unperturbed state. Conduction bands are assumed to be $\varepsilon_k=-\frac{1}{3}(\cos k_x+\cos k_y+\cos k_z)$ for simple cubic lattice and $\varepsilon_k=-\cos k$ for one-dimensional lattice. V=0.2.

For the case with a small number of localized electrons, we can adopt the low-density approximation. In the followings, we confine ourselves to Ne/N=2. In this approximation, we have

$$\phi_q(\nu) = T\frac{1}{N}\sum_{\omega,k_1} g^0_{q-k_1}(\nu-\omega)\; g^0_{k_1}(\omega) \quad , \tag{4.1}$$

$$t_q(\nu) = \frac{U}{1+U\phi_q(\nu)} \quad , \tag{4.2}$$

$$\Sigma_k(\omega) = -T\frac{1}{N}\sum_{q,\nu} g^0_{q-k}(\nu-\omega)\; t_q(\nu) \quad . \tag{4.3}$$

We show $\phi_q(0)$ calculated for the simple cubic lattice in Fig.2 and for one-dimensional case ($\varepsilon_k= -\cos k$) in Fig.6. From Fig.2, we can see that ϕ_q becomes larger as E_0 increases and that q-dependence of ϕ_q is small in the three-dimensional case.

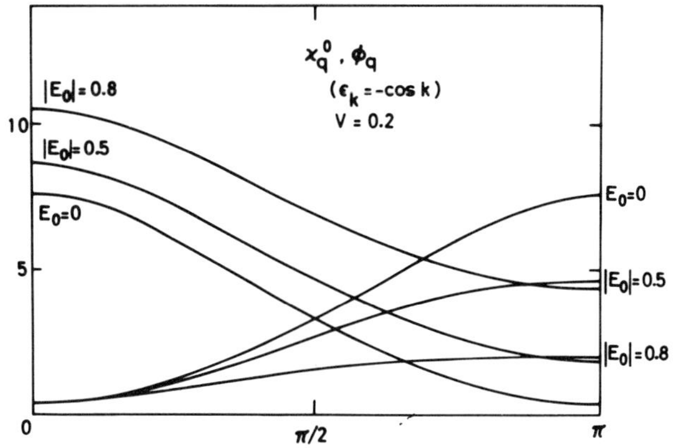

Fig.6 The wave-vector dependence of $\chi_q^0(0)$ and $\phi_q(0)$ for one-dimensional case.

In three-dimensional case, we have a magnetic instability when $U_{eff}\ \chi_q^0$ is larger than unity where U_{eff} is given by $U/[1+U\phi_q(0)]$. As U_{eff} can be replaced by $[\phi_q(0)]^{-1}$ for large U, we have no magnetic instability as far as $\phi_q(0)$ is larger than χ_q^0. In this case we have a paramagnetic ground state.

On the other hand, the q dependence of ϕ_q is so large for one-dimensional case that we must solve (4.1) to (4.3). We show the selfenergy shift $\mathrm{Re}\Sigma_k(\omega)$ for one-dimensional case in Fig.7. Since the selfenergy shift is almost

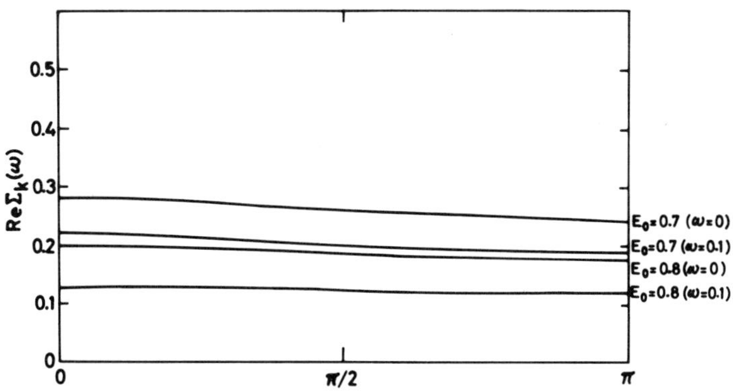

Fig.7 The k dependence of the selfenergy shift, $\mathrm{Re}\Sigma_k(\omega)$ for one-dimensional case (ε_k=-cosk). U=10, V=0.2.

independent of wave vector k in contrast with $\phi_q(0)$, we can renormalize the result of the low-density approximation by replacing E_0 with $E_0+\Sigma_{k=0}(0)$. That is, in the renormalized case we have only to replace E_0 in the unrenormalized case by that reduced by $\Sigma_{k=0}(0)$. We show E_0 dependence and U dependence of the selfenergy shift $\Sigma_k(0)$ for one-dimensional case in Figs.8 and 9. From Fig.9 we can see that the selfenergy shift is proportional to logU in the range of $\Delta<<U<D$, where Δ is given by $\rho_0 V^2$.

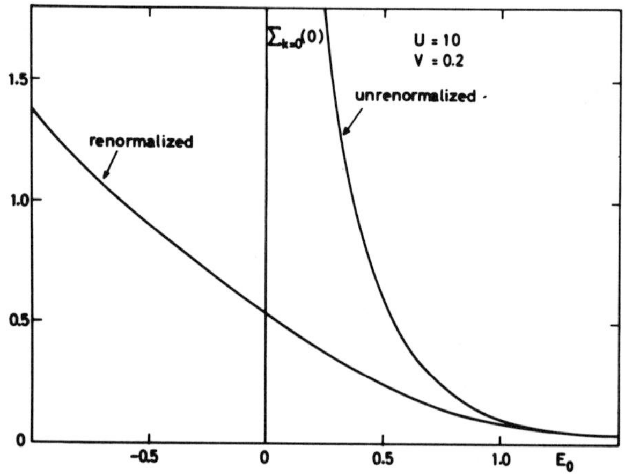

Fig.8 The E_0 dependence of the selfenergy shift $\Sigma_k(0)$ for one-dimensional case. One is unrenormalized and the other renormalized by replacing E_0 with $E_0+\Sigma_{k=0}(0)$. U=10, V=0.2.

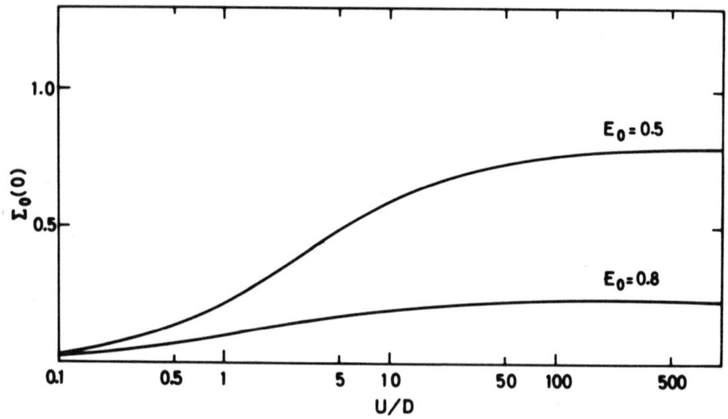

Fig.9 The U dependence of the selfenergy shift $\Sigma_k(0)$ for one-dimensional case (unrenormalized). V=0.2.

This U dependence of the selfenergy for $\Delta<<U<<D$ generally holds for all the dimensions [2]. Eq.(4.1) can be calculated at the absolute zero of temperature as

$$\phi_q(\nu\pm i\delta) = \frac{1}{N}\sum_{k_1}\{\frac{\alpha^-_{q-k_1}\ \alpha^-_{k_1}}{\nu\pm i\delta - E^-_{q-k_1} - E^-_{k_1} + 2\mu} + \frac{\alpha^+_{q-k_1}\ \alpha^+_{k_1}}{-(\nu\pm i\delta) + E^+_{q-k_1} + E^+_{k_1} - 2\mu}\} \quad . \tag{4.4}$$

Using $\alpha^\pm_k \ll 1$ for $|E^\pm_k - \mu| \gg |V|$, we can approximate ϕ_q for $|\nu| \gg |V|$ as

$$\phi_q(\nu\pm i\delta) = \frac{1}{N}\ \sum_{k_1} \frac{1}{\nu\pm i\delta}\ \{\alpha^-_{q-k_1}\ \alpha^-_{k_1} - \alpha^+_{q-k_1}\ \alpha^+_{k_1}\}$$

$$= \frac{1}{\nu\pm i\delta}\frac{1}{N}\ \sum_{k_1}\ \{\alpha^-_{q-k_1} + \alpha^-_{k_1} - 1\}$$

$$= \frac{1}{\nu\pm i\delta}\ (2n_\ell - 1) = -\ \frac{A}{\nu\pm i\delta} \quad , \tag{4.5}$$

where

$$A = 1 - 2n_\ell \quad ,$$

and

$$n_\ell = \frac{1}{N}\sum_k \alpha^-_{k_1} = n_{\ell\sigma} \quad . \tag{4.6}$$

From (4.3), the selfenergy shift can be written as

$$\mathrm{Re}\Sigma_k\ (0+i\delta) = -\ \frac{1}{N}\sum_q \int_{-\infty}^{0} \frac{dx}{\pi}\ t^R_q(x+i\delta)\ \mathrm{Im}\ g^R_{q-k}(x)$$

$$= \frac{1}{N}\sum_q\ t^R_q(E^-_{q-k} - \mu + i\delta)\ \alpha^-_{q-k} \quad . \tag{4.7}$$

Using here (4.5) and putting $z = E^-_{q-k} - \mu$, we obtain

$$\mathrm{Re}\Sigma_k\ (0+i\delta) \simeq \frac{1}{N}\sum_q \frac{\alpha^-_{q-k}\ U(E^-_{q-k}-\mu)}{E^-_{q-k}-\mu-UA} = \rho\int_{-D}^{\Delta} \frac{Uz\alpha^-_{q-k}}{z-UA}\ \left[\frac{dE^-_{q-k}}{d\varepsilon_{q-k}}\right]^{-1} \cdot dz$$

$$= \rho\int_{-D}^{\Delta} \frac{Uz}{z-UA}\ \frac{\alpha^-_{q-k}}{\alpha^+_{q-k}}\ dz \quad . \tag{4.8}$$

Here we use $\alpha^+_{q-k} = z^2/(z^2+V^2)$, and $\alpha^-_{q-k} = V^2/(z^2+V^2)$ and obtain

$$\mathrm{Re}\Sigma_k\ (0+i\delta) = \frac{\Delta}{A}\ \log\frac{UA}{\Delta} \simeq \Delta\log\frac{U}{\Delta} \quad , \quad \text{for}\quad n_\ell \ll 1 \quad . \tag{4.9}$$

5. Perturbation Expansion for the Symmetric Case

In this section we investigate the ground-state properties of the Anderson lattice for the symmetric case. The symmetric case means that the following relations are fulfilled;

$$E_0 = -\frac{U}{2} \quad , \tag{5.1}$$

$$\varepsilon_{-k} = \varepsilon_k \quad , \tag{5.2}$$

$$\varepsilon_{k+G} = -\varepsilon_k \; , \quad G = \frac{K}{2} \quad , \tag{5.3}$$

and $Ne/N = 2$.

When E_0 is equal to -U/2, we can rewrite the d-electron part of the Hamiltonian as [3]

$$E_0 \sum_{i,\sigma} n_{i\sigma} + U \sum_i n_{i\uparrow} n_{i\downarrow} = U \sum_i \{(n_{i\uparrow}-\frac{1}{2})(n_{i\downarrow}-\frac{1}{2}) - \frac{1}{4}\} \quad . \tag{5.4}$$

Using this relation we divide the Hamiltonian of (2.1) into the unperturbed part H_0, and perturbation, H' as

$$H_0 = \sum_{k,\sigma} \varepsilon_k n_{k\sigma} + V \sum_{k,\sigma} (a^+_{k\sigma} c_{k\sigma} + c^+_{k\sigma} a_{k\sigma}) - \frac{U}{4} N \quad , \tag{5.5}$$

where

$$a_{k\sigma} = \frac{1}{\sqrt{N}} \sum_i e^{ikR_i} a_{i\sigma} \quad ,$$

and

$$H' = \frac{U}{N} \sum_{k_1,k_2,q\neq 0} a^+_{k_1+q\uparrow} a^+_{k_2-q\downarrow} a_{k_2\downarrow} a_{k_1\uparrow} \quad . \tag{5.6}$$

The Green's function of the localized electron in the unperturbed state is given by (2.2) and (2.3) in which E_0=0. From symmetry relation (5.3), the following relations are derived:

$$E^{\pm}_{k+G} = - E^{\mp}_k \quad , \tag{5.7}$$

$$\alpha^{\pm}_{k+G} = \alpha^{\mp}_k \quad , \tag{5.8}$$

$$g^0_{k+G}(\omega) = - g^0_k(-\omega) \quad , \tag{5.9}$$

$$g_k^o(\tau) = T \sum_{\omega_n} g_k^o(\omega_n)\, e^{i\omega_n\tau} \quad , \tag{5.10}$$

$$g_{k+G}^o(\tau) = -\, g_k^o(-\tau) \quad . \tag{5.11}$$

With the use of (5.9), we can show that the free energy and the selfenergy part of the localized electron have only the even order terms with respect to U [3], as follows:

1) We choose an arbitary one of the free-energy or selfenergy diagrams for up spin. As $g_{-k}^o(\omega)=g_k^o(\omega)$, we can change the sign of wave vectors for all the Green's functions in the closed loops with down spin.
2) Then, we shift the wave vector by G for all the Green's functions in all the closed loops with down spin.
3) Since $g_{G-k}^o(\omega)=-g_{-k}^o(-\omega)$, the contribution of the above n-th order diagram is equal to $(-1)^n$ times the contribution from the diagram in which all the closed loops with down spin are replaced by those with the inverse direction of rotation.
4) In the n-th order-diagrams, we have always those diagrams in which closed loops rotate in the inverse direction. On account of the factor $(-1)^n$ in 3), two diagrams in which closed loops rotate in the opposite direction cancel out in each other for odd-order terms.

By the similar reasoning, it is also shown that $\chi_{\uparrow\uparrow}$ and $\chi_{\uparrow\downarrow}$ have only even-order terms and odd-order terms, respectively. These results just correspond to the symmetric case of the single impurity [3].

The lowest order term of the selfenergy part is the second-order term which is shown diagrammatically in Fig.10. This term can be calculated as follows:

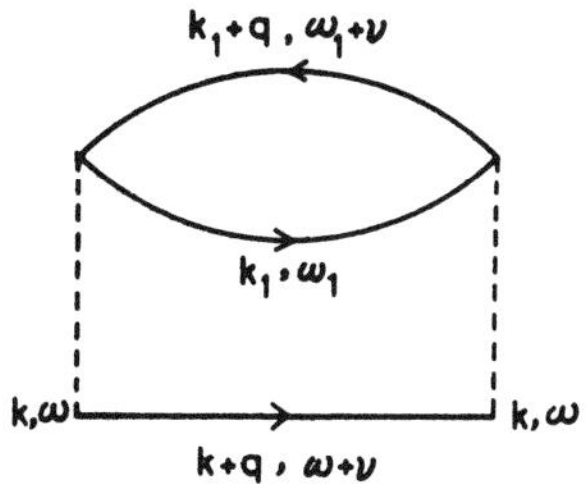

Fig.10 The second-order diagram of the self-energy.

$$\Sigma_k^{(2)}(\omega) = U^2\, T \frac{1}{N} \sum_{\nu,q} g_{k+q}^o(\omega+\nu)\, \chi_q^o(\nu) \quad , \tag{5.12}$$

$$\chi_q^o(\nu) = -\, T \frac{1}{N} \sum_{\substack{\omega_1 \\ k_1}} g_{k_1}^o(\omega_1)\, g_{k_1+q}^o(\omega_1+\nu) \quad . \tag{5.13}$$

At the zero of temperature, we can easily integrate this with respect to ν and ω_1 to obtain the following result:

$$\Sigma_k^{(2)}(\omega+i\delta) = (\frac{U}{N})^2 \sum_{q,k_1} \alpha_{k_1}^- \alpha_{k_1+q}^+ \{\frac{\alpha_{k+q}^+}{\omega+i\delta+E_k^- -E_{k_1+q}^+ -E_{k+q}^+ +\mu} + \frac{\alpha_{k+q}^-}{\omega+i\delta+E_{k_1+q}^+ -E_{k_1}^- -E_{k+q}^- +\mu}\} , \quad (5.14)$$

where we can put $\mu=0$.

As $\alpha_k^\pm \ll 1$ for $|E_k^\pm| \gg |V|$, we can put as $|\omega| \gg |E_{k_1}^- - E_{k_1+q}^+ - E_{k+q}^+|$ and $|E_{k_1+q}^+ - E_{k_1}^- - E_{k+q}^-|$ for $|\omega| \gg |V|$. Therefore, if $|\omega| \gg |V|$, $\Sigma_k^{(2)}(\omega+i\delta)$ can be approximated as [3]

$$\Sigma_k^{(2)}(\omega+i\delta) \simeq (\frac{U}{N})^2 \sum_{q,k_1} \alpha_{k_1}^- \alpha_{k_1+q}^+ \{\frac{\alpha_{k+q}^+}{\omega+i\delta} + \frac{\alpha_{k+q}^-}{\omega+i\delta}\} = \frac{1}{4}\frac{U^2}{(\omega+i\delta)} . \quad (5.15)$$

The density of states for the localized electron is given by

$$\rho_\ell(\omega) = -\frac{1}{\pi}\frac{1}{N}\sum_k \text{Im}\, g_k(\omega+i\delta) , \quad (5.16)$$

$$g_k(\omega+i\delta) = [\omega + i\delta - \Sigma_k^{(2)}(\omega) - \frac{V^2}{\omega+i\delta-\varepsilon_k}]^{-1} . \quad (5.17)$$

We show the density of states of the localized electron, $\rho_\ell(\omega)$, obtained for typical values of U and V in Fig.11. There, calculations have been done for

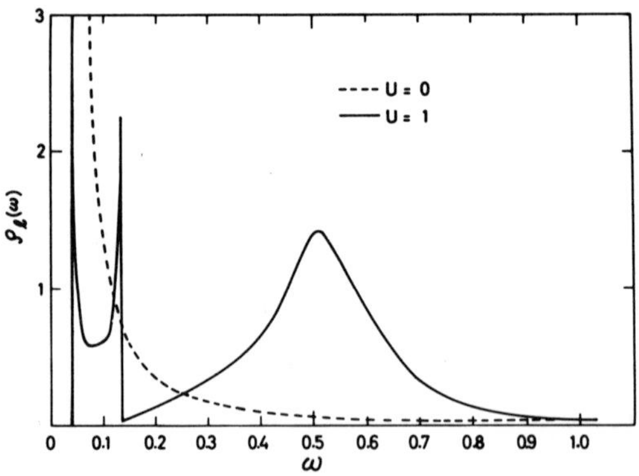

Fig.11 Density of states of the localized electrons, $\rho_\ell(\omega)$, as a function of ω, for one-dimensional case (ε_k=-cosk). U=0 and U=1, V=0.2, Ne/N=2.

one-dimensional conduction band. For U>>V, $g_k(\omega)$ has two poles at $\omega=\pm U/2$ as (5.15) shows. Therefore, $\rho_\ell(\omega)$ has two peaks at $\omega=\pm U/2$. This situation is quite similar to the d-electron state density for the symmetric Anderson Hamiltonian of a single impurity [3].

It should be noted that the other peaks appear near the gap. These peaks are caused by the hybridization effects. The reason for their appearance can be found from Fig.12. In the case of U=0, the lowest value of E_k^+ is $E_{k=0}^+=(-1+\sqrt{1+4V^2})/2$ and the highest value of E_k^- is $E_{k=G}^-=-E_{k=0}^+$. Therefore, the two peaks

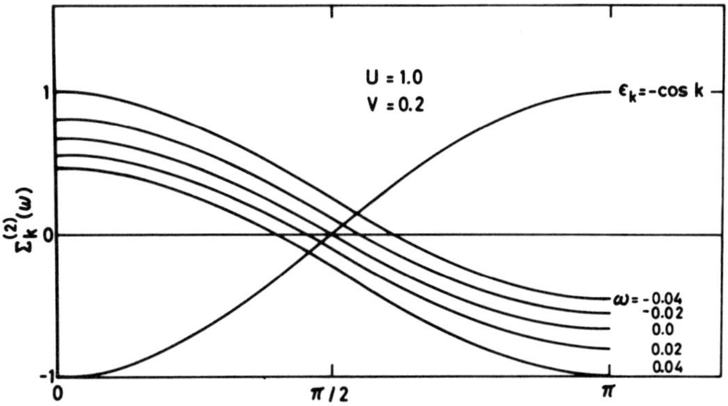

Fig.12 The second order selfenergy $\mathrm{Re}\Sigma_k^{(2)}(\omega)$ for one-dimensional case. U=1.0, V=0.2.

near the gap for U=0 are made of the states near k=0 and k=G, respectively. On the other hand, in the case of a large value of U we have a gap between the states near $E_{k=G/2}^+$ and $E_{k=G/2}^-$, as shown in Fig.12. The change of the minimum- and maximum-energy states in the Brillouin zone is originated from the k dependence of the selfenergy. In our case, we have $\partial\Sigma_k(\omega)/\partial k<0$.

The similar k dependence of the selfenergy is also obtained for the simple cubic lattice. The ω dependence of the selfenergy is generally obtained as

$$\frac{\partial\Sigma_k(\omega)}{\partial\omega} < 0 \quad . \tag{5.18}$$

This property of the selfenergy makes the gap narrower. In the following, we elucidate the density of states near the gap by using the expansion of the selfenergy part around the Fermi surface. $\Sigma_k(\omega)$ is expanded around $k=\pi/2$ and $\omega=0$ as

$$\Sigma_k(\omega) \simeq \alpha\omega + \beta k' \quad , \tag{5.19}$$

where

$$\alpha = \left.\frac{\partial\Sigma_k(\omega)}{\partial\omega}\right|_{\omega=0} , \qquad \beta = \left.\frac{\partial\Sigma_k(\omega=0)}{\partial k}\right|_{k=\frac{\pi}{2}} , \tag{5.20}$$

and $k'=k-\pi/2$.

ε_k is also expanded as

$$\varepsilon_k \simeq \gamma k' , \quad \text{for} \quad |k'|<<1 , \tag{5.21}$$

γ^{-1} being the density of states for conduction electrons at the Fermi energy. Here, we assume $\alpha<0$, $\beta<0$, $\gamma>0$. Since the density of states near the gap is given by the contributions of single poles, $\rho_\ell(\omega)$ is written as

$$\rho_\ell(\omega) = \frac{1}{N}\sum_k \alpha_k^+(\omega)\ \delta(\omega-E_k^+(\omega)) , \quad \text{for } 1>>\omega>0 , \tag{5.22}$$

where

$$E_k^+(\omega) = \frac{1}{2}\left(\varepsilon_k + \Sigma_k(\omega) + \sqrt{(\varepsilon_k-\Sigma_k(\omega))^2 + 4|V|^2}\ \right) , \tag{5.23}$$

and

$$\alpha_k^+(\omega) = \frac{1}{2}\left\{1 + (\Sigma_k(\omega) - \varepsilon_k) \Big/ \sqrt{(\varepsilon_k - \Sigma_k(\omega))^2+|4|V^2}\right\} . \tag{5.24}$$

Now, we use the following relation:

$$\delta(\omega - E_k^+(\omega)) = \delta(k - k_c)\left(\frac{\partial\omega}{\partial k}\right)^{-1} , \tag{5.25}$$

where k_c is determined by

$$\omega - E_k^+(\omega) = 0 . \tag{5.26}$$

In general, we have four solutions of (5.26). $d\omega/dk$ is obtained from (5.26) as follows:

$$\frac{d\omega}{dk} - \frac{E_k^+(\omega)}{\partial\omega}\frac{d\omega}{dk} - \frac{\partial E_k^+}{\partial k} = 0 , \tag{5.27}$$

$$\frac{\partial E_k^+(\omega)}{\partial\omega} = \frac{\partial E_k^+(\omega)}{\partial\Sigma_k(\omega)}\frac{\partial\Sigma_k(\omega)}{\partial\omega} \simeq \alpha_k^+\alpha , \tag{5.28}$$

$$\frac{\partial E_k^+}{\partial k} = \frac{\partial E_k^+}{\partial \Sigma_k}\frac{\partial \Sigma_k}{\partial k} + \frac{\partial E_k^+}{\partial \varepsilon_k}\frac{\partial \varepsilon_k}{\partial k} \simeq \alpha_k^+\beta + \alpha_k^-\gamma \quad . \tag{5.29}$$

Thus, we obtain $\rho_\ell(\omega)$ near the gap as

$$\rho_\ell(\omega) = \frac{1}{2\pi}\sum_{k_c}\alpha_{k_c}^+ \frac{(1-\alpha_{k_c}^+\alpha)}{|\alpha_{k_c}^+\beta+\alpha_{k_c}^-\gamma|} \quad . \tag{5.30}$$

The density of states near the gap is enhanced by a factor of $1-\alpha_{k_c}^+\alpha$. At k_c satisfying $\alpha_{k_c}^+\beta+\alpha_{k_c}^-\gamma=0$, $\rho_\ell(\omega)$ diverges for one-dimensional case.

As an example, we calculate the contributions of typical points satisfying $\varepsilon_k=\Sigma_k(\omega)$ to the density of states. By the following relations

$$\gamma k' = \alpha\omega + \beta k' \quad , \tag{5.31}$$

and

$$\omega = E_{k'}^+(\omega) = \varepsilon_k + |V| \quad , \tag{5.32}$$

ω and k' are determined as

$$\omega_0 = \frac{(\gamma-\beta)}{(1-\alpha)\gamma-\beta}V \quad , \tag{5.33}$$

and

$$k_0' = \frac{\alpha}{(1-\alpha)\gamma-\beta}V \quad . \tag{5.34}$$

Using the relation $\alpha_k^+=\alpha_k^-=\frac{1}{2}$ for $\varepsilon_k=\Sigma_k(\omega)$, we obtain the contribution of k_0' to $\rho_\ell(\omega_0)$ as

$$\frac{1}{2\pi}\left(1-\frac{\alpha}{2}\right) / |\beta+\gamma| \quad . \tag{5.35}$$

Now, we consider the dependence of the gap on parameters α, β and γ. The lowest value of $E_k^+(\omega)$ is determined by

$$\frac{dE_k^+(\omega)}{dk} = 0 \quad , \tag{5.36}$$

and

$$\omega = E_k^+(\omega) \quad . \tag{5.37}$$

Eq.(5.36) is rewritten as

$$\alpha_k^+\beta + \alpha_k^-\gamma = 0 \quad . \tag{5.38}$$

From (5.37) and (5.38), we obtain the band edge as

$$\omega = \frac{2V}{(1-\alpha)\gamma-\beta} \sqrt{-\gamma\beta} \quad . \tag{5.39}$$

From (5.39), we can see that the band gap is reduced by the electron interaction.

The ω dependence of the selfenergy makes the gap narrower, but does not destroy it. The gap survives under the finite Coulomb repulsion. This can be proved for the symmetric bands in one- and two-dimensional cases.

For the symmetric case defined by (5.1) to (5.3), the unperturbed Green's function $g_k^o(\omega)$ satisfies relation (5.9). From this relation, the following relation which the selfenergy satisfies is derived:

$$\Sigma_{k+G}(\omega) = - \Sigma_k(-\omega) \quad . \tag{5.40}$$

Since $\Sigma_k(\omega)=\Sigma_{-k}(\omega)$ holds, for k_o on the Fermi surface we have $\Sigma_{k_o}(0)=0$ if $\Sigma_{G-k_o}(0)=\Sigma_{k_o}(0)$ is satisfied, because from (5.40) we have $\Sigma_{G-k_o}(0)=-\Sigma_{k_o}(0)$.

For one-dimensional case, $G=\pi$ and the Fermi point is $k_o=\pi/2$. So we have the relation of $G-k_o=k_o$ for k_o. For two-dimensional case, $G=(\pi,\pi)$ and the Fermi line is given by $k_y^o=\pi-k_x^o$ etc. Therefore, for (k_x^o, k_y^o) on the Fermi line we have

$$\Sigma_{\pi-k_x^o,\pi-k_y^o}(0) = \Sigma_{k_y^o,k_x^o}(0) = \Sigma_{k_x^o,k_y^o}(0) \quad .$$

For three-dimensional case, we can show that $\Sigma_{k_o}(0)$ vanishes on each side of six squares determined by $k_y^o=\pi-k_x^o$, $k_z=\pm\pi/2$ and the relations obtained by exchanging x, y and z in a cyclic way. On the other parts of the Fermi surface, $\Sigma_{k_o}(0)$ may presumably not vanish exactly, but numerical calculations show that the selfenergy shifts on the Fermi surface are very small for reasonable values of V and U. So, it may be considered that even for three-dimensional case, the gap would survive up to a considerable value of U.

For a large value of U, the s-d model may be a good approximation for the system having two electrons per atom. On the basis of this s-d model, the localized spin situated at each lattice site is coupled with the conduction electrons by the effective s-d exchange interaction, whose exchange integral is given by V^2/U. For this system, the susceptibility of the conduction electrons, χ_q^o, diverges at q=G for the symmetric half-filled band in which $\varepsilon_{G+k}=-\varepsilon_k$ is satisfied. Therefore, the localized spins are ordered in the antiferromagnetic arrangement. This ordering makes a gap at the Fermi surface of the conduction electrons. So, this s-d picture also leads to the insulating ground state.

The problem then we have is which one of the two ground states is more stable for a large value of U, one without any antiferromagnetic ordering we have considered in this section by the perturbation method and the other with antiferromagnetic long-range order derived from the s-d model. In the one- and two-dimensional cases, it is natural to suppose that the state with a gap due to mixing at the Fermi surface changes smoothly to the short-range antiferromagnetic state with a gap caused by the nesting in the s-d exchange model. However, in the three-dimensional case, we expect a transition from the paramagnetic state without long-range order to the antiferromagnetic state with long-range order.

6. Discussion

In this paper we considered the ground-state properties of the periodic Anderson Hamiltonian from several sides.

Particularly, we investigated the ground state of this Hamiltonian for a symmetric conduction band and for Ne/N=2. The calculated density of states of the localized electrons shows some structures near the gap at the Fermi surface and also large peaks at $\omega=\pm U/2$. The appearance of these peaks at $\omega=\pm U/2$ is quite similar to the case of the symmetric Anderson Hamiltonian for a single impurity. This seems to indicate the essential nature of the dense Kondo state.

In the one- and two-dimensional cases, the gap is shown to survive even when Coulomb repulsion U gets large. It seems likely that as U increases the state moves smoothly to the short-range antiferromagnetic state stabilized by the nesting effect due to the s-d exchange interaction in the case of a large value of U. This can be seen from the movement of the position of the gap towards G/2.

In the three-dimensional case also, the gap initially produced by the s-d mixing does not seem to be destroyed by introducing the Coulomb repulsion between d electrons for a symmetric band, although we can not give an exact proof of it in this paper. However, we expect that the state without long-range order makes a transition to the state with antiferromagnetic long-range order in the three-dimensional cases. Detailed consideration about this transition will be an important problem of future studies.

In this paper, we considered mainly the symmetric bands. The consideration for asymmetric bands is also left to the future studies.

References

1. H.J. Leder, B.Mühlschlegel: Z.Physik B29, 341 (1978)
2. H. Fukuyama, A. Sakurai: Prog. Theor. Phys. 62, 595 (1979)
3. K. Yosida, K. Yamada: Prog. Theor. Phys. Suppl. No.46, 244 (1970),
   K. Yamada, Prog. Theor. Phys. 53, 970 (1975),
   K. Yosida, K. Yamada: Prog. Theor. Phys. 53, 1286 (1975)

f Bands in Lanthanium Compounds

Akira Yanase and Akira Hasegawa*

Department of Physics, Tohoku University, Sendai, 980, Japan

*General Education Department, Niigata University,
Niigata, 950-21, Japan

1. Introduction

In order to discuss the valence fluctuating states in the rare earth compounds quantitatively, it is necessary to have the information on the conduction or valence bands, namely the shape of the Fermi surface and the density of states etc.. Moreover the interaction between 4f states on the rare-earth ions and the valence states plays an important role in the valence fluctuating phenomena. One of the ways to obtain these informations is the band calculation of the lanthanium compounds. In the La compounds, the 4f bands lie a few eV. above the Fermi level. The hybridization with the valence states causes an appreciable band width of the 4f bands and an appreciable distortion of the conduction and valence bands. Recently, we have calculated the electronic energy band structure of various La compounds with the 4f states of the La atoms [1]-[4]. Calculated results for the Fermi surfaces account reasonably well for the experimental results for the de Haas-van Alphen effect [5]-[8]. The calculated density of states for the valence bands agrees well with X-ray photoemission measurements[9]-[12]. In this paper, we will review these results and report the values of the mixing integrals which are directly derived from the band structures.

2. LaB6

The unit cell of the LaB6 crystal lattice consists of a La atom and a regular octahedron of six B atoms [13][14]. The La atoms and the B octahedra form two interpenetrating simple cubic lattices. In our earlier calculation, the 4f states of the La atoms were neglected completely [1]. We used the Xα method for the exchange term in the one-electron potential of the self-consistent APW calculation and neglected the relativistic effects. The calculated band structure is characterized by wide B sp bands and La sd bands. The La d bands appear near the Fermi energy, EF, and a strong hybridization of La d and B sp bands occurs. The strong hybridization causes a large wave-vector dependence of the d bands and EF cuts one such d band. There are three equivalent electron Fermi surfaces centred at the X points of the Brillouin zone, which are connected by necks at the axes. The number of electrons in these Fermi surfaces is one electron per cell. The calculated results for the Fermi surface agree well with the de Haas-van Alphen experiments, except the fact that the necks are too large [5][6].

To make a quantitative comparison with the experiments, both areas and cyclotron masses from some selected extremal cross sections were calculated. In table 1, the results are shown. The calculated results for the extremal areas agree well with the experimental results. However, the calculated cyclotron masses are much smaller than the observed ones. The disagreements are about 50%. A similar tendency was obtained between the calculated density of states, N(EF), at the Fermi energy and the observed ones. The calculated values for N(EF) are 9.57 states/Ryd.cell for α=1.0 and 8.78 for α=2/3, where α is the parameter in the $X\alpha$ potential. The recent experimental value is, on the other hand, 13.8 states/Ryd.cell [15]. We have ascribed an origin of this disagreement between the calculated and experimental density of states to the complete neglect of the 4f states of the La atoms.

Table 1 Areas and cyclotron masses for some selected extremal cross sections of the Fermi surface of LaB6. The calculated values are taken from ref.[1].

Center of orbit Direction of magnetic field	X <100>	Γ <100>	M <100>
Area ($2\pi/a$)			
Experiment			
Arko et al.(1976)	0.330	0.036	0.13
Ishizawa et al.(1977)	0.330	0.0353	0.135
Band calculation(1977)			
α=1.0	0.312	0.0372	0.141
α=2/3	0.313	0.0317	0.149
Cyclotron mass (free-electron mass)			
Experiment			
Arko et al.(1976)	0.610		
Ishizawa et al.(1977)	0.64		
Band calculation(1977)			
α=1.0	0.44	0.23	0.46
α=2/3	0.40	0.22	0.42

Recently, we have recalculated the energy band structure for LaB6 with the 4f states of La by a self-consistent APW method [2]. The method of this calculation is similar to that used in the previous work. Here, however, we use the local-spin-density approximation for the exchange term in the one-electron potential, and take the relativistic effects (except the spin-orbit interaction) into account. Fig. 1 shows the energy band along the symmetry axes and table 2 lists the values of the energies at the symmetry points in the Brillouin zone. The band structure obtained in this calculation agrees well with the angle resolved photoemission measurement [16], and the agreement is better than the previous calculation. Narrow 4f bands appear about 3 eV. above the Fermi energy. In an unpublished calculation which was carried out without the relativistic effects, the 4f bands appear about 1 eV. above the Fermi energy. This indicates the importance of the relativistic effects in the band calculation including 4f states.

The La 4f states interact with the B 2p states. This B 2p-4f hybridization effect causes an appreciable band width for the 4f bands and distortions of the conduction bands. For instance, at the Γ point of the Brillouin zone, the B 2p states form a Γ25 state just above the Fermi energy. As a result of the hybridization with the Γ25 state of the 4f bands, this Γ25 state of the conduction band goes down toward EF. Therefore, the width of the conduction band becomes smaller, and the density of states at EF larger than that of the conduction band without the effect of the 4f bands. The calculated results for N(EF) is 13.3 states/Ryd.cell, in good agreement with the experimental value mentioned above. The Fermi surface of this calculation remains about the same as in the previous calculation. From the energies of the two Γ25 states, we can estimate the mixing matrix element between these two states as 0.11 Ryd.. Using the other energies of the 4f bands, we can estimate the two-center integrals as (pfσ)=-0.037 and (pfπ)=0.010 Ryd.. The definition of the two-center integrals is given in Ref.[26].

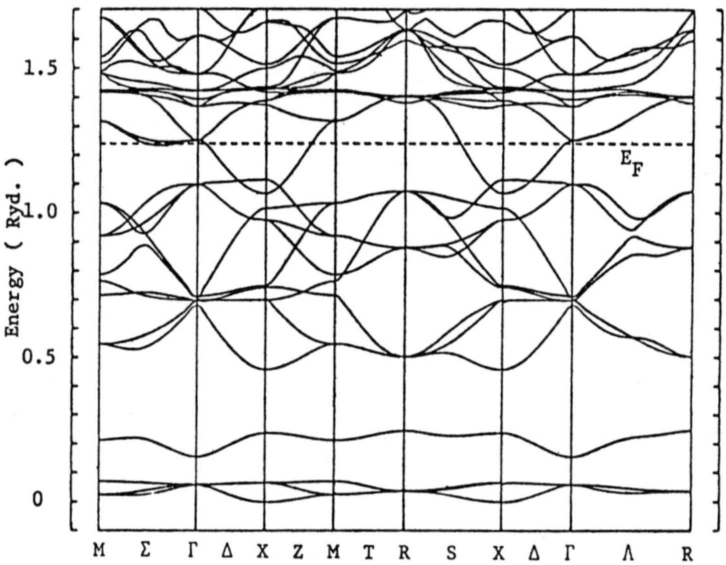

Fig.1 Self-consistent relativistic APW energy band structure for LaB6 with 4f bands. EF shows the Fermi energy.

We have also calculated the ferromagnetic band structure for EuB6 [17]. Because the Eu ion is divalent, EuB6 is expected to be semiconductive or semimetallic as CaB6 is [10]. Both of the top of the valence band and the bottom of the conduction band are located at three X points. The narrow bands due to the Eu 4f states lie 0.03 Ryd. below and 0.3 Ryd. above the Fermi energy for the majority and minority spin electrons, respectively. Near EF at the X point the 4f bands distort the valence band of the majority spin state appreciably. Here, the electron state at the top of the valence band, X3, consists mainly of B 2p states, and its wave function has the f symmetry about the Eu nucleus. Therefore, this states can mix strongly with the state with the same symmetry in the 4f bands. As a result of this strong 2p-4f hybridization effect, one X3 state goes up above EF, and another X3 state down below EF. The same values of the two-center integrals as in LaB6 could explain the separation of these states obtained by the APW calculation. In contrast to the valence band, the electron state of the conduction band bottom for majority spin state, X3', consists mainly of the B 2p and Eu d states, and its wave function has the d symmetry about the Eu nucleus. Therefore, the 4f bands distort scarcely the conduction band near EF. Consequently, the top of the valence band is higher by about 0.03 Ryd. than the bottom of the conduction band and then the ferromagnetic EuB6 becomes semimetallic. Experimentally EuB6 is semiconductive in the paramagnetic state and semimetallic in the ferromagnetic state [18]. We can conclude that the main mechanism of this observed transition is due to the B 2p-4f hybridization effect in the ferromagnetic state. It should be noticed that the selective nature of the mixing is important in this mechanism.

Table 2 Energies at the symmetry points in the Brillouin zone of the self-consistent relativistic energy band structure for LaB6.

Label	Energy(Ryd.)			Atomic character
Γ1	0.1543	0.6805	1.7050	B-s B-p La-s
Γ12	0.7081	1.3666		B-s B-p La-d
Γ15	0.0584	0.6972	1.4202	B-s B-p B-p La-p La-f
Γ25	1.2486	1.4774		B-p La-f
Γ2'	1.3915			La-f
Γ25'	1.0965	1.6088		B-p La-d
R1	0.2450	1.3791	1.5937	B-s B-p La-d
R15	0.5015	1.0736	1.6281	B-s B-p B-p La-d
R12'	1.6318			La-d
R15'	1.4026	1.7010		B-p La-f
R25'	0.0360	0.8797	1.4016	B-p La-p La-f
X1	-0.0011	0.2375	0.7415	B-s B-s B-p B-p
	1.3715	1.6591		La-p La-f
X2	0.7484	1.4241	1.7160	B-s B-p La-f
X3	1.1139	1.4345		B-p La-f
X5	0.0646	0.9747	1.4140	B-p B-p La-p
	1.4290			La-f La-f
X2'	1.5110			La-d
X3'	1.0653			B-p La-d
X4'	0.4576	1.0141	1.4970	B-s B-p B-p
	1.6562			La-s La-d
X5'	0.6985	1.3863	1.7450	B-s B-p B-p B-p La-d
M1	0.2120	0.7149	1.5143	B-s B-s B-p B-p La-d
M3	0.7854	1.4876		B-p La-s La-d
M4	1.5379			B-p La-d
M5	0.9205	1.6703		B-p B-p La-d
M1'	1.4169			La-f
M2'	0.0695	1.4195		La-p La-f
M4'	0.7637	1.4229		B-s B-p B-p La-f
M5'	0.0240	0.5455	1.0330	B-s B-p B-p B-p
	1.3179	1.4177	1.4811	La-p La-f La-f

3. LaAl2

The intermetallic compound LaAl2 belongs to a large group of rare-earth metal compounds in the cubic Laves phase, and has recently attracted particular interest, because it serves as a proper reference material for the study of the Kondo problem in its alloys with Ce and the valence fluctuating states in CeAl2 [19][20]. The space group of the cubic Laves phase crystal is identical to that of the diamond lattice, i.e. Fd3m. The method of the APW calculation is the same as that used in the recent calculation on LaB6[3]. The Fermi surface consists of various sheets in this compound. The eighth band has small hole pockets around L points in the Brillouin zone. The ninth band has complex mulitiply connected hole sheet distributed along the Brillouin zone boundary. A remarkable feature of this hole sheet is big arms connecting nearest-neighbour L points. Since L points form a simple cubic lattice in the repeated zone scheme, the topology of these arms is a junglegym. In addition, there exist small hole sheets surrounding X and W points. These three kinds of sheet connect with each other by slender arms. The density of states at EF of the ninth band amounts to about 80% of the total.

The tenth band has a large closed electron sheet centred at the Γ point. This electron sheet contains 0.43 electrons per unit cell. It is essentially a sphere but has bumps in <100> directions. The calculated results for the extremal cross-sectional area of the tenth band Fermi surface agree within 3% with the experimental results [7][8].

To investigate an actual effect of the 4f bands on the energy band structure, we have recalculated the band structure with modified logarithmic derivatives in which we have smoothed out the singularity of the l=3 component for the La ion. But, in a new band structure in which the 4f bands are removed in this way, the band structure near EF remains about the same. Therefore, it is rather difficult to find any experimental evidences from which the mixing parameters can be estimated.

Both of CeB6 and CeAl2 are believed to show typical examples of the dense Kondo system [19]-[22]. The difference in the transport properties of these compounds may reflect the difference in the band structure. There is one sheet of the Fermi surface with a light effective mass in CeB6, while, in CeAl2, there are one light electron sheet and one heavy hole sheet of the Fermi surfaces. The numbers of the electrons and holes are the same in this compensating material.

4. LaSb

The compound LaSb with NaCl structure is semimetallic and serves as a proper reference material to study a new type of the valence fluctuating states found in CeSb and CeBi. Very strong magnetic anisotropy was observed in the ordered phase of CeSb and CeBi, while the effective crystal field splitting in the paramagnetic phase is unusually small compared with other rare-earth pnictides [23][24]. To explain this anisotropy, a model of the anisotropic p-f virtual mixing mechanism or the anisotropic virtual valence fluctuating process is proposed recently [25]. The calculated results agree quantitatively well with the observed anomalous behaviour of the anisotropy and the crystal field splitting. The mechanism may be understood in the following way. The 4f states coexist with the valence bands and, through large p-f mixing integrals, make virtual bound states with fairly large width. When holes appear in the valence bands, the holes of the virtual bound state also appear. The amount of the holes in the virtual bound state depends sensitively on the magnetic structure and temperature and thus induce various unusual magnetic properties. Strong anisotropy observed in the ferromagnetic state is one of such unusual properties.

In order to estimate the p-f mixing parameters, we have used the band structure of LaSb [26], where the top of the valence bands is located at Γ point and the bottom of the conduction bands at three X points. The carrier density is 0.013 electrons and a compensating number of holes per unit cell. The two $\Gamma 15$ states of the La 4f and the Sb 5p bands repel each other due to the hybridization between these states. A similar situation occurs in X5' states. Using the shifts of these two 4f states, the two-center integrals between the nearest neighbour La 4f and Sb p states are estimated as $(pf\sigma)=-0.029$ and $(pf\pi)=0.015$ Ryd.. While the calculated band structure below the 0.6 Ryd. can be reproduced by the LCAO interpolation formula having 64 parameters with the maximum error of 0.01 Ryd.. From this interpolation, the two-center integrals are determined as $(pf6\sigma)=-0.041$ Ryd., $(pf6\pi)=0.006$ Ryd. and $(pf6\pi)=0.010$ Ryd., where f6 demotes the T1u type state and f7 the T2u type state. The other two-center integrals have no contribution to the mixing between the nearest-neighbour La 4f and Sb p states.

5. Conclusions

When the symmetry of the valence states is favourable, the absolute magnitude of its mixing matrix element with the 4f state of the rare—earth ions becomes about 1eV.. The strongest mixing of the 4f states occurs with the p states of B in LaB6 and p states of Sb in LaSb, because of the shortest atomic distance in crystal lattice. A similar tendency of the 4f mixing may appear in almost all the rare-earth compounds, because usually the metalloid or metallic ions other than the rare-earth ions tend to come at the nearest neighbour of the rare-earth ions. The order of magnitude of the matrix elements derived from the results of the APW band calculations are supported by the agreement with the experiments of the strong anisotropy in the CeSb and CeBi. However, this mixing depends very much on the symmetry of the states and then depends on the wave vector even in the same band. Since the 4f states have very anisotropic wave function, the mixing is strongly anisotropic and strongly depends on the ordering of the 4f angular momentum. Therefore, when the effect of the mixing with 4f states in the valence fluctuating materials is considered quantitatively, it might be essential to take into account the respective band structure and 4f configuration explicitly, because the effective mixing matrix element takes a value in a wide range from 0 to 1 eV. depending on the situation, even though the two-center integrals between the 4f and p states of the nearest-neighbour atoms in the compounds might be always of the order of a few tenth of eV..

References

1. A.Hasegawa and A.Yanase, J.Phys.F 7(1977) 1205
2. A.Hasegawa and A.Yanase, J.Mag.Mag.Mat. 15-18(1980) 887

3. A.Hasegawa ans A.Yanase, J.Phys.F 10(1980) 847
4. A.Hasegawa, J.Phys.C 13(1980) 6147
5. A.J.Arko, G.Crabtree, D.Karim, F.M.Mueller, L.R.Windmiller, J.B.Ketterson and Z.Fisk, Phys.Rev. B13(1976) 5240
6. Y.Ishizawa, T.Tanaka, E.Bannai and S.Kawai, J.Phys.Soc.Japan 42(1977) 112
7. J.Reichelt and K.Winzer, Phys.Stat.Solidi b89(1978) 487
8. E.Seitz, B.Lengeler, G.Kamm and J.Kopp, J.Physique Colloq. 40(1979) C5-76
9. M.Aono, S.Kawai, S.Kono, M.Okusawa, T.Sagawa and Y.Takehana, J.Phys.Chem.Solids 37(1976) 215
10. A.Hasegawa and A.Yanase, J.Phys.C 12(1979) 5431
11. M.Aono, T.C.Chiang, J.A.Knapp, T.Tanaka and D.E.Eastman, Phys.Rev. B21(1980) 2661
12. Y.Baer, R.Hauger, Ch.Zurcher, M.Campagna and G.K.Wertheim, Phys.Rev. B18(1978) 4433
13. P.Blum and F.Bertaut Acta Crystallogr. 7(1954) 81
14. R.Naslein, J.Etourneau and P.Hagenmuller: In "Boron and Refractory Borides, ed. by V.I.Matkovich (Springer, Berlin, Heidelberg, New York 1968) pp 262
15. T.Kasuya, K.Takegahara, T.Fujita, T.Tanaka and E.Bannai, J.Physique Colloq. 40(1979) C5-308
16. M.Aono,T.C.Chiang, J.A.Knapp, T.Tanaka and D.E.Eastman, Solid State Comm. 32(1979) 271
17. A.Hasegawa and A.Yanase, J.Physique Colloq. 41(1980) C5-377
18. T.Kasuya, K.Takegahara, M.Kasaya and T.Fujita, J.Physique Colloq. 41(1980) C5-161
19. C.D.Bredl, F.Steglich and K.D.Schotte, Z.Phys. B29(1978)327
20. F.Steglich, C.D.Bredl, M.Loewenhaupt and K.D.Schotte J.Physique Colloq. 40(1979) C5-76
21. A.Takase, K.Kojima, T.Komatsubara and T.Kasuya, Solid State Comm. 36(1980) 461
22. M.Kawakami, S.Kunii, T.Komatsubara and T.Kasuya, Solid State Comm. 36(1980) 435
23. R.J.Birgeneau, E.Bucher, J.P.Maita, L.Passell and K.C.Turberfield, Phys. Rev. B8(1973) 5345
24. P.Burlet, J.Rossat-Mignod, H.Bartholin and O.Vogt, J.Physique 40(1979) 47
25. K.Takegahara, H.Takahashi,A.Yanase and T.Kasuya, J.Phys. C13(1980) in press.
26. K.Takegahara, Y.Aoki and A.Yanase, J.Phys. C13(1980) 583

Valence Fluctuating States in Ce Compounds: Survey of Our Group Work

T. Kasuya
Department of Physics, Tohoku University, Sendai, 980, Japan

1. Introduction

Recently the valence fluctuating state in some rare-earth compounds is attracting much attention of both the theoretical and the experimental physicists [1][2]. Here we use the word of the valence fluctuating state in a wide sense that the 4f electronic configurations fluctuate in space and time in a wide sense.

As shown in Fig.1 [3], the 4f electron in the rare-earth atoms is well shielded by the $5s^2$ and $5p^6$ closed shells and thus keeps its atomic character in the crystals. However, when the 4f levels sit just on the Fermi level and two different valence states can coexist, it is possible even for the 4f electron to move through the crystal either by the quantum tunneling making a very narrow quasi-particle band or by the thermal hopping. Various unusual properties are observed which we call the typical, in a narrow sense, valence fluctuating phenomena. In some cases, the above 4f band may make a Wigner crystal such as in SmB_6 [4][5][6][7], in which an amorphous Wigner crystal is believed to be formed, still moving around on some rare-earth sites, but in some cases with increasing amount of interaction it may form atomic order, in which the 4f electron can not move freely through the crystal and thus we call it as a mixed valence material distinguishing from the valence fluctuating state in the above sense. The definition that the 4f level is at the Fermi energy is also vague because the 4f level has width by various mechanisms. In particular in Ce compounds the width can be fairly large, about one eV. As the center of the 4f level moves down, the valence fluctuation occurs in the tail part. Most remarkable feature of this situation is the dense Kondo-like behavior typically seen in various Ce compounds [1][2]. However, when such fluctuation occurs with the valence band, the actual feature changes drastically.

We have studied the present topic for the recent several years both theoretically and experimentally. In this article, we restrict our attention mostly on Ce compounds and make a brief survey of our group work on this subject. At first a general consideration for the valence fluctuating phenomena is done concerning important fundamental interactions and expected physical phenomena with very brief descriptions of the related our works. Then survey of Ce compounds is done by classifying them into three categories, the 4f-bands materials accompanied with the delocalized-localized 4f transition typically shown by CeN, the dense Kondo system with the d-f mixing typically shown by CeB_6 and the anomalous materials typically given by CeBi and CeSb with the p-f mixing.

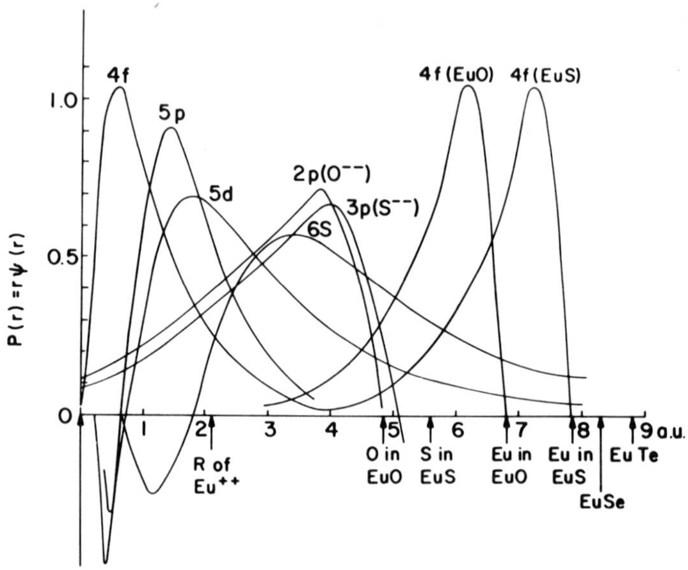

Fig.1 Atomic radial wave functions $\psi(r)$ multiplied by r, P(r) for 4f, 5p, 5d and 6s in Eu^{++} are shown as functions of distance r in a.u. for Eu^{++} at the original site. P(r) for 5s is nearly equal to that of 5p. For comparison, 2p and 3p functions of O^{--} and S^{--} for the nearest neighbour sites in EuO and EuS, as well as 4f wave functions for the nearest neighbour Eu sites are also shown.

2. General Consideration

When we call valence of a rare-earth atom, it means $Z'-n_f$ in which Z' means the nuclear charge subtracted by the electron charge up to the $5p^6$ closed shell and n_f the number of the 4f electrons. The most rare-earth atoms prefer trivalent state in the crystals. An energy to take out one 4f electron from a trivalent rare-earth atom and to add it on the 4f state of another trivalent atom requires about 20eV [8]. This is the so called correlation energy, the most important energy. This means that the valence fluctuation is difficult to occur in the atomic system. In the crystals this correlation energy is reduced in various ways. In the ionic crystals, both the electronic and the ionic polarizations can reduce the above value to a half-value. In metals, the conduction 5d electrons screen the charge more effectively and thus the correlation energy can be reduced to nearly one third, that is, about 7eV [9]. This is still much larger than the 4f-band width, at most leV in Ce compounds with the metallic 4f bands [10], and thus the 4f band is described by the limit of strong correlation in a good approximation.

In the rare-earth atoms, the well-known Russell-Saunders coupling, or the so called L-S coupling scheme, is well applicable. Then, the splitting between the ground and the first excited configurations is usually much larger than 0.1eV, at least around 0.04eV in Eu^{+++} and Sm^{++} between the singlet and the triplet states. Therefore, in the metallic 4f bands, the L-S coupling energy is comparable to the band width and thus the situation is complicated, but in the typical valence fluctuating 4f quasi-particle band such as in SmB_6

the width is about 100K or less and then the atomic configuration is well kept even for the transfer process. This is the origin of the word of valence fluctuation. This means that the 4f band is not a simple one but a quasi-particle one with many body effect. On the other hand the crystal field splitting is at most a couple of hundred K and thus comparable to the quasi-4f-band width. Therefore the treatment of the crystal field is complicated.

As seen in Fig.1, direct overlapping of the 4f states on the nearest-neighbour sites in the crystal is very small and thus the actual transfer occurs through the mixing with the conduction 5d electrons and/or the valence p electrons. In the usual compounds, the anion sits the nearest neighbour of the cation and thus the p-f mixing is most important. In the 4f-band model, which is obtained naturally by the usual band calculation, the 4f-band width of Ce compounds can be the order of one eV. In the heavier rare earths, the 4f wave function shrinks as the increasing number of 4f electron due to the increasing effective charge for the 4f electron and thus the band width decreases. For example, in EuO shown in Fig.1, the 4f-band width is a few tenth of eV which mostly comes from the wave-number dependence of the p-f mixing matrix, in the present case the antibonding effect for the 4f electron [11]. This bare band width becomes much narrower in the actual valence fluctuating quasi-4f-particle band due to various mechanisms. To treat this problem, we should consider the screening effect first because it has strong correlation with the band reduction mechanism.

As mentioned before, in some materials, the 4f electrons make bands in the similar sense as the usual 3d bands but in some materials the 4f electrons behave like atomic localized states making charge ordering. One of the most important force to make such a localization is of course the Coulomb interaction between the 4f electrons on different lattice sites. In the crystals, the bare Coulomb interaction is screened in various ways. In ionic crystals, the electronic and the ionic polarizations screen the Coulomb interaction. The former effect is typically represented by the optical dielectric constant ε_∞ and the latter $\varepsilon_p = (\varepsilon_\infty^{-1}-\varepsilon_0^{-1})^{-1}$, in which ε_0 is the static dielectric constant representing the total screening effect due to the above two mechanisms. Usually the typical frequency for the optical dielectric constant, which may correspond to the atomic optical excitation energy or a typical band gap, is high and thus always follows the 4f electron transfer fairly well. On the other hand the typical frequency for the ionic dielectric constant is the optical phonon frequency which is comparable to the 4f-quasi-band width and thus makes the situation complicated. In EuO mentioned before, for example, the gain of the energy to screen an extra charge on a lattice site by the ionic polarization was calculated to be about 0.7eV [12], which is also equal to the strain energy Δ of the lattice system in the harmonic approximation. When the 4f-band width is smaller than Δ, it is better for the 4f electron to localize on a lattice site to get full merit of the screening effect. Then the effective transfer matrix of the 4f electron is reduced by the overlapping integral of the polarization cloud which follows the tunneling transfer of the 4f electron. This reduction factor is estimated to be the order of $\exp[-\Delta/h\omega_{LO}]$ in which ω_{LO} is the typical LO phonon frequency, which is observed to be 0.05 eV in EuO giving an extremely small value for the reduction factor [3][13]. The above consideratión indicates that the ionic polarization has two effects, to screen the charge reducing the interaction and thus localization tendency but in the same time reducing the transfer matrix and thus the band width encouraging the localization. This duality is the very general feature for any screening mechanism. For example, the electronic screening mentioned before also contributes to the reduction factor in the same form as in the ionic screening but now $\hbar\omega_{LO}$ is replaced by a typical band gap or a typical atomic excitation energy which is usually the same order of or larger than Δ, which means now the polarization energy due to the electronic polarization. Therefore the reduction factor is not so much less than unity. In some special cases, for example in a narrow gap material, the reduction factor due to the electronic polarization can become substantially smaller than unity. In this case, however, ε_∞ becomes larger and thus ε_p smaller reducing the lattice polaron effect. Therefore the total band reduction becomes rather weaker. This is thought to be the case of Sm_3X_4, in which X=S, Se and Te, where Sm^{++} and Sm^{+++} coexist in the ratio of 1:2 showing the thermal valence fluctuation in higher temperature with the activation energy of about 0.12eV, which originates the lattice polaron [14][15][16]. Note that at low temperature the polaron moves by the tunneling with the reduced effective transfer but in higher temperature it is easier to move diffusively by mounting the potential barrier of the local polarization energy thermally. The tunneling motion at low temperature is not yet definitely seen even though there are some evidence in AC conductivity and in Mössbauer effect [16][17]. The screened Coulomb interaction is reduced to the order of 0.1eV in the nearest-neighbour sites but the long-range charge ordering temperature can be reduced to much lower temperature in the case of an incommensurable number of charge carrier per lattice site. Actually no long-range charge ordering is observed in Sm_3X_4 down to 1K.

In metals, the long-range character of the LO phonon interaction with the charged carrier is reduced by the screening of the conduction electron. Therefore, the polaron effect is also reduced and becomes the same order of magnitude as that of the acoustic phonon, in which Δ is of the order of 0.1eV. Note that in the latter the deformable potential originating from the shift of energy level by distortion is usually important for the electron-phonon coupling. Furthermore, the Coulomb interaction between 4f electrons on the nearest-neighbour sites is also reduced to the order of 0.1eV due to the screening of the conduction electrons. Of course, this screening also contributes to the reduction of the transfer matrix. Different from the earlier calculations taking into account only the leading term [18][19], oun numerical calculation shows much larger reduction giving a value of about 0.3 [20]. Nevertheless, in metals, compared with insulators, the effective reduction factor for the transfer integral is smaller while the screening of the Coulomb interaction is larger in favor of the existing of the delocalized narrow 4f-quasi-particle band. For the actual band width, the following reduction factors mentioned before should be also taken into account. (1) The band narrowing effect due to the strong correlation limit in the sense of the Hubbard model [15]. (2) Reduction of the transfer matrix because of the atomic many-body state given by the L-S coupling scheme instead of the individual 4f state treated in the simple band calculation. By these effects, for example in SmB_6 known as the most typical 4f-quasi-particle band material, the band width is reduced from the bare band width of a few tenths of eV mostly induced by the p-f mixing mechanism to the order of 100K in good agreement with experimental value of about 200K [7]. On the other hand, the Coulomb interaction of the 4f electrons in the near neighbour is estimated to be one order of magnitude larger. However, because the number of the charged carrier is not commensurable, the ratio of Sm^{++} and Sm^{+++} is about 4:6, the atomic localized charge ordering is not observed but the Wigner crystal is believed to be formed [5][7]. In actual crystal, the Wigner crystal is easily distorted by crystal imperfections which we call amorphous Wigner crystal. Note that the existence of the crystal lattice and the incommensurability are essentially important for the amorphous Wigner crystal formation. For detailed evidence of the Wigner crystal formation in SmB_6, see detailed papers [5][7]. The similar situation is expected also in TmSe [21]. To explain an important experimental fact of a gap formation in the one-particle excitation, the d-f mixing model was also proposed based on a simple model [22][23][24]. In actual SmB_6, due to many multiplicity of the 4f-quasi-particle, that is Sm^{+++}, and due to larger p-f mixing effect, the gap cannot be formed by the d-f mixing effect. See the feature of the real 4f-band structure in LaB_6 given in Yanase's paper in this volume [10].

So far we have treated the general polaron effect and the reduction factor in the real 4f transfer. The situation is different in the virtual valence fluctuation in which a 4f electron is virtually transfered to a conduction or a valence band electron and then comes back again to the initial site. In this virtual charge transfer process, the polarization is not necessary to follow the process as is known in the optical transition as the Frank-Condon principle. This is particularly important to consider the dense Kondo-like behavior in Ce compounds because the single-site virtual valence fluctuation is important for the single-site Kondo state which is much favorable than the real charge transfer process. When the 4f level is near the Fermi level, however, the accompanied energy shift effect becomes important. This problem is treated again in the following sections. In anyway, the polaron effect is complicated showing very different features depending sensitively on phenomena. This makes the valence fluctuation problem very attractive giving rich variety.

In the following, we restrict ourselves in the Ce compounds and treat three typical cases as typical examples of valence fluctuating problem.

3. Localized-Delocalized Transition

Among Ce compounds, Ce metal itself has been studied from the earliest date and the α-γ phase transition, in which the crystal structure is same and only the volume decreases in the low-temperature and high-pressure α phase, has been the most controversial problem [25]. Two models were proposed. The first one is the valence transition from the trivalent γ phase to the tetravalent α phase accompanied by the $4f \rightarrow 5d$ promotion. This model was criticized by Johanson [26] because α-Ce is substantially different from the typical tetravalent metals. Then he proposed the Mott transition of the 4f state, that is, the localized to delocalized transition of the 4f state. These two have been considered to be contradictory models but more close look shows that actually they are very similar indicating a common feature. Many Ce compounds show the similar α-γ phase transitions [1] in which the situations seem to be essentially same as that in Ce metal.

For the γ phase, both the models coincide, the well-localized atomic 4f state. The picture is shown in Fig.2a in which the lower 4f levels mean the occupied $4f^1$ state, that is, the energy required to take out one 4f electron and put it on the reservoir with the same Fermi energy as the compound. The initial state has fourteen multiplicity per atom due to J=5/2 and 7/2 multiplets. Note that because of the many-body character only one level is possible to be occupied. The energy is measured from the Fermi level. The upper 4f levels mean the energy required to take out one electron from the reservoir and put it on a 4f state at a Ce^{+++} atom. Therefore there are thirteen possible ways to add one extra 4f electron per atom for each initial $4f^1$ state, or thirty three multiplets in total for the final $4f^2$ states. Therefore these are not one-electron levels but typical many-body levels. The difference between the above two levels is the effective correlation energy and thus the order of 7eV mentioned before. Note that the Madelung energy and the all polarization effects considered in §2 are included in the present level scheme. When the f-d and f-p mixing effects are taken into accout, the

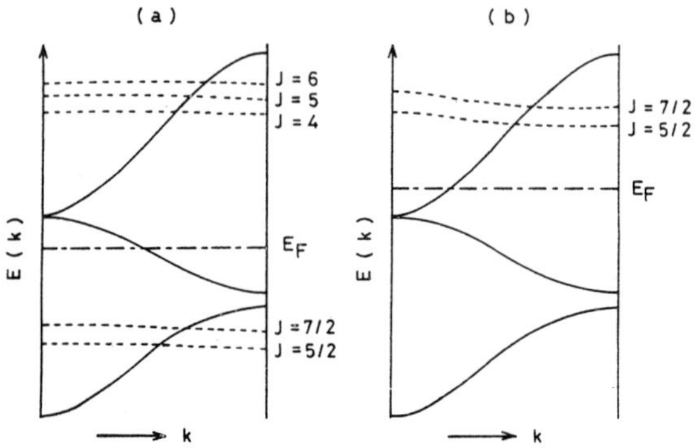

Fig.2(a) Schematic energy levels for $Ce^{+++}(4f^1)$ compounds. The solid lines mean usual one-particle band dispersion for conduction and valence bands while dotted lines the many-body 4f levels measured from Fermi energy E_F shown by dot-dashed line. The lower dotted lines are for the $4f^1$ levels and the upper dotted lines for the $4f^2$ levels corresponding the initial state of $4f^1(J=5/2)$. The $4f^2$ levels for the initial $4f^1(J=7/2)$ are shifted up by the energy difference of J=7/2 and J=5/2 states.
(b) Schematic energy levels for Ce^{++++} compounds. The scheme for the 4f-band picture is essentially same as those in La-compounds typically shown in LaB_6 [10]. Here the atomic picture for narrow band width is shown. The Fermi level E_F is shifted from (a) due to increase number of conduction electrons.

above levels are shifted and broadened. As is clear from the preceding consideration, the broadening occurs in two steps. One is a broader one with the bare mixing matrix corresponding to the Frank-Condon mechanism and another is a much narrower one due to the reduced mixing matrix mentioned before. When the second-order perturbation is applicable, the energy shift of the occupied state is given by the two terms, one corresponding to the $4f^1 \rightarrow 4f^0 5d$ virtual transition and another to the $4f^1 5d \rightarrow 4f^2$ virtual transition. Because of the large multiplicity number, the number of the path for the one-particle picture, thirteen in the latter process is much larger than the former, one, and thus the latter mechanism is usually more important even for a large denominator due to the correlation energy. On the other hand, for the final state, the number of the path for one-particle picture is equal, fourteen, but the denominators are smaller than the second process for the initial state causing shift up of the occupied level. This means in another view point that due to the mixing of the second 4f electron the binding energy of the first 4f electron decreases with increasing extent of the 4f wave function. When the lattice shrinks either due to decreasing temperature or applied pressure the mixing matrix increases which may cause the γ-α transition. The most remarkable feature of the γ phase is the dense Kondo behavior, which is treated in the next section.

For the α phase, too, there are no fundamental difference between the two models. In the α phase, it is necessary to give energy to put an electron to a 4f state from the reservoir. In the 4f-band model, the situation is essentially same as in La compounds, for example LaB_6 in this volume [10]. Even in LaB_6, which is one of the most typical monovalent metals, by the mixing effect, the total expectation value of the 4f atomic state $\langle n_f \rangle$ is expected to be several tenths. In Ce compounds, the 4f levels are shifted down a little bit to give $\langle n_f \rangle$ about unity or a little bit more but still expected to be a few eV above Fermi level. The situation is similar in the atomic picture, too. As shown in Fig.2b, only difference is that now the 4f bands are described by the strongly correlated polarons accompanied by various types of polarization clouds. Even so, the bare mixing matrix is effective for the major mixing process and thus the level of the 4f state is determined so as to give appropriate value of $\langle n_f \rangle$ similar to that of the band model. These polarization features effective particularly in the atomic 4f model are believed also to be included at least partly by the effective potential in the usual 4f-band model. Actually the levels of the 4f states obtained by the band calculation, in which the potential is obtained without any adjustable parameter, agree fairly well with the experimental values [10]. In this sense, there are no essential differences between two pictures. In some compounds, the 4f band may be a narrow quasi-particle small polaron band but in most Ce compounds the 4f bands seem to be sufficently wide as in LaB_6. There is, however, one big difference between La^{+++} and Ce^{++++} compounds. In Ce compounds, because the energy difference between tri- and tetra-valent states is small, any lattice imperfection has possibility to produce trivalent state or a complex mixture around the defect. Therefore the observed properties depend sensitively on samples exhibiting extrinsic unusual properties. In this sense, experiment using well-controlled sample is essentially important. Note that, as is clear in the above considereation, non-integral value of $\langle n_f \rangle$ does not mean anomalous valence fluctuating properties.

If the γ-α transition occurs continuously, really anomalous properties may be expected in the midway of the transition. It is, however, difficult problem to follow the transition theoretically, because of the difficulty to describe the mixture of the one-particle and the many-body states. Even for the simplest model for one impurity problem, it is not so easy to follow the transition, which is known as a limit of the asymmetric Anderson model [27]. In this process, too, in Ce compounds, different from Sm compounds, the narrow quasi-particle polaron band has not yet been observed but rather broad 4f band seems to be common. In this case, as in the asymmetric Anderson model [27], nothing particularly anomalous phenomena seem to occur. Here, again, experiments with more well-controlled samples are required.

4. Dense Kondo System, CeB_6

CeB_6 is the most typical example of the dense Kondo system. It crystalizes in the CsCl type in which Ce atoms form a simple cubic lattice and a octahedral B_6 molecule replaces Cl [28]. The band structure is expected to be similar to that of LaB_6 except the 4f-band part. The occupied 4f level is expected to be 1-2eV below the Fermi level from the resonant photoemission experiment [29]. In Fig.3, temperature dependences of resistivity ρ and thermoelectric power Q are shown as most typical dense Kondo properties [7]. The resistivity has a peak at T_1=3.15K and a sharp shoulder at T_2=2.2K. These correspond to magnetic phase transitions as seen later. The temperature dependence of resistivity above 5K is fitted very well by the well-known form applicable for the dilute Kondo system giving the Kondo temperature T_K=7.7K, the unitarity limit resistivity ρ_u=121$\mu\Omega$cm for s=0.4 [30]. The resistivity of $Ce_{.75}La_{.25}B_6$ is also fitted very well with the parameters, T_K=6.5K, ρ_u=93$\mu\Omega$cm for s=0.4 [31]. When ρ per Ce atom is plotted by T/T_K, both the resistivity curves coincide very well indicating that any interference effect among individual Kondo state on each Ce atom is not seen even at $T\sim0.5T_K$ [31]. The thermoelectric power also shows clear logT dependence from 6K up to 80K, then with a kink changes to another logT slope up to room temperature.

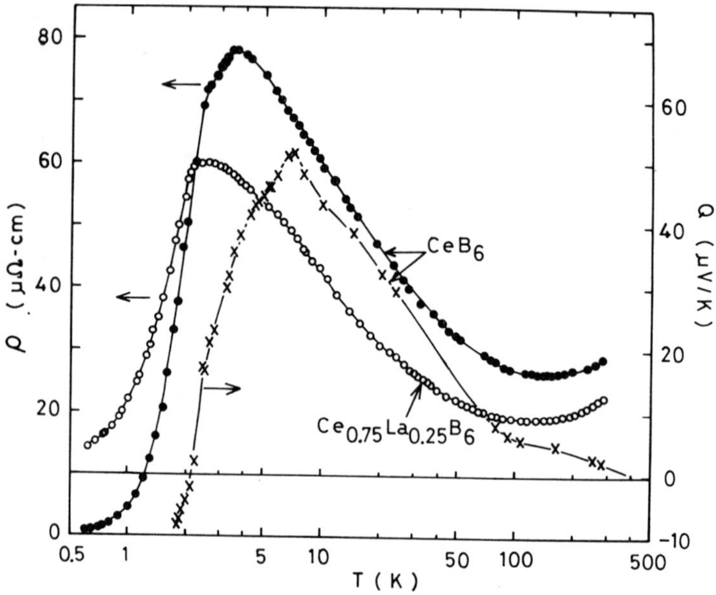

Fig.3 The temperature dependences of the electrical resistivity ρ for CeB_6 and $Ce_{.75}La_{.25}B_6$ are shown as functions of logT in the left-hand-side scale. The thermoelectric power Q of CeB_6 is also shown as a function of logT in the right-hand-side scale. Note the zero point in the scale.

Magnetic measurements [7][31][32] and their analyses [33][34] indicate that the spin-orbit split ground multiplet J=5/2 splits further by the cubic crystal field in which the Γ_7 doublet is lower than the Γ_8 quartet because of a large Coulomb interaction of the Γ_8 quartet with the 5d(e_g) conduction band electrons. This situation also causes a larger d-f Coulomb exchange interaction in Γ_8 which kills the Kondo state of Γ_7. Therefore, the dense Kondo behavior is believed to come from the Γ_7 doublet. Estimation of ρ_u has some difficulties. The simplest model may be as follows. The conduction bands are formed by the three branches with the minimums at the three X points of the k space. When we simply assume that the Γ_7 Kondo states are formed independently by these three branches, it may be possible to apply the usual s-band model with S=1/2 impurity. Then

$$\rho_u = \frac{4\pi\hbar}{\nu^2 e^2 k_F} = 112\mu\Omega\text{cm} \qquad (1)$$

where ν=3 is the number of valleys, k_F is the Fermi wave vector for each valley and thus $5.15\times10^7\ \text{cm}^{-1}$. The agreement with the value obtained before is very good. When we assume instead of the above model that each Kondo state is constructed by the three branches equally, a factor $\nu\sin^2(\pi/6)$=3/4 should be multiplied to (1) giving ρ_u=84μΩcm, still not so bad agreement. Most ambiguity comes from the fact that the three branches are not independent but their Fermi surfaces touch each other making twelve necks [10]. Because of the saddle point character of the neck, the total density of state around the necks is about three times larger in LaB_6 than that of the main spherical regions we have considered so far. Fermi surface of CeB_6 is not known but the low-temperature transverse magnetoresistance indicates [31] that in CeB_6, too, the neck should exist even if it may be smaller than that in LaB_6. See Fig.6. It is difficult to evaluate the neck effect but it may be an important factor to explain the difference from ρ_u in dilute Ce in LaB_6, ρ_u=300K for T_K=1K [35]. Actually, when we multiply ρ_u simply by a factor four corresponding to the actual density of state including the necks, we obtain reasonable agreement with ρ_u of dilute Ce-LaB_6 system. Note that the neck part contributes a little to the conductivity because of small velocity. It should be also noticed that we can not use the simple s-band model with the $\ell=3$ scattering in the present system as was done by [35]. Of course, there is a big possibility that, due to increasing interference effect with lowering temperature, the low-temperature resistivity is suppressed giving spuriously higher T_K and smaller ρ_u. More detailed investigation is necessary on this point. It is interesting to notice that the entropy at T_1=3.15K, at which the resistivity shows maximum, is 0.88R≡1.27Rln2, in which R=8.31joul/moleK, and becomes 1.15R=1.66Rln2 at T=8K∿T_K [36]. While the specific-heat measurement on the dilute alloy of Ce on $LaAl_2$ [37][38] indicates that the entropy of the doublet dilute Kondo state is quenched about 30% at T∿0.5T_K and about 25% at T∿T_K. This means that in CeB_6 a substantial amount of quartet is still coexisting even in low temperature even though the crystal field splitting is expected to be about 50K from the analysis of the magnetic measurement [33]. We interpreted the T_1 phase transition as follows. The dispersion of Γ_8 due to the mixing type exchange interaction J_{QD} increases with decreasing temperature because of increasing number of the Γ_7 doublet and the bottom of the Γ_8 band touches the doublet level at $T=T_1$, similar to the phase transition in the singlet ground state system [33]. Then, to explain the above-evaluated large Γ_8 population, an unusually sharp large peak of density of state should exist at the bottom of the Γ_8 band [39]. In this sense, too, a smaller T_K value seems to be reasonable.

An unusual logT dependence of the inverse life time T_1^{-1} in the NMR measurement on Boron was also found by Yasuoka group in CeB_6 from 100K up to 800K [40]. This seems to have relation with the persistence of the dense Kondo state but still no detailed theoretical calculation exists. Temperature dependence of the Hall constant is weak. The evaluated effective number of the conduction electrons per Ce atom is 0.99 at room temperature in good agreement with that of LaB_6 and 0.90 at 4.2K. This change is also consistent with the dense Kondo picture due to large distribution of energy-dependent life time [7]. Note also a small effect of the necks on the Hall constant.

Recently the neutron diffraction experiment for zero magnetic field has been performed in Grenoble revealing interesting properties [41]. Nearly equal intensities for two wave vectors, [1/4, 1/4, 0] and [1/4, 1/4, 1/2], appear rather gradually at T_1=2.7K and increase continuously with decreasing temperature without any visible change at T_2=2.2K. As shown in Fig.4, a large sharp peak is observed at T_2 in the specific-heat measurement [42]. Ultrasonic measurement also showed a peak of absorption at T_2 but no peak at T_1 [43]. These experiments indicate rather stronger magnetic phase transition at T_2 than at T_1. This strange fact is considered to be due to the interference effect of the dense Kondo system. The neutron diffraction experiment determines the ordered wave vectors but can not determine the ordered phase factors. Actually, in the present case, we can choose the phase factors so as to create a homogeneous strength of local moment on each site as well as to create three quartets on non-magnetic site. From various magnetic properties obtained so far, we think the following model most probable. Below T_1, the antiferromagnetic sheets and the Kondo non-magnetic sheets are considered to coexist in the sandwich-like way. For the pure exchange interacting system,

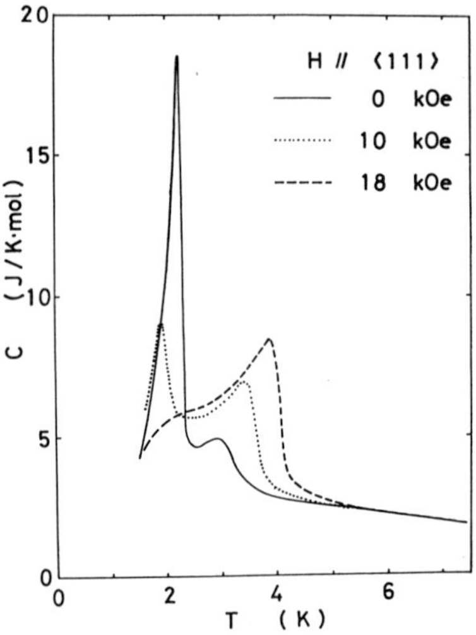

Fig.4 Specific heat of CeB_6 under various strengths of magnetic field applied along [111] axis.

this kind of ordering is impossible to be stable because there is no reason for the non-magnetic sheets to exist. The underlying principle is a kind of conservation of the Kondo character. Because of a fairly extended character of each Kondo state, the interference effect should be of long range. Therefore the total interference effect is roughly proportional to the total amount of the Kondo character but does not depend sensitively on the detailed local structure. When both the entropy and the energy of each Kondo state under an effective field decrease sharply at first and then rather gradually with increase of the induced moment, the above described structure is stable at higher temperature in which the entropy effect is dominating. With decreasing temperature, the energy term becomes more important and at T_1 the energy term is thought to overcome the entropy term, then the state with the homogeneously induced moment becomes more stable. Therefore, the T_2 transition is defined as the phase factor phase transition due to the interference effect of the dense Kondo system, or the induced moment Kondo lattice system. To confirm this picture, more detailed investigation is necessary.

To study the characteristic properties of the Kondo lattice system, temperature dependences of the specific heat [44] and resistivity [31] have been measured in low temperature. As shown in Fig.5, the resistivity of CeB_6 is given by the form, $\rho=(0.5+4.2T^4)$ μΩcm for 0.53K<T<1.7K. This is different from the well-known T^2 law observed in $CeAl_2$ and $CeAl_3$, other typical dense Kondo systems, and also expected in the Fermi liquid theory. However, in $Ce_{.75}La_{.25}B_6$, we observed clear T^2 law as shown also in Fig.5, that is, $\rho=(9.2+12.8T^2)$ μΩcm for 0.56K<T<1.8K. To check the nature of the residual resistivity, the magnetoresistance was measured as shown in Fig.6 at the lowest temperature. In $Ce_{.75}La_{.25}B_6$, the resistivity decreases from 13.7μΩcm at zero field to 2.7μΩcm at 95kOe and is still decreasing gradually. On the other hand, in CeB_6, the longitudinal magnetoresistance is nearly constant, 0.35μΩcm, for the field value between 30 to 95kOe. Note that below 30kOe the system shows some antiferromagnetic phase transitions including domain rearrangement. Note also that the transverse magnetoresistance shows clearly the cyclotron orbital effect indicating the neck effect. As shown in Fig.7, even in CeB_6, the temperature-dependent part of the resistivity shows large negative magnetoresistance approaching the value of the residual resistivity

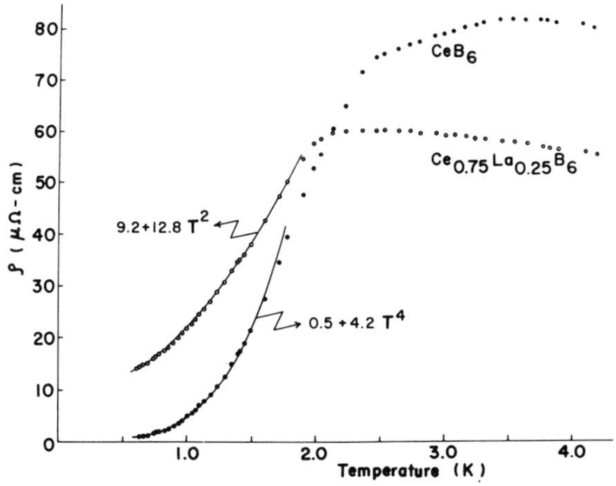

Fig.5 Resistivities of CeB_6 and $Ce_{0.75}La_{0.25}B_6$ are shown as functions of temperature. The solid lines indicate the fitting curves using the forms shown in the text.

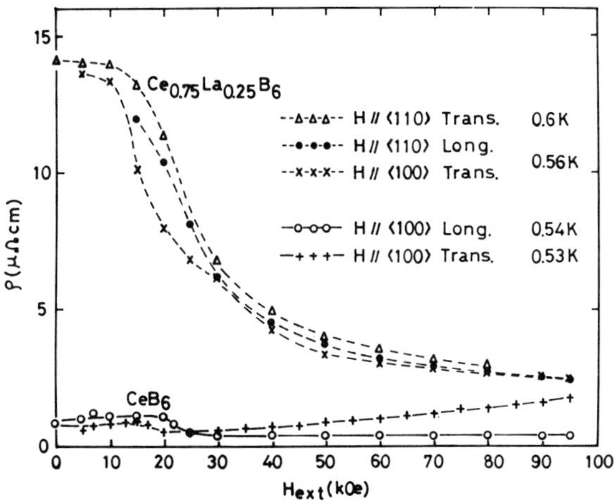

Fig.6 Resistivities of CeB_6 and $Ce_{0.75}La_{0.25}B_6$ are shown as functions of external magnetic field in the lowest available temperatures.

at high field. These features seem to indicate the following physical picture. In CeB_6, the residual resistivity is mostly due to non-magnetic impurities or defects probably on boron sites. The thermal elementally excitation on the perfect Kondo lattice seems to be a wave like as is common in any ordered system giving a higher T power, T^4, in the temperature dependence of the resistivity. Specific-heat measurement done recently by Furuno [44] shows that in the same temperature range the T^3 law exists consistently with the resistivity result. Note that the phonon-like, and the usual antiferro-magnon-like, q-linear dispersion gives the T^3 specific heat and the T^5 resistivity in the simple metals. In the present system, due to the extra-

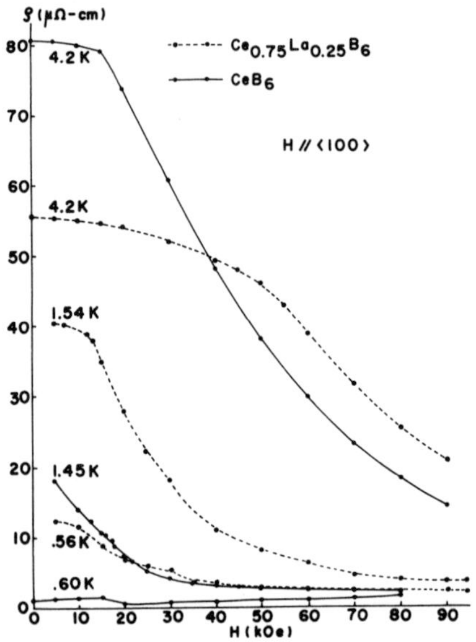

Fig.7 Resistivities of CeB_6 and $Ce_{0.75}La_{0.25}B_6$ are shown as functions of external field for various fixed temperatures.

boundaries produced by the antiferromagnetic ordering, the temperature dependence of the resistivity is more complicated. But anyhow the large value of resistivity indicates that the main scattering mechanism is due to the broken coherence of the Kondo lattice due to the thermal fluctuation. Note that even at room temperature, at which the spins are nearly perfectly at random dominated by the stronger scatterer Γ_8 states, the magnetic resistivity is about 20μΩcm because of weakened Kondo effect. In $Ce_{.75}La_{.25}B_6$, the coherent Kondo lattice character is broken locally around each La impurity, or each La cluster, causing large individual Kondo scattering. The large negative magnetoresistance is due to weakened Kondo scattering. Then the observed T^2 dependence may be simply interpreted as the broken wave character due to impurities. In this respect, it is interesting to notice that below 0.6K down to 0.1K the temperature dependence of the specific heat on CeB_6 changes to $T^{1.3\pm0.2}$ law. This weak T dependence may be also due to impurity effect. Note that in $CeAl_3$ and in $CeAl_2$ the T-linear law of the specific heat had been reported [45][46]. However, it should be noticed that it is difficult to obtain good single crystals of these materials. More detailed careful experiments are needed.

5. Ce Pnictides, p-f Mixing System

Ce pnictides crystallize in the rock-salt-type structure. CeN is a typical valence fluctuating material as was described in §3. In other Ce pnictides, Ce is mostly trivalent. The inelastic neutron scattering done on various rare earth pnictides indicates that except Ce pnictides the point charge model is well applicable [47]. Ce pnictides are anomalous and in particular in CeSb and in CeBi the crystal field splittings between Γ_7 and Γ_8 are very small, 3.5meV and 0.8meV, respectively, one order of magnitude smaller than those expected from the extrapolation of other pnictides, about 260K for both materials. Contrary to such small crystal fields in the paramagnetic region, the magnetic anisotropies in the ordered phases are extraordinarily large, too large to get accurate values but substantially larger than one million Oe [48]. Ordered phases are also very unusual including many phases which are consisted with the strong Ising-like moments insisting [001] direction and furthermore coexisting of zero-moment sites in CeSb [49]. This zero-moment state is, however, not the singlet state but is due to the usual thermal fluctuation of the degenerated doublet due to small effective field [50]. Various properties indicate that Γ_7 is the lower level in zero field in the paramagnetic region but the polarized neutron scattering under 47kOe revealed that Γ_8 is the dominant character [51].

The band calculation on La and Y pnictides [52] shows that the materials are typical indirect narrow gap semiconductors or semimetals with the bottom of the conduction band at the three X points and the top of the valence band at the Γ point, in agreement with various experiments. The gap becomes narrower in heavier pnictogen, seems to close in near As. Therefore CeBi and CeSb are semimentals in the paramagnetic region with equal number of holes and electrons, about a few percents per Ce atom. As is seen in Fig.1, because the pnictogen atoms sit on the nearest-neighbour sites of Ce, the p-f mixing is much more important than the d-f mixing. We have enphasized this point for more than ten years but still many physicists do not recognize its importance. The observed anomalous properties in CeBi and CeSb should be attributed in most part to the p-f mixing and the small number of holes in the valence band near at the Γ point.

In the Γ point, the valence bands are split into the spin-orbit coupled J=3/2 quartet, P(3/2), and the J=1/2 doublet, P(1/2). It is easily shown that each of the quartet couples through the p-f mixing with a particular Γ_8 quartet of the 4f states but Γ_7 and P(1/2) have no mixing matrixes [53]. Because the 4f levels sit about one or two eV below the Fermi level [54], P(3/2) levels are pushed up and Γ_8 pulled down by the bonding-antibonding effect. When both the states are occupied, no real energy gain is obtained when a Γ_7 state changes to a Γ_8 state. Therefore, the energy shift of the P(3/2) holes is observed as the real energy gain of Γ_8 relative to Γ_7. It is possible to evaluate the p-f mixing matrix from the band calculation of LaSb [52][53] as follows, (pfσ)=0.5eV and (pfπ)=-(pfσ)/2. Then assuming that the band structure of CeSb without the p-f mixing is the same as that of YSb, in which the number of holes is 3% per Y, we can evaluate the effective Γ_8-Γ_7 splitting due to the mixing to be near 100K in the dilute Ce in YSb and 240K in CeSb with the number of holes of 1.26% for each P(3/2) state, in good agreement with experimental value. In CeBi, because of the increase of the p-f mixing matrix, the cancellation is more perfect and in CeAs and in CeP the cancellation is less, again consistent with experiments [56]. The temperature dependence of the Γ_8-Γ_7 splitting is complicated because the numbers of P(3/2) and P(1/2) change with temperature. Furthermore, our preliminary calculation shows that in the intermediate temperature region a state in which the Γ_8 quartet states are further split into two doublets becomes stable [56]. Finally, a quartet ferromagnet in which one quartet state is split becomes most stable. In this situation, however, the corresponding P(3/2) state is pushed up so large that the holes can not be localized around the Γ point. Therefore, it is necessary to solve the whole valence bands exactly for a particular configuration of the 4f state. Then, of course, P(3/2) and the Γ_8 and Γ_7 are not the eigen states. We solved the problem for five different states of the 4f electron; two kinds of Γ_8, three different directions, [100], [110] and [111], of the moment with the maximum moment value J=5/2. Some results for the dispersion in the valence bands and the resultant Fermi surfaces are shown in Figs.8 and 9. The numerical results are shown in Table I [56][57]. It is seen that concerning the mixing energy the quartet ferro and the J=5/2 fully saturated ferro have nearly equal energy

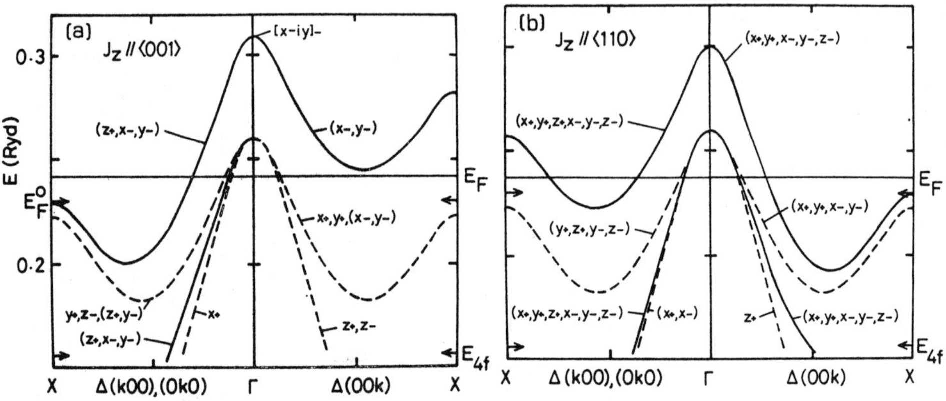

Fig.8 Dispersion curves of the perturbed valence bands along three Δ axes are shown. Shifted bands are shown by solid lines and unshifted bands by broken lines. Each state is constructed by a linear combination of states shown in the parentheses. Positions of the unperturbed Fermi energy E_F^0 and the 4f level E_{4f} are indicated by arrows. (a) is the case for the perfect ferromagnetism with moment along <001> and (b) along <110>.

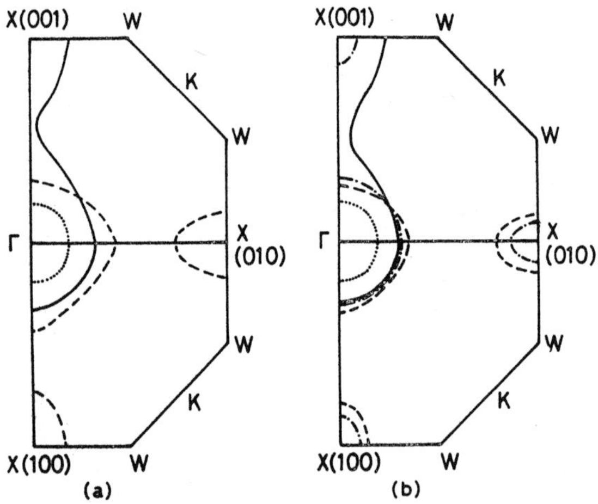

Fig.9 Calculated Fermi surfaces for the highest valence band are shown in various cases. (a) The solid line for the |Qν> ordering and the broken lines for the |Qλ> ordering. (b) The solid line for J_z//<001>, the broken lines for J_z//<110> and the dot-dashed lines for J_z//<111>. Unperturbed Fermi surface is shown by the dotted line. See Table I.

gain but the latter can save the point charge crystal field energy much more than the former and thus the J=5/2 ferromagnetic state is most stable in agreement with experimental result. Experimentally, the ground state without external field is not ferromagnetic but a complicated antiferromagnetic state with [001] ferromagnetic sheets. With some applied field, the ferromagnet with nearly fully saturated J=5/2 ferromagnetic state becomes stable. So

Table I Expectation value of J_z, $\langle J_z \rangle$, Fermi energy E_F, total number of holes n_h, spin polarization of the conduction bands $n_{\downarrow}-n_{\uparrow}$, calculated energy shift, $U=U_p+U_d+U_{PC}-U_D$, in which U_p is the energy gain through the valence band, U_d through the conduction band, U_{PC} the point charge crystal field energy and U_D the energy of the doublet, and reduction of the 4f character Δn_f. The value of the crystalline field splitting by the point charge model is taken to the 230K and thus U_D=-13.21meV.

Ferromagnetic ordered state		$\lvert Q\nu \rangle$	$\lvert Q\lambda \rangle$	$J_z=5/2$		
				J_z//<001>	J_z//<110>	J_z//<111>
$\langle J_z \rangle$		11/6	3/6	5/2	5/2	5/2
E_F	(Ryd)	0.242	0.242	0.242	0.238	0.237
n_h	(per Sb)	0.082	0.082	0.082	0.064	0.059
$n_{\downarrow}-n_{\uparrow}$	(per Sb)	0.014	0.003	0.019	0.017	0.017
U_p	(meV)	-16.74	-16.33	-15.17	-8.64	-7.21
U_d	(meV)	-4.10	-3.85	-4.20	-1.96	-1.57
U_{PC}	(meV)	6.61	6.61	3.30	-0.83	-2.20
U	(meV)	-1.02	-0.32	-2.86	1.78	2.23
Δn_f	(% per Ce)	2.20	2.20	2.23	1.25	0.89

far, we have investigated only the ferromagnetic configuration. The strong ferromagnetic interaction exists mostly within the [001] plane because the J_z=J state mixes mostly with the p states on the pnictogens in the same [001] plane. In this sense, the interplane interaction is weak in agreement with the experimental fact. It is interesting to note that in CeSb the energy difference between the J=5/2 ferromagnetic state and the perfect doublet state is surprizingly small. This corresponds to the fact that in CeSb the perfect doublet non-magnetic layers coexist mixing with the perfect J=5/2 ferromagnetic layers even at low temperature because of the entropy gain. In this sense, CeSb is situated near at the critical condition. In CeBi, because the p-f mixing is larger, the energy difference is larger and thus no mixing of the doublet layers is observed. In CeAs and in CeP, the p-f mixing is smaller and thus the perfect doublet state is most stable as is observed experimentally. On the other hand, the anisotropy energy corresponding to the different orientations of the perfect ferromagnetic moment is large. The anisotropy field may be defined such as, for example for [110],

$$g_J \mu J H(1-\cos\tfrac{\pi}{4}) = \Delta U(110) \qquad (2)$$

in which g_J is the usual g factor and thus 6/7, μ the Bohr magneton, J=5/2 and $\Delta U(110)$ is the energy difference between [110] and [100] ferromagnetisms, and thus 4.64meV. Then H is evaluated to be 1280kOe, consistent with the experimental fact mentioned before. It is also clear that the first excited state observed in the ordered CeBi and CeSb phases by neutron scattering should be to the doublet [58][59] because the energy difference to other unoccupied quartets is very large because no corresponding P(3/2) valence holes exist. This is, of course, a very crude description and a more detailed calculation for the excitation spectra is in progress.

The calculation under applied field is now in progress but it is clear from the above consideration that even in the paramagnetic region, when a sufficiently large external field is applied to induce substantial amount of the moment, population of a particular type of the Γ_8 quartet increases with a non-linear enhancement factor changing the main character of the induced moment to Γ_8 as was observed by the polarized neutron scattering mentioned before. It is also expected that with increasing amount of the induced mo-

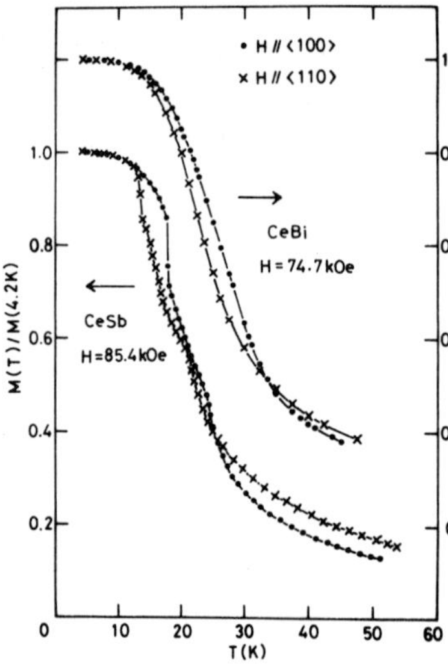

Fig.10 Temperature dependences of the normalized magnetization of CeSb and CeBi under fixed magnetic fields along [100] and [110].

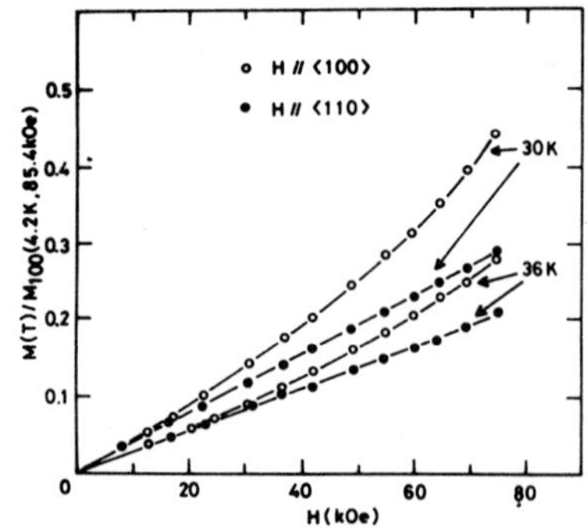

Fig.11 Field dependences of the normalized magnetization of CeBi for two different directions of the applied field, [100] and [110].

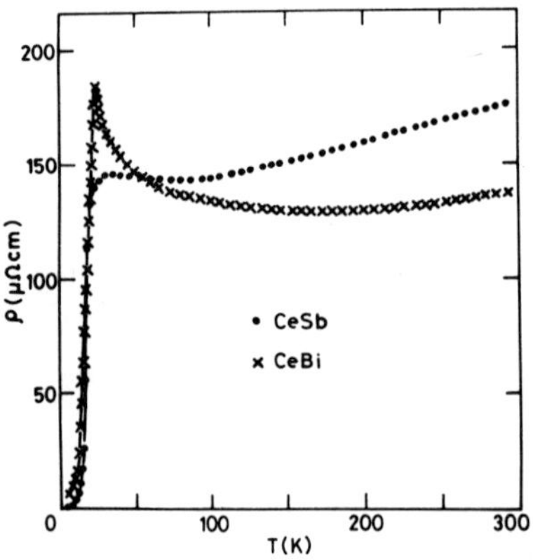

Fig.12 Temperature dependences of the electrical resistivity for CeBi and CeSb.

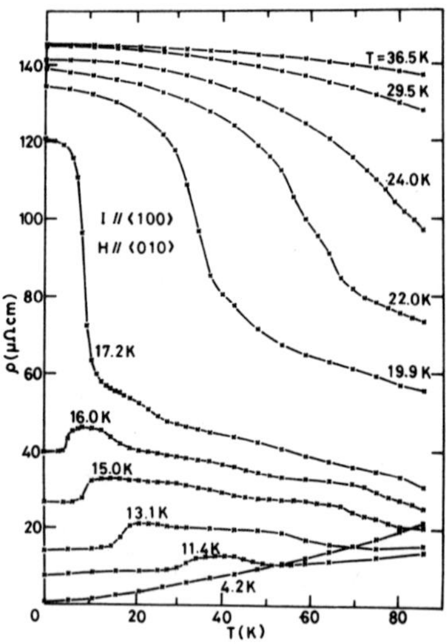

Fig.13 Magnetic field dependences of the electrical resistivity of CeSb at various fixed temperatures.

ment, a similar anisotropy as was observed in the ordered phases should be also induced. This was directly checked by measuring the induced moments in various directions and the results are shown in Figs.10 and 11 [60]. Because of the cubic crystal, the initial susceptibility should be isotropic but with increasing amount of the applied field the non-linear increase appears in the [100] direction. As a result, for a field strength of about 80kOe, nearly the same amount of the anisotropy as that in the ordered phase was observed even at the temperature as high as 50K, about $3T_N$ in CeSb.

It is interesting to check the transport properties in connection with the dense Kondo behavior observed in the f-d mixing system, because now the f-p mixing is much larger which should causes higher T_K. The results for the temperature dependences of the resistivities of CeSb and CeBi are shown in Fig.12. There are some similarities with those of $CeAl_2$ and $CeAl_3$ as well as other so called valence fluctuating Ce compounds but further detailed investigations are necessary to solve the present point accurately. It is also interesting to check how the degenerated doublet non-magnetic sheet behaves in the scattering process. The results are shown in Fig.13. A sudden small increase of the resistivity at a low-field value in each curve for T=16-11.4K means the beginning of the mixed phase with a large amount of non-magnetic sheets. The change is smaller than that we expected. This may have a relation with the fact that in the ordered phase the main current carrier seems to be the highly shifted valence holes which have, however, a little mixing with the doublet and thus suffer only weak scattering.

6. Conclusions

Various anomalous properties of the valence fluctuating, in the wide sense, Ce compounds have been surveyed by classifying them into three groups based mostly on our group's works. We could partly understand their unusual behaviors but still many mysterious properties are unsolved awaiting our further attacking.

Finally, the members of our group contributed to the present problem are listed. In Tohoku University, A. Yanase, A. Hasegawa, now in Niigata University, K. Takegahara, Y. Aoki, K. Hanzawa, M. Takeshige and Miss H. Takahashi for theoretical work including band calculation, T. Komatsubara, T. Ishii, both in Tsukuba University now, T. Suzuki, M. Kasaya, S. Kunii, T. Fujita, M. Kawakami, now in Kagoshima University, H. Shida, Y. Isikawa, A. Takase, T. Nakajima, M. Sera, N. Sato, H. Kimura and M. Sugita for experimental work. In other institutes, the collaborated works with Prof. M. Date and Mr. T. Sakakibara in Osaka University, Prof. H. Yasuoka and Mr. H. Takigawa, Prof. S. Tanuma and Mr. I. Oguro, Prof. W. Sasaki and Mr. T. Furuno in University of Tokyo and Dr. J. Rossat-Mignod and Dr. P. Burlet in Grenoble are greatly appreciated.

References

1. Valence Instabilities and Related Narrow-Band Phenomena, ed. by R.D. Parks (Plenum Press, New York 1977).
2. Proc. of Int. Conf. on Valence Fluctuations in Solid, Santa Barbara, 1981.
3. T. Kasuya, CRC Critical Reviews in Solid State Sciences 3 (1972) 131.
4. T. Kasuya, K. Kojima and M. Kasaya, in R.D. Parks (ed.) Valence Instabilities and Related Narrow-Band Phenomena (Plenum Press, N.Y., 1977). p.137.
5. T. Kasuya, K. Takegahara, T. Fujita, T. Tanaka and E. Bannai, J. de Phys. 40 (1979) C5 308.

6. M. Kasaya, H. Kimura, Y. Isikawa, T. Fujita and T. Kasuya, to be published in [2].
7. T. Kasuya, K. Takegahara, Y. Aoki, K. Hanzawa, M. Kasaya, S. Kunii, T. Fujita, N. Sato, H. Kimura, T. Komatsubara, T. Furuno and J. Rossat-Mignod, to be published [2].
8. C.E. Moore, Atomic Energy Levels, NBS circular No.467 (1958).
9. These values are obtained from experiments of photoemission for occupied state and bremsstahlung isochromat for unoccupied state. J.K. Lang, Y. Baer and P.A. Cox, Phys. Rev. Lett. 42 (1979) 74.
10. A. Yanase and A. Hasegawa, in this volume.
11. T. Kasuya, IBM J. Res. Dev. 14 (1970) 214.
12. T. Kasuya, J. Appl. Phys. 41 (1970) 1090.
13. More exact form has a distant dependent part.
14. F. Holtzberg, J.M.D. Coey, S. von Molnar and B. Cornut, J. Appl. Phys. 49 (1978) 2098.
15. P. Wachter, Phys. Lett. 58A (1976) 484.
16. A. Ochiai, T. Suzuki and T. Kasuya, J. de Physique Colloq. 41 (1980) C5-71.
17. M. Sugita, S. Kunii, M. Fujioka and T. Kasuya, to be published.
18. J. Kondo, Physica 84B (1976) 40, 207.
19. K. Yamada and K. Yosida, Progr. Theor. Phys. 62 (1979) 363.
20. M. Takeshige, Master Thesis, Tohoku University, 1981, and to be published.
21. G.K. Wertheim, W. Eib, E. Kaldis and M. Campagna, Phys. Rev. 22 (1980) 6240, and see its references.
22. N.F. Mott, Metal-Insulator Transitions, Taylor and Francis, London 1974, p.27.
23. J.W. Allen, B. Batlogg and P. Wachter, Phys. Rev. B20 (1979) 4807.
24. J.W. Allen and R.M. Martin, J. de Physique 41 (1979) C5-171.
25. K.A. Gschneider Jr. [1] 89.
26. B. Johansson, J. Phys. Chem. Solids 39 (1978) 467.
27. K. Yosida and A. Sakurai, Prog. Theor. Phys. 61 (1979) 1597; J. Magn. Magn. Mat. 15-18 (1980) 949, K. Yamada, Prog. Theor. Phys. 62 (1979) 354, H. Fukuyama and A. Sakurai, Prog. Theor. Phys. 62 (1979) 595, S. Inagaki, Prog. Theor. Phys. 62 (1979) 1441.
28. J.Etourneau, J.P. Mercurio and P. Hagenmuller: In *Boron and Refractory Borides*, ed. by V.I. Matkovich (Springer, Berlin, Heidelberg, New York 1977) p. 115.
29. H. Sugawara, A. Kakitani, I. Nagakura, T. Ishii, T. Komatsubara and T. Kasuya, ISSP Activity Report of Synchrotron Radiation Laboratory, 1979, p.57.
30. A. Takase, K. Kojima, T. Komatsubara and T. Kasuya, Solid State Commun. 36 (1980) 461.
31. N. Sato, T. Komatsubara, S. Kunii, T. Suzuki and T. Kasuya, to be published in [2].
32. M. Kawakami, S. Kunii, T. Komatsubara and T. Kasuya, Solid State Commun. 36 (1980) 435.
33. Y. Aoki and T. Kasuya, Solid State Commun. 36 (1980) 317.
34. K. Hanzawa et. al., to be published.
35. K. Samwer and K. Winger, Z. Physik B25 (1976) 269.
36. T. Fujita, M. Suzuki, T. Komatsubara, S. Kunii, T. Kasuya and T. Ohtsuka, Solid State Commun. 35 (1980) 569.
37. S.D. Bader, N.E. Phillips, M.B. Maple and C.A. Luengo, Solid State Commun. 16 (1975) 1263.
38. F. Steglich, Z. Physik B23 (1976) 331.
39. Y. Aoki, et. al., to be published.
40. H. Takigawa, Master Thesis, ISSP, University of Tokyo 1980, and private communication of H. Takigawa and H. Yasuoka.

41. J. Rossat-Mignod, P. Burlet, T. Komatsubara, S. Kunii and T. Kasuya, to be published.
42. T. Fujita, M. Suzuki, T. Komatsubara, S. Kunii, T. Kasuya and T. Ohtsuka, Solid State Commun. 35 (1980) 569.
43. T. Nakajima, S. Kunii, T. Komatsubara and T. Kasuya, J. Mag. Mag. Mater. 15-18 (1980) 967, T. Komatsubara et. al., to be published.
44. T. Furuno et. al., to be published.
45. F. Steglich, C.D. Bredl, M. Loewenhaupt and K.D. Schotte, J. de Phys. 40 (1979) C5 301.
46. K. Andres, J.E. Graebner and H.R. Ott, Phys. Rev. Letters 35 (1975) 1779.
47. R.J. Birgeneau, E. Bucher, J.P. Maita, L. Passell and K.C. Turberfield, Phys. Rev. B8 (1973) 5345.
48. G. Busch and O. Vogt, Phys. Letter 25A (1967) 449, P. Burlet, J. Rossat-Mignod, H. Bartolin and O. Vogt, J. de Physique 40 (1979) 47.
49. J. Rossat-Mignod, P. Burlet, J. Villain, H. Bartholin, W. Tcheng-Si, D. Florence and O. Vogt, Phys. Rev. B16 (1977) 440.
50. J. Rossat-Mignod, P. Burlet, H. Bartholin, O. Vogt and R. Lagnier, J. Phys. C: Solid State Phys. 13 (1980) 6381.
51. J.X. Boucherle, A. Delapalme, C.J. Howard, J. Rossat-Mignod and O. Vogt, Proc. of Intern. Symp. on Physics of Actinides and Related 4f Materials, held at Zürich, 1980.
52. A. Hasegawa, J. Phys. C: Solid State Phys. 13 (1980) 6147.
53. T. Kasuya, K. Takegahara, M. Kasaya, Y. Isikawa, H. Takahashi, T.Sakakibara and M. Date "Anomalous Anisotropics in Rare-Earth Magnetic Compounds and Their Behavior and High Magnetic Field", in *Physics in High Magnetic Fields*, ed. by S. Chikazumi and N. Miura (Springer, Berlin, Heidelberg, New York 1981) p. 150.
54. Y. Baer, R. Hauger, Ch. Zürcher, M. Campagna and G.K. Wertheim, Phys. Rev. B18 (1978) 4433.
55. K. Takegahara, A. Yanase and T. Kasuya, J. de Physique 41 (1980) C5-327 Erratum J. de Physique 41 (1980) 1231, K. Takegahara, H. Takahashi, A. Yanase and T. Kasuya, J. Phys. C: Solid State Phys. 13 (1980) in press.
56. H. Takahashi, Master thesis for Faculty of Science, Tohoku University, 1981.
57. K. Takegahara, H. Takahashi, A. Yanase and T. Kasuya, submitted to Solid State Commun.
58. H. Heer, A. Furrer, W. Hälg and O. Vogt, J. Phys. C: Solid State Phys. 12 (1979) 5207.
59. A. Furrer, W. Halg, H. Heer and O. Vogt, J. Appl. Phys. 50 (1979) 2040.
60. T. Suzuki, M. Sera, H. Shida, K. Takegahara, A. Yanase and T. Kasuya, to be published in [2]

Index of Contributors

Dynamics of Solids and Liquids by Neutron Scattering

Editors: S. W. Lovesey, T. Springer
1977. 156 figures, 15 tables. XI, 379 pages
(Topics in Current Physics, Volume 3)
ISBN 3-540-08156-9

Contents:
S. W. Lovesey: Introduction. – *H. G. Smith, N. Wakabayashi:* Phonons. – *B. Dorner, R. Comès:* Phonons and Structural Phase Transformations. – *J. W. White:* Dynamics of Molecular Crystals, Polymers, and Adsorbed Species. – *T. Springer:* Molecular Rotations and Diffusion in Solids, in Particular Hydrogen in Metals. – *R. D. Mountain:* Collective Modes in Classical Monoatomic Liquids. – *S. W. Lovesey, J. M. Loveluck:* Magnetic Scattering.

Monte Carlo Methods in Statistical Physics

Editor: K. Binder
1979. 91 figures, 10 tables. XV, 376 pages
(Topics in Current Physics, Volume 7)
ISBN 3-540-09018-5

Contents:
K. Binder: Introduction: Theory and "Technical" Aspects of Monte Carlo Simulations. – *D. Levesque, J. J. Weis, J. P. Hansen:* Simulation of Classical Fluids. – *D. P. Landau:* Phase Diagrams of Mixtures and Magnetic Systems. – *D. M. Ceperley, M. H. Kalos:* Quantum Many-Body Problems. – *H. Müller-Krumbhaar:* Simulation of Small Systems. – *K. Binder, M. H. Kalos:* Monte Carlo Studies of Relaxation Phenomena: Kinetics of Phase Changes and Critical Slowing Down. – *H. Müller-Krumbhaar:* Monte Carlo Simulation of Crystal Growth. – *K. Binder, D. Stauffer:* Monte Carlo Studies of Systems with Disorder. – *D. P. Landau:* Applications in Surface Physics.

Neutron Diffraction

Editor: H. Dachs
1978. 138 figures, 32 tables. XIII, 357 pages
(Topics in Current Physics, Volume 6)
ISBN 3-540-08710-9

Contents:
H. Dachs: Principles of Neutron Diffraction. – *J. B. Hayter:* Polarized Neutrons. – *P. Coppens:* Combining X-Ray and Neutron Diffraction: The Study of Charge Density Distributions in Solids. – *W. Prandl:* The Determination of Magnetic Structures. – *W. Schmatz:* Disordered Structures. – *P.-A. Lindgård:* Phase-Transitions and Critical Phenomena. – *G. Zaccai:* Application of Neutron Diffraction to Biological Problems. – *P. Chieux:* Liquid Structure Investigation by Neutron Scattering. – *H. Rauch, D. Petraschek:* Dynamical Neutron Diffraction and Its Application.

B. Saleh

Photoelectron Statistics

With Applications to Spectroscopy and Optical Communication
1978. 85 figures, 8 tables. XV, 441 pages
(Springer Series in Optical Sciences, Volume 6)
ISBN 3-540-08295-6

Contents:
Tools from Mathematical Statistics: Statistical Description of Random Variables and Stochastic Processes. Point Processes. – Theory: The Optical Field: A Stochastic Vector Field or, Classical Theory of Optical Coherence. Photoelectron Events: A Doubly Stochastic Poisson Process or Theory of Photoelectron Statistics. – Applications: Applications to Optical Communication. Applications to Spectroscopy.